Earthworks in Europe

Geological Society books refereeing procedures

The Society makes every effort to ensure that the scientific and production quality of its books matches that of its journals. Since 1997, all book proposals have been refereed by specialist reviewers as well as by the Society's Books Editorial Committee. If the referees identify weaknesses in the proposal, these must be addressed before the proposal is accepted.

Once the book is accepted, the Society Book Editors ensure that the volume editors follow strict guidelines on refereeing and quality control. We insist that individual papers can only be accepted after satisfactory review by two independent referees. The questions on the review forms are similar to those for *Journal of the Geological Society*. The referees' forms and comments must be available to the Society's Book Editors on request.

Although many of the books result from meetings, the editors are expected to commission papers that were not presented at the meeting to ensure that the book provides a balanced coverage of the subject. Being accepted for presentation at the meeting does not guarantee inclusion in the book.

More information about submitting a proposal and producing a book for the Society can be found on its web site: www.geolsoc.org.uk.

It is recommended that reference to all or part of this book should be made in one of the following ways:

RADFORD, T. A. (ed.) 2012. *Earthworks in Europe*. Geological Society, London, Engineering Geology Special Publications, **26**.

HODGSON, I. F., BEALES, S. P & CURD, M. J. 2012. Use of tyre bales as lightweight fill for the A421 improvements scheme near Bedford, UK. *In*: RADFORD, T. A. (ed.) *Earthworks in Europe*. Geological Society, London, Engineering Geology Special Publications, **26**, 101–108. http://dx.doi.org/10.1144/EGSP26.12

GEOLOGICAL SOCIETY ENGINEERING GEOLOGY SPECIAL PUBLICATION NO. 26

Earthworks in Europe

EDITED BY

T. A. RADFORD
Atkins Ltd, Epsom, UK

2012
Published by
The Geological Society
London

THE GEOLOGICAL SOCIETY

The Geological Society of London (GSL) was founded in 1807. It is the oldest national geological society in the world and the largest in Europe. It was incorporated under Royal Charter in 1825 and is Registered Charity 210161.

The Society is the UK national learned and professional society for geology with a worldwide Fellowship (FGS) of over 10 000. The Society has the power to confer Chartered status on suitably qualified Fellows, and about 2000 of the Fellowship carry the title (CGeol). Chartered Geologists may also obtain the equivalent European title, European Geologist (EurGeol). One fifth of the Society's fellowship resides outside the UK. To find out more about the Society, log on to www.geolsoc.org.uk.

The Geological Society Publishing House (Bath, UK) produces the Society's international journals and books, and acts as European distributor for selected publications of the American Association of Petroleum Geologists (AAPG), the Indonesian Petroleum Association (IPA), the Geological Society of America (GSA), the Society for Sedimentary Geology (SEPM) and the Geologists' Association (GA). Joint marketing agreements ensure that GSL Fellows may purchase these societies' publications at a discount. The Society's online bookshop (accessible from www.geolsoc.org.uk) offers secure book purchasing with your credit or debit card.

To find out about joining the Society and benefiting from substantial discounts on publications of GSL and other societies worldwide, consult www.geolsoc.org.uk, or contact the Fellowship Department at: The Geological Society, Burlington House, Piccadilly, London W1J 0BG: Tel. + 44 (0)20 7434 9944; Fax + 44 (0)20 7439 8975; E-mail: enquiries@geolsoc.org.uk.

For information about the Society's meetings, consult *Events* on www.geolsoc.org.uk. To find out more about the Society's Corporate Affiliates Scheme, write to enquiries@geolsoc.org.uk.

Published by The Geological Society from:
The Geological Society Publishing House, Unit 7, Brassmill Enterprise Centre, Brassmill Lane, Bath BA1 3JN, UK

The Lyell Collection: www.lyellcollection.org
Online bookshop: www.geolsoc.org.uk/bookshop
Orders: Tel. + 44 (0)1225 445046, Fax + 44 (0)1225 442836

British Library Cataloguing in Publication Data

A catalogue record for this book is available from the British Library.
ISBN 978-1-86239-352-3
ISSN 0267-9914

Distributors
For details of international agents and distributors see:
www.geolsoc.org.uk/agentsdistributors

Typeset by Techset Composition Ltd, Salisbury, UK
Printed by Cambrian Printers Ltd, Aberystwyth, UK

Contents

Preface

The design and construction of earthworks has a significant impact on a project both in terms of cost as well as environmental consequences, including setting and achievement of sustainability targets. This volume focuses on recent developments in: the suitability of materials, including the reuse of arising, for engineering or landscape purposes; soil improvement techniques; geotechnical risk and performance, including the design response to natural and man-made hazards; measurement and mitigation of any adverse environmental impact; and asset management in the face of climate change.

The introduction of European Standards and the national annexes has brought about a review of the design codes used for earthworks and highlighted the variations in practice between countries. The specifications for the execution of earthworks vary considerably from one country to another; in particular because they were primarily derived from US specifications established in the 1950s, or are based on national experience, which is often difficult to export. However the enhancement and enlargement of the European Union has increased the need to consider professional activities beyond national boundaries.

The 2nd International Seminar on Earthworks in Europe (2009) followed on from the inaugural event held in France in 2005. The first seminar was organized primarily to promote discussions across the European countries. Consequently it was a natural progression to ensure that the 2nd seminar built upon these discussions to facilitate the promotion of technical progress and to aid understanding of recent developments/issues in the design and construction of earthworks in member states of the European Union (including new and candidate countries). It is hoped that the 2nd seminar and future events will promote the professions of engineering geology and geotechnical engineering in the design of earthworks for transport infrastructure Hopefully it will also encourage dialogue between earthworks professionals throughout Europe and enhance pan-European linkages and collaboration, facilitate discourse on national practices and the establishment of European standards and encourage an increase in coordinated earthworks research in Europe.

The editor wishes to thank all those involved in the organization of the 2nd International Seminar on Earthworks in Europe. This includes all members of the organizational and technical committees for giving their time; the thanks are also extended to the employers of those professionals who without their backing and commitment to the profession such events would not be possible.

Particular thanks are due to the Sessions Chairs; Alex Kidd (PIARC), Ursula Lawrence (seminar Publicity), Peter Gilbert (BGA), Ivan Hodgson (EGGS) and Anna Morley (seminar Secretary) who aided the smooth running of the seminar, as well as to the treasurer, Ian Duncan. To the presenters who, without them there would not have been a seminar. The authors are thanked for showing patience during the editorial process. Recognition is due to Georgina Worrall and her staff in the Geological Society Conference Office for their assistance running the event. The editor would also like to thank Angharad Hills at the Geological Society Publishing House and Jim Griffiths for their help in getting this Special Publication into print.

Tracey Radford

Chair, 2nd International Seminar on Earthworks in Europe

An introduction to earthworks in Europe

James S. Griffiths[1]* & Tracey Radford[2]

[1] School of Geography, Earth & Environmental Sciences, University of Plymouth, Drake Circus, Plymouth, Devon, PL4 8AA, UK

[2] Atkins Ltd, Woodcote Grove, Ashley Road, Epsom, KT18 5BW, UK

*Corresponding author (e-mail: jim.griffiths@plymouth.ac.uk)

Earthworks encompasses a range of engineering activities that goes well beyond the conventional digging out of soil and rock and movement of the material to another part of a site. Although these activities may be amongst the most fundamental of civil engineering construction activities, and take place whenever and wherever a development takes place, earthworks also include embankment dams, foundations (especially those for linear infrastructure) such as capping layers for road construction, structural fill for retaining walls, and for reinforced and anchored soil and earth. Conventional cut and fill operations are often somewhat disparagingly known in the UK as 'muckshift', the processes of excavation to form temporary and permanent cuttings and trenches, controlled placement of material for structural backfill, creation of embankments including earth and rockfill dams, and preparation of foundations for all forms of structure is a highly skilled geotechnical design practice. To undertake this work safely and economically requires an evaluation of the engineering geology of the site where the works are to take place, accurate descriptions of the nature and properties of the materials involved, detailed material specifications, assessment of earthwork quantities, settlement calculations for structures and foundations, analysis of short and long term slope and foundation stability, and the design of appropriate construction practices. The methods for excavating materials must be determined, the short and long term stability of excavations must be evaluated, and the suitability of any materials to be used in construction needs to be established. During any fill placement monitoring is necessary to ensure the materials meet specification and which may define pre-determined methods. It is also necessary to ensure that post-compaction properties meet performance specification, rates and amounts of settlement fall within expected limits, and that the foundations and side slopes of all structures meet design requirements.

Once construction is finished the long term behaviour of excavations (i.e. cuttings) and structures (embankments, backfill, reinforced earth, etc.) requires inspection and, in some instances, monitoring for signs of distress to ensure remedial measures can be designed and implemented should failures occur. Given that the largest embankment in the world is presently the Nurek Dam in Tajikistan at over 300 m in height and 730 m long holding back over 10 km^3 of water (Stephens 1976; Ghasimi 1994), the critical importance of safe design, construction and long term monitoring is readily apparent. However, large scale earthworks do fail and examples of embankment dam failures are to be found in many countries, including the USA. (e.g. Hebgen Dam in Montana in 1959; Sherard 1959) and the UK (e.g. Carsington Dam in Derbyshire in 1984; Skempton & Vaughan 1993). Similarly a study of the M1 motorway earthworks in Bedfordshire in England identified a considerable acceleration in the rate of failure of cuttings and embankments on the heavily overconsolidated Gault Clay after a period of about 15 years (Andrews 1990; Perry 1995). Indeed, in England assessing the probability of failure is seen as part of the maintenance of highway earthworks as geotechnical assets (Highways Agency 2003, 2008).

There are many large earth and rockfill embankment dams; however, a substantial proportion of earthworks around the world are either associated with the construction of transport infrastructure or loose tipped waste (non-engineered fill). At the 2nd International Seminar on Earthworks in Europe that took place in the London in 2009, the main focus of the meeting was the design and construction of cuttings and engineered fill for transport infrastructure. This is reflected in the papers compiled for this present volume as outlined below.

Earthworks associated with transport infrastructure are essentially linear in nature, comprising cuttings and embankments, and are either built from the arisings from adjacent cuttings or, in certain circumstances, specially imported materials. Soil and rock from cuttings are used directly on site (reused), used after treatment (recycled) or transported off-site as waste (potentially a costly process) for either recovery or disposal. This activity has occurred since prehistoric times when the first earthworks were built as defensive ditches and ramparts around settlements, drainage

From: Radford T. A. (ed.) 2012. *Earthworks in Europe*. Geological Society, London, Engineering Geology Special Publications, **26**, 1–4, http://dx.doi.org/10.1144/EGSP26.1

ditches and the first roadways. Such pre-historic activities are found throughout the world, such the Pre-Columbian earthworks discovered in western Amazonia (Pärssinen, *et al.* 2009), with canals being dated to 4000 BC in Iran and earth dams at 3200 BC in Jordan (McFarlan 1989). With the industrial revolution there was a need for cuttings to remain stable in the long term and for embankments to carry the heavy loads associated firstly with waterways, then railways, and, in the 20th century, highways. These long linear combinations of cuttings and embankments are structures that require a far more sophisticated approach to design and construction (e.g. see Ekins *et al.* 1993; Perry 1995; Vaughan *et al.* 2004) and form a substantial component of the construction, associated construction duration, and cost of modern transport infrastructure. The papers in this volume provide an insight into the development of modern earthworks construction practice, mainly in the context of highways. These studies build upon previous notable conferences and symposia on the subject that in the UK include Clays Fills (Institution of Civil Engineers 1979), Failures in Earthworks (Institution of Civil Engineers 1985), Compaction Technology (Institution of Civil Engineers 1988) and Engineered Fills (Clarke *et al.* 1993).

The critical importance of the safe construction of earthworks is recognized globally by a series of important publications (e.g. U.S. Bureau of Reclamation 1974; Geotechnical Engineering Office 1984; Canadian Geotechnical Society 1985; Horner 1988; Geotechnical Control Office 1990; Halcrow 1994; Trenter 2001). In the UK this importance was demonstrated by the production of the Code of Practice for Earthworks CP2003 in 1959 (BSI 1959). This was revised in 1981 as BS6031 (BSI 1981) and then again in 2009 (BSI 2009) when it was incorporated in Eurocode 7 (BSI 1997; Simpson & Driscoll 1998). The history and development of BS6031 is reviewed by **Lamont** whilst **Harris** assesses the design of stable earthworks according to the requirements of Eurocode 7. Design and construction practice for major highways in the UK is controlled by the *Design Manual for Roads and Bridges* (DMRB 1994; see also Department for Transport 2012). Advice Note HA70/94 established that highway embankments can be built according to a method specification (defined material placed and compacted in a standard way) or an end-product specification (testing of the completed embankment to ensure it meets design requirements). **Perry** examines the management and contractual risk associated with these construction methods and goes on to evaluate the long term risk of cutting and embankment failures in the context of asset management. The concept of asset management for the English Highways Agency is a theme also discussed by **Power *et al.*** who recognize that a key tool in the process is use of the Geotechnical Data Management System (Highways Agency 2003, 2008). Both of these papers build upon the earlier work reported by the Construction Industry Research and Information Association (Perry *et al.* 2003*a*, *b*) and the research undertaken by the Transport Research Laboratory (e.g. Perry 1989). Because their network was established in the 60s and 70s another facet of the UK Highways Agency concern has been the potential for deep-seated delayed failure of cuttings, as experienced on the London Underground (Skempton 1964, 1996), and this work is presented by **Parry**.

Most countries have their own national codes of practice, such as the Australian Standard AS3798 (1990). In Spain, design work is carried out according to the Technical Regulation for the Construction of Roads, Motorways and Masonry Bridges (PG3 2002) as discussed by **Dapena & Parilla**. In the Czech Republic there are a range of well-defined standards based on Eurocodes and this includes the newly revised system CSN 73 6133 '*Design and Execution of Earthworks on Highways*' which has recently been introduced and is described by **Herle**. There are also learned societies that promote co-operation and exchange of knowledge with respect to earthworks design and **Gomes Correia *et al.*** present the work of the Technical Committee TC3 'Geotechnics of Pavements' which is part of the International Society for Soil Mechanics and Geotechnical Engineering (ISSMGE). They highlight progress towards the performance specification of earthworks and the development of design, specification and construction techniques for tyre bales (Winter *et al.* 2006).

Whilst codes of practice and technical committees can provide the general standards of work required for safe and economic earthworks construction, the use of particular materials, some associated with certain regions, may need a more localized study. **Rutty & Johnston** discuss the situation in Ireland where the material used as engineered fill is derived primarily from glacial soils and some excavated rock and they note the widespread use of the Moisture Condition Value as a control (Parsons & Boden 1979; Parsons 1992; Matheson & Winter 1997; Winter 2004). **Petkovšek & Majes** report on the situation in Slovenia where over 500 km of motorways have been constructed since 1994 over bedrock of stiff clays and argillaceous sedimentary rock. Because of the decrease in suction following construction that led to wetting and softening of the embankments, for much of this work use has been made of Proctor compaction diagrams complemented with the corresponding soil water characteristic curve. In the Irish and Slovenian examples the materials used in construction of the embankments were derived from arisings from cuttings. However, some situations require the use of completely different materials for the engineered fill and **Hodgson *et al.*** present a case study of the use of tyre bales as a lightweight fill for a road improvement scheme near Bedford in the UK, which draws upon the methods developed by Winter *et al.* (2006). More conventional alternative fill materials that have been widely used are the waste material from heavy industries, including mining. This is the situation in the Czech Republic and **Kresta** describes the use of colliery spoil, fly ash and blast furnace slag for fill.

Despite careful design and construction, cuttings and embankments do still fail; indeed one could argue that any engineer who has not had an earthworks failure is likely to have been over-conservative in his/her design and cost the client money (Stephens, pers. comm. 1984). At the Earthworks Seminar this issue was addressed by three papers, each looking at a different facet of failure. **Macdonald *et al.*** examined the well-established practice of installing deep counterfort drains into cut slopes to reduce pore water pressures and act as a reinforcing agent. They recognized that Hutchinson (1977) still represents the standard empirical technique for the design of such drains but point out that there is little published data on drain performance. **Lamont-Black *et al.*** reviewed a completely different technique for stabilizing an old Victorian railway embankment built out of London Clay, namely the use of electrokinetic geosynthetics and electrokinetic soil nails. The reported success of the technique suggests that this is a method that warrants further research. The third paper in this sequence, by **Macdonald *et al.***, looked at how seasonal effective stress changes can result in the progressive failure of clay embankments and how this correlates with soil moisture deficit values. The paper goes onto to discuss how rainfall and temperature forecasts may be used as a predictor of changes in the soil moisture deficit.

Another theme that emerged from the Earthworks Seminar was the need to take account of environmental issues, including climate change. The effects of climate change on the stability of embankment slopes was addressed by **Toll *et al.*** who describe the building and monitoring of a full-scale trial embankment. **O'Riordan & Phear** take an overview of the environmental impact and look at the reuse of excavated tunnel spoil, using ground treatment to minimize earthwork impact and combining performance-based design with carbon footprint and cost analyses. The issue of carbon footprint and energy emissions as a measure of the environmental impact and how this could assist in engineering design selection is addressed in the paper by **Pantelidou *et al.*** which brings the book to its conclusion.

The 'Earthworks in Europe' seminars have been established to discuss and promote the technical progress and understanding of recent developments and issues in the design and construction of earthworks in European countries. This volume of papers derived from the 2nd international seminar held in the UK represents an important part of the dissemination of good practice.

The authors wish to thank Mike Winter of TRL for his support and advice in bringing this paper to completion.

References

ANDREWS, A. D. 1990. *Determining the Age of Failure of Motorway Earthworks from Aerial Survey Photographs*. Research Report 257, Transport Research Laboratory, Crowthorne.

AUSTRALIAN STANDARD, AS3798. 1990. Guidelines on earthworks for commercial and residential developments. Standards Australia, Sydney, New South Wales.

BSI. 1959. *Code of Practice for Earthworks*. CP2003, British Standards Institute, London.

BSI. 1981. *Code of Practice for Earthworks*. BS6031:1981, British Standards Institute, London.

BSI. 1997. *Eurocode 7: Geotechnical Design – Part 1 General Rules*. DD ENV 1997-1, British Standards Institute, London.

BSI. 2009. *Code of Practice for Earthworks*. BS6031:2009, British Standards Institute, London.

CANADIAN GEOTECHNICAL SOCIETY. 1985. *Canadian Foundation Engineering Manual*. 2nd edn. Canadian Geotechnical Society, Ottawa.

CLARKE, B. G., JONES, C. J. F. P. & MOFFAT, A. I. B. (eds) 1993. *Engineered Fills*. Thomas Telford, London.

DEPARTMENT FOR TRANSPORT., 2012. *Design Manual for Road and Bridgeworks*. Website accessed 27 April 2012 at http://www.dft.gov.uk/ha/standards/dmrb/index.htm

DESIGN MANUAL FOR ROADS AND BRIDGES. 1994. Volume 4 Geotechnics and Drainage, Section 1 Earthworks, Part 5 HA 70/94, *Construction of Highway Earthworks*. Highways Agency.

EKINS, J. D. K., CATER, R. & HOUNSHAM, A. D. 1993. Highway embankments – their design, construction and performance. In: CLARKE, B. G., JONES, C. J. F. P. & MOFFAT, A. I. B. (eds) *Engineered Fills*. Thomas Telford, London.

GEOTECHNICAL CONTROL OFFICE. 1990. *Review of Design Methods for Excavations*. Civil Engineering Services Department, Hong Kong.

GEOTECHNICAL ENGINEERING OFFICE. 1984. *Geotechnical Manual for Slopes*. 2nd edn, Engineering Development Department, Hong Kong.

GHASIMI, R., 1994. *Tajikistan*. A World Bank Country Study, World Bank, Washington, D.C.

HALCROW, 1994. *Study of Engineered Fills for Building Purposes*. Report for the Building Research Establishment.

HIGHWAYS AGENCY. 2003. HD41/03: Maintenance of Highway Geotechnical Assets. *Design Manual for Roads and Bridges*, Volume 4, Part 3.

HIGHWAYS AGENCY. 2008. HD22/08: Managing Geotechnical Risk. Maintenance of Highway Geotechnical Assets. Design Manual for Roads and Bridges, Volume 4, Part 2.

HORNER, P. C. 1988. *Earthworks*. Institution of Civil Engineers Works Construction Guide, 2nd edn, Thomas Telford, London.

HUTCHINSON, J. N. 1977. Assessment of the effectiveness of correction measures in relation to geological conditions and type of slope movement. *Bulletin of the International Association of Engineering Geology and the Environment*, **16**, 131–155.

INSTITUTION OF CIVIL ENGINEERS. 1979. *Clay Fills*. Proceedings of the Conference held at the ICE 14–15 November 1978. Thomas Telford, London.

INSTITUTION OF CIVIL ENGINEERS. 1985. *Failures in Earthworks*. Proceedings of the Symposium held at the ICE March 6–7 1985. Thomas Telford, London.

INSTITUTION OF CIVIL ENGINEERS. 1988. *Compaction Technology*. Proceedings of the Conference held at the ICE 29th October 1987. Thomas Telford, London.

MATHESON, G. D. & WINTER, M. G. 1997. *Use and application of the MCA with particular reference to glacial tills*. TRL Report 273, Transport Research Laboratory, Crowthorne.

MCFARLAN, D. 1989. *The Guinness Book of Records*. Guinness Superlatives Ltd, Enfield.

PARSONS, A. W. 1992. *Compaction of Soils and Granular Materials*. A review of research performed at the Transport Research Laboratory. HMSO, London.

Parsons, A. W. & Boden, J. B. 1979. *The Moisture Condition Test and its Potential Application in Earthworks*. Report SR522, Transport and Road Research Laboratory, Crowthorne.

Pärssinen, M., Schaan, D. & Ranzi, A. 2009. Pre-Columbian geometric earthworks in the upper Purús: a complex society in western Amazonia. *Antiquity*, **83**, 1084–1095.

Perry, J. 1989. A Survey of Slope Conditions on Motorway Earthworks in England and Wales. TRRL Research Report 199, Transport Research Laboratory, Crowthorne.

Perry, J. 1995. Engineering geology of soils in highway construction: a general overview. In Eddleston, M., Walthall, S., Cripps, J. C. & Culshaw, M. G. (eds) *Engineering Geology of Construction*. Geological Society Engineering Geology, Special Publication, **10**, 189–203.

Perry, J., Pedley, M. & Bradley, K. 2003*a*. Infrastructure cuttings: condition, appraisal and remedial measures. CIRIA Report C591. Construction Industry Research and Information Association, London.

Perry, J., Pedley, M. & Reid, M. 2003*b*. Infrastructure embankments: condition, appraisal and remedial measures. CIRIA Report C592. 2nd edn, Construction Industry Research and Information Association, London.

PG3. 2002. *Pliego de Prescripciones Téchnicas Generales para Obras de Catteras y Puentas*. Ministerio de Fomento, España.

Sherard, J. L. 1959. What the earthquake did to Hebgen Dam. *Engineering News Record*, **163**, 26.

Skempton, A. W. 1964. Long term stability of clay slopes. *Geotechnique*, **11**, 77–101.

Skempton, A. W. 1996. Embankments and cuttings on the early railways. *Construction News*, **11**, 33–39.

Skempton, A. W. & Vaughan, P. R. 1993. The failure of Carsington Dam. *Geotechnique*, **43**, 151–173.

Simpson, B. & Driscoll, R. 1998. *Eurocode 7: A Commentary*. BRE Publications, Construction Research Communications Ltd, London.

Stephens, J. H. 1976. *The Guinness Book of Structures*. Guinness Superlatives Ltd, Enfield.

Trenter, N. A. 2001. *Earthworks: A Guide*. Thomas Telford Ltd, London.

U.S. BUREAU OF RECLAMATION. 1974. *Earth Manual: A Water Resource Technical Publication*. 2nd edn, U.S. Department of the Interior, Washington, D.C.

Vaughan, P. R., Kovacevic, N. & Potts, D. M. 2004. Then and now: some comments on the design and analysis of slopes and embankments. *In*: Jardine, R. J., Potts, D. M. & Higgins, K. G. (eds) *Advances in Geotechnical Engineering: The Skempton Conference*. Thomas Telford, London, 241–290.

Winter, M. G. 2004. Determination of the acceptability of glacial tills for earthworks. *Quarterly Journal of Engineering Geology and Hydrogeology*, **37**, 187–204.

Winter, M. G., Watts, G. R. A. & Johnson, P. E. 2006. *Tyre Bales in Construction*. TRL Published Project Report PPR080. Transport Research Laboratory, Crowthorne.

Revision of BS 6031: Code of practice for earthworks

Donald Lamont

Health & Safety Executive, Bootle, UK (e-mail: donald@hyperbaricandtunnelsafety.co.uk)

Abstract: The paper summarizes the work of BSI committee B/526/0-/0-/DG1 which is responsible for the current revision of BS 6031:1981 'Code of Practice for Earthworks'.
Explanatory note: This paper was prepared in advance of the Earthworks Seminar whilst BS 6031 was in the process of public consultation and final revision prior to it being published in December 2009.

History of BS 6031

BS 6031 was first published as CP 2003 in 1959. It was revised and republished as BS 6031 in 1981 which remains the published version to this day, despite major changes in earthworks practice over the intervening decades. A second revision commenced in 2006 and was issued by BSI as a Draft for Public Comment in June 2009.

BS 6031:1981

The existing Code is in 3 sections:-

Section 1 – General – that is scope, references, definitions, safety procedures
Section 2 – Cuttings and embankments, grading and levelling
Section 3 – Trenches, pits and shafts

It has been recognized for some time that there were problems with the text of BS6031: 1981. It is long winded and descriptive and not a stimulating or informative document to read. It is ignored by major infrastructure clients such as Highways Agency and Network Rail hence is generally disregarded by other less prominent clients seeking to have earthworks undertaken. There is no reference in the current text to the EN 474 'Earthmoving machinery – safety' series of standards. Only health and safety legislation enacted in the 1960s is covered. BS 6031 is incompatible with EC7 which could make it attractive to some! More importantly it does not reflect current practice in the design of earthworks or in their construction on site. As a result BS 6031:1981 is irrelevant to most earthworks engineers.

The problems with BS 6031 are summarized in the foreword to the 1981 edition which states 'The construction methods described in the code are necessarily quite basic in conception. Some of them, especially those used for supporting the sides of excavations, are traditional in character, and show few changes from the methods described in the 1959 edition of this code'. Given that the 1959 edition probably took at least 2–3 years to draft, this is hardly a good advertisement for its use in current projects. There are few rope-operated face shovels used in earthworks today but hydraulic backacters and articulated dump trucks were not around in the late 1950s.

BS 6031:2009

So what of the revision? The first and most important change from BS 6031:1981 is that major infrastructure owners and authorities specifying earthworks are supporting the revised standard. Once again BS 6031 will become the national standard for the design and construction of earthworks. It will be applicable to road and rail projects and can be applied to the construction of all engineered cuts and embankments. Although earth dams are excluded from the scope, much of the guidance in the standard can be applied there too. For example, the guidance on safety in earthworks construction is applicable to any earthworks operation.

Committee B/526/0-/0-/DG1 first met in early 2006. It was felt to be particularly important that membership was drawn from a wide cross section of organizations with an interest in earthworks. Currently members representing major infrastructure owners, designers, local authorities, contractors, trench support suppliers and industry regulators sit on the committee. Work started as a revision of the existing text. Very soon it became apparent that nothing short of a full rewrite of Section 2 would suffice if a text fit for the 21st century was to be produced. Fortunately funding was made available by Highways Agency and Network Rail

From: Radford T. A. (ed.) 2012. *Earthworks in Europe*. Geological Society, London, Engineering Geology Special Publications, **26**, 5–7, http://dx.doi.org/10.1144/EGSP26.2

and a technical author produced a first draft of the rewritten text of Section 2. This was discussed at length in committee. Section 3 was revised by the committee and then revised again to match the style of the new Section 3.

The code has been completely rewritten. It is still in three sections but is now set out in a more concise format with extensive use made of bulleted lists. The order of the text is based on 'Earthworks lifecycle'. This is discussed below.

Extensive use has been made of authoritative references in the field of earthworks and geotechnics. To match current social and political policy, there is guidance on sustainability and the safety of earthworks plant and process. At an early stage the committee decided that the 1981 edition should remain extant as it contains useful guidance on timber support for excavations.

The 'Earthworks lifecycle'

The earthworks lifecycle is shown diagrammatically in Figure 1. It illustrates the various stages from initial conception of an earthwork through design and construction to asset management. The latter leads to the identification of defects, the remediation of which require planning, design, etc. thus the cycle begins again.

The revised text is also in three sections which correspond broadly to those in the current version. These are:

- Section 1: General: this includes the scope, normative references and terms and definitions along with chapters on the control of risk, sustainability in planning earthworks, site conditions and investigations.
- Section 2: entitled 'Earthworks that are designed to be self-supporting' and has chapters on the design of earthworks, the specification of earthworks, the construction of earthworks, handover, post-construction testing and monitoring, earthworks asset management along with guidance on the decommissioning and disposal of earthworks assets.
- Section 3: entitled 'Temporary excavations' includes chapters on 'Temporary excavations', 'Construction procedure', 'Trenches, pits and shafts' and 'Excavation in rock'.

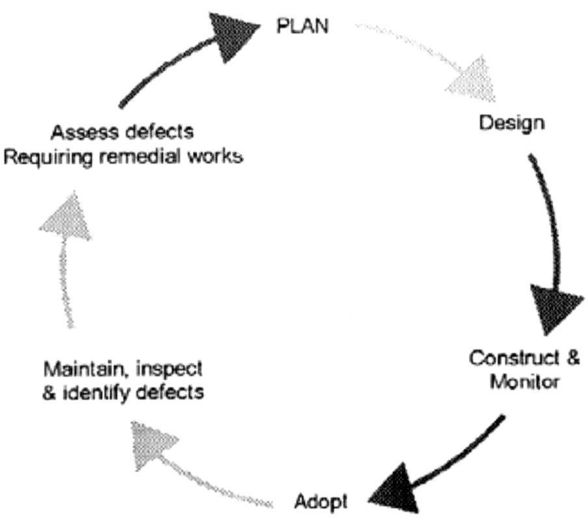

Fig. 1. Earthworks lifecycle.

The primary reference in the revised BS 6031 is the Highways Agency '*Specification for Highway Works*'. In addition, the text draws heavily on BS EN 1997-1:2004 (Eurocode 7). In this respect BS 6031 forms part of the 'non-conflicting complementary information' supporting Eurocode 7 in the UK.

The BSI committee has adopted a particular format for the text. Part of a typical clause from the text, selected more or less at random to illustrate aspects of the format, is set out below. The use of **bold underlined** text in the part-clause below is purely for illustrative purposes.

Part of Cl 7.2.4.1 – General.

When selecting the characteristic values, due account should be taken of the items listed in **BS EN 1997-1:2004, 2.4.5.2**, the following items are of particular relevance to slope design:

- geological, historical and other background data;
- the amount of measured data relating to the parameter value under consideration;
- the variability of the measured data and the degree of confidence in the data;
- **the extent of the zone of ground governing the limit state under consideration**;
- the ability of the ground to transfer load from weak to strong zones; and
- the consequences of failure at the limit state under consideration.

Note: 3 Statistical methods may be used to select characteristic values for the ground properties and procedures for the application of statistical methods for this purpose are described in detail by Frank *et al.* (2004). The use of statistical methods implies that sufficient data are available to permit a meaningful evaluation to be undertaken and, where statistical methods are used, BS EN 1997-1:2004, **2.4.5.2** recommends that the calculated probability of a worse value governing the occurrence of the limit state under consideration should be no greater than 5%.

As can be seen above there are references to BS EN 1997-1:2004 that is Eurocode 7, an example of a bulleted list and a reference to an authoritative reference text.

The next step

Proposals exist for one or more CEN standards on the topic of 'Earthworks'. CEN/BT WG203 concluded that earthworks are a well defined and economically important field of activity that had not been covered by CEN standardization works. It proposed that a new Technical Committee for earthworks be created which it believes would not conflict with existing Technical Committees. Given the comprehensive and up to date content of the revised BS 6031, might this form the basis for the CEN standard(s)? (Note: CEN TC396 'Earthworks' was created in September 2009 to undertake the work described above.)

Conclusion

The purpose of this presentation was to highlight the work carried out to revise BS 6031 to bring it into line with BS EN 1997-1 (Eurocode 7, part 1) and the Highways Agency Design Manual for Road and Bridge Works (http://www.dft.gov.uk/ha/standards/dmrb/index.htm), whilst retaining those elements of the Code of Practice that remain valid. Since the paper was presented at the UK Earthworks Conference the revised code of practice has been published (http://shop.bsigroup.com/en/ProductDetail/?pid=000000000030234058).

Stability of earthworks to Eurocode 7

A. J. Harris* & A. J. Bond

Tony Gee and Partners LLP, Hardy House, 140 High Street, Esher, KT10 9QJ
Geocentrix Ltd, Scenic House, 54 Wilmot Way, Banstead, SM7 2PY

**Corresponding author (e-mail: andrew.harris@tonygee.com)*

Abstract: The introduction of Eurocodes to engineering practice represents an importantchange in the approach to geotechnical engineering, as for the first time there is a unified set of codes dealing with the main structural materials including the ground. All of the Eurocodes adopt a limit state design philosophy which is different from the lumped factor of safety approach adopted by traditional geotechnical design practice. This paper examines the implications of the introduction of the Eurocodes, in particular EN1997, when assessing the stability of earthworks. It outlines the various committees responsible for the development of the codes and discusses the application of EN1997 to the assessment of the stability of slopes highlighting the difficulties in applying the various design approaches. Examples from practice are used to highlight where the code is different from traditional practice. A survey of practice in other European countries is outlined which indicates the differing approaches to ensuring adequate reliability being taken across Europe highlighting options that could be considered by the geotechnical community. The paper concludes that the introduction of the Eurocodes does not require a significant change to the way stability analysis is carried out.

In April 2010, the integrated suite of Eurocodes for structural engineering replaced equivalent national standards. This presents particular challenges to those dealing with geotechnical engineering as, for the first time, the same limit state approach to design was adopted for all structural materials including the ground.

EN1997-1 and EN1997-2 (EC7) present the principles for carrying out geotechnical design adopting a limit state philosophy where uncertainty is accounted for by the use of partial factors. Although there are codes of practice currently in use that adopt limit state principles for some aspects of geotechnical design (e.g. retaining structures BS8002 1994) such principles have not been adopted universally. EC7 represents the first attempt to provide a limit state framework for all geotechnical design which can be readily applied to both the geotechnical and the structural materials (concrete, steel, timber, and aluminium) aspects of the design.

The design of earthworks is a particular facet of geotechnical engineering where the ground does not generally interact with other structural materials except where they provide surcharge loading or where the ground is reinforced. This paper considers the application of EC7 to embankments and cuttings but does not discuss reinforced soils or the design of piled embankments. For guidance on reinforced soils the reader is referred to BS8006 which has been revised to make it compliant with EC7 where appropriate.

European documentation for geotechnics

A suite of technical committees has been set up by the European Committee for Standardization (CEN) to generate and maintain the documentation that governs structural design. The relevant committees for geotechnical engineering are: TC250/SC7 responsible for EN1997; TC288 responsible for execution standards; TC341 responsible for investigation and testing; and TC369 responsible for earthworks. Figure 1 represents the interrelationship and scope of TC250, TC288 and TC341 and Figure 2 shows how TC369 draws on the results of the other technical committees.

The output from these committees is a series of documents which either have already replaced or will eventually replace equivalent British Standards, that is BSs 1377, 8002, 8004, and 8081. Some British Standards remain after revision (e.g. BSs 6031, 8006, and 5930). Many of the European Standards (ENs) have now been approved and have been published as current British Standards. In addition National Annexes (NAs) have been approved which allow for country specific application of the standards. These NAs cannot modify the main body and principles of the standards but provide country specific values for the partial factors that are required to be used in practice.

From: Radford T. A. (ed.) 2012. *Earthworks in Europe*. Geological Society, London,
Engineering Geology Special Publications, **26**, 9–19, http://dx.doi.org/10.1144/EGSP26.3

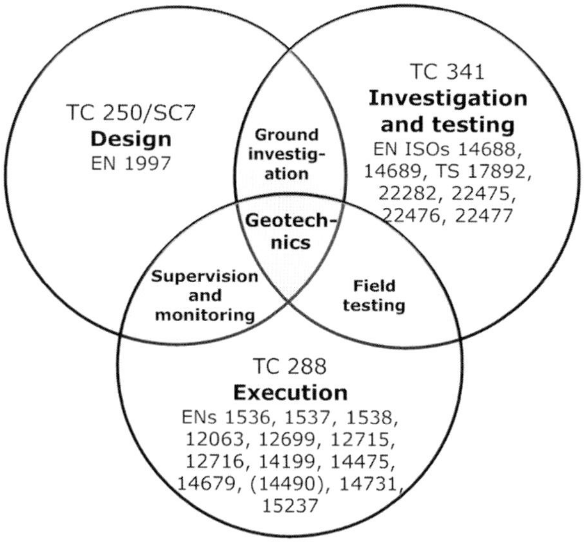

Fig. 1. CEN technical committees responsible for geotechnical engineering.

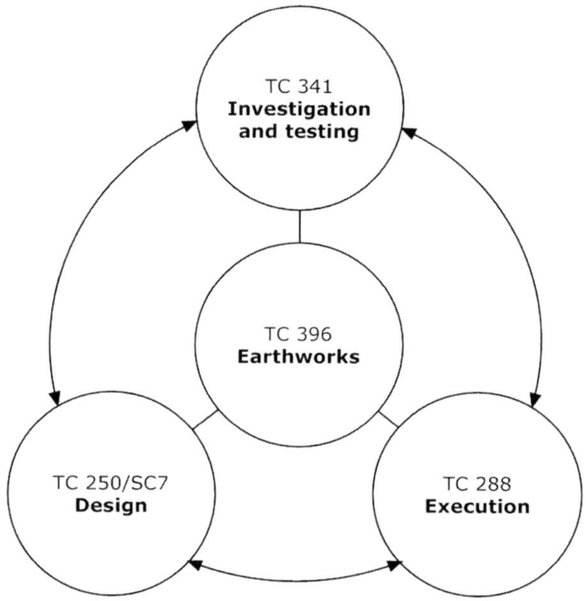

Fig. 2. Earthworks sits at the junction of investigation and testing, design, and execution.

Principles of EC7 for slope stability analysis

The assessment of the stability of slopes is governed by the general requirements of EC7 for the verification of strength as expressed below:

$$E_d \leq R_d$$

E_d = design effect of actions
R_d = design resistance corresponding to that effect

This requirement applies to limit state GEO:

> *'Failure or excessive deformation of the ground, in which the strength of soil or rock is significant in providing resistance'*

EN 1997-1 §2.4.7.1(1)P

> ...and to ultimate limit state STR
>
> *'Internal failure or excessive deformation of the structure or structural elements ... in which the strength of structural materials is significant in providing resistance'*

EN 1997-1 §2.4.7.1(1)P

When assessing the stability of an unreinforced slope formed of soil or rock materials, limit state GEO is the more relevant of the two limit states. There are two sections in EN1997-1 which deal directly with slope stability: Section 11 'Overall Stability' (which covers overall stability of the ground and movements in the ground related to foundations, retaining structures, natural slopes, embankments, and excavations) and Section 12 'Embankments' (which covers embankments for small dams and infrastructure). The contents for each section are presented under similar headings, as given below:

- §x.1 General (2 paragraphs in S11/2 paragraphs in S12)
- §x.2 Limit states (2/2)
- §x.3 Actions and design situations (6/8)
- §x.4 Design and construction considerations (11/13)
- §x.5 Ultimate limit state design (26/7)
- §x.6 Serviceability limit state design (3/4)
- §11.7 Monitoring (2)/§12.7 Supervision and Monitoring (5)

The general methodology for applying partial factors for the verification of strength is given in Figure 3, from which it should be noted that partial factors, and thus design values, are assessed independently for actions (loads), geometry, and material parameters. The notation adopted for the various factors used by EC7 is given in the Appendix at the end of this paper. The partial factors recommended in Annex A to EN1997-1 and given in Table 1 have been adopted in the UK National Annex for slope stability assessment.

EC7 presents three different design approaches to cater for different philosophies in the application of partial factors. These approaches are summarized in Table 2. Design Approach 1 is sub-divided into two combinations: Combination 1 applies partial factors greater than 1.0 to actions

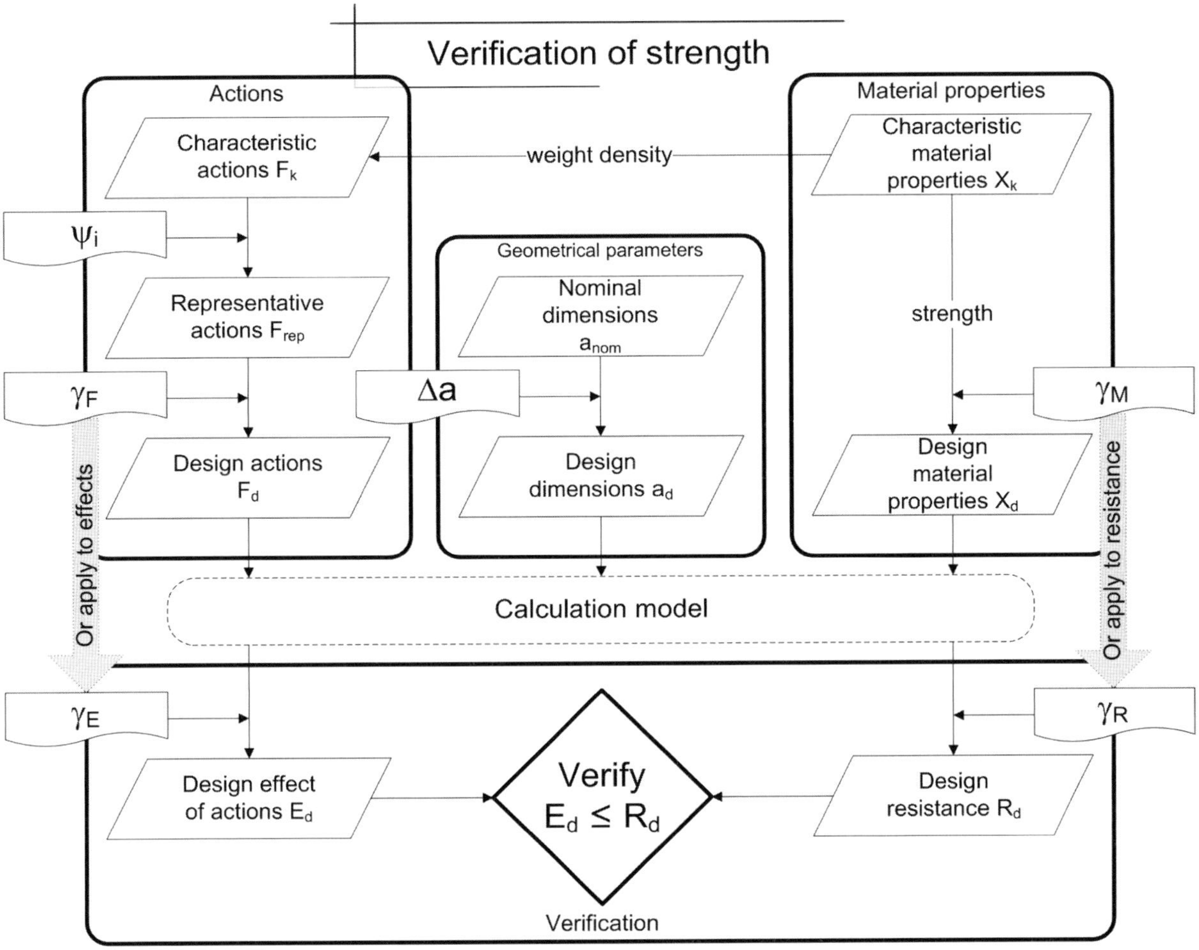

Fig. 3. Application of partial factors (after Bond & Harris 2008), for definition of symbols see Appendix: notation.

while leaving material properties unchanged; Combination 2 applies partial factors to material properties while leaving permanent actions unchanged (except a small partial factor is applied to variable actions). In Design Approach 1, no resistance factors are applied when considering slopes; Design Approach 2 applies partial factors to action effects and resistances at the same time; and Design Approach 3 applies partial factors to material properties while leaving permanent actions unchanged (a small partial factor is applied to variable actions). For slopes, Design Approach 3 and Design Approach 1 Combination 2 are essentially the same.

The following clauses in EN 1997-1 and the UK National Annex govern the selection of design approaches:

- The manner in which equations are applied shall be determined using one of three design approaches (EN 1997-1 §2.4.7.3.4.1(1) P)
  - design approaches apply only to STR and GEO limit states
  - each nation can choose which one (or more) to allow
- . . . only Design Approach 1 is to be used in the UK (UK NA to BS EN 1997-1 §NA.4)

Factors of safety and partial factors

Traditional limit equilibrium calculations express the factor of safety against failure along a given slip plane in two main ways:

(1) for circular shaped slip surfaces in the undrained condition the factor of safety is defined as the ratio of the restoring moment to the overturning moment

(2) for all shapes of sliding surface in total or effective stress terms the factor of safety may be defined as the

Table 1. *Partial factors from EN 1997-1 Annex A (after Bond & Harris 2008)*

Parameter		*Symbol*	*Action factors*		*Material factors*		*Resistance factors*			
			A1	*A2*	*M1*	*M2*	*R1*	*R2*	*R3*	*R4*
Permanent action (G)	Unfavourable	γ_G	1.35	1.0						
	Favourable	($\gamma_{G,fav}$)	1.0							
Variable action (Q)	Unfavourable	γ_Q	1.5	1.3						
	Favourable	–	(0)	(0)						
Shearing resistance (tan φ)		γ_{φ}			1.0	1.25				
Effective cohesion (c')		γ_c								
Undrained shear strength (cu)		γ_{cu}				1.4				
Unconfined compressive strength (qu)		γ_{qu}								
Weight density (γ)		γ_{γ}				1.0				
Bearing resistance (Rv)		γ_{Rv}					1.0	1.4	1.0	(1.0)
Sliding resistance (Rh)		γ_{Rh}						1.1		
Earth resistance (Re)	Walls	γ_{Re}						1.4		
	Slopes	γ_{Re}						1.1		
Pile resistance							Varies			

Table 2. *Application of partial factor sets for different design approaches (after Bond & Harris 2008)*

Structure	*Design Approach*			
	1		*2*	*3*
	Combination 1	*Combination 2*		
General structures	A1 + M1 + R1	A2 + M2 + R1	A1 + M1 + R2	(A1*/A2†) + M2 + R3
Slopes			[A1] + M1 + R2	[A2]*† + M2 + R3
Axially loaded piles & anchors		A2 + (M1‡/M2§) + R4	as for general structures	

Partial factors on: Actions; M, material factors; R, resistances.
Partial factors underlined provide main source of safety (i.e. have values >1.0).
Partial factors [in square brackets] applied to action effects.
**On structural actions;* †*On geotechnical actions;* ‡*For calculating resistances*; §*For calculating unfavourable actions* (e.g. *down-drag).*

ratio of the average available shear strength to the average mobilized shear strength along the sliding surface (Bromhead 1992)

The first approach is akin to applying a partial factor to the effect of an action or to the resistance. The second is the same as applying a partial factor to the material properties. However, for translational slides and undrained circular slips the effect of applying a factor of safety to the effect of the resistance is the same as applying it to the shear strength.

In an EC7 design the partial factors applied to actions are different depending on whether the action is permanent, variable, or accidental and also whether it is favourable or unfavourable. Within standard slope stability calculations it is relatively simple to account for variable or accidental actions separately and thus apply appropriate partial factors as necessary. However, for permanent actions this is more difficult as, in effective stress slope stability analysis, both the actions (or their effects) and the resistances are a function of the weight density of the soil.

If a design approach requires that a partial factor greater than 1.0 is applied to unfavourable permanent actions or their effects then by definition it should only be applied to those actions which are unfavourable. Within a slope stability calculation this would be the overturning moment in a circular slip analysis, the downslope component of force in a translational slide analysis or the force parallel to the base of the slice for positive values of α (slope of base of slice) or the slope angle β for a translational slide. However, EC7 introduces the Single Source Principle (EN 1997-1 §2.4.2(9)P NOTE). This states that if both favourable and unfavourable actions are derived from the same source then they may be treated as the same when applying partial factors.

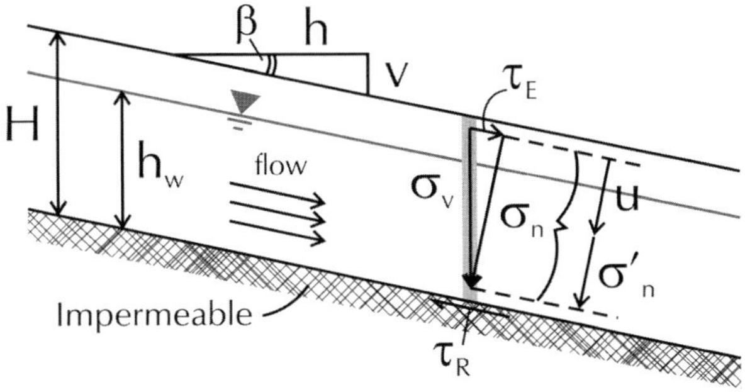

Fig. 4. Translational slide (after Bond & Harris 2008), for definition of symbols see Appendix: notation. H, depth of soil; h_w, depth of water; β, inclination of slope to horizontal; σ_v, total vertical stress; σ_n, total normal stress; σ'_n, effective vertical stress; τ_E, shear stress effect; τ_R, shear stress resistance.

Thus considering the translation slide shown in Figure 4 the traditional formula for factor of safety F is:

$$F = \frac{c' + (1 - r_u)\sigma_v \cos^2 \beta \tan \varphi}{\sigma_v \sin \beta \cos \beta}$$

where $r_u = (\gamma_w h_w/\sigma_v)$ and $\sigma_v = \gamma H$.

In terms of EC7's partial factors, the above formula can be re-written as:

design action

$$E_d = \tau_{Ed} = \sigma_{vd} \sin \beta \cos \beta = \gamma_G\ \sigma_{vk} \sin \beta \cos \beta,$$

design resistance

$$R_d = \tau_{Rd} = c'_d + (1 - r_u)\sigma_{vd} \cos^2 \beta \tan \phi_d$$

$$R_d = \tau_{Rd} = \frac{(c'_k/\gamma_c) + (1 - r_u)\gamma_G \sigma_{vk} \cos^2 \beta(\tan \phi_k/\gamma_\phi)}{\gamma_{Re}}$$

The above equations are valid if the single source principle is adopted, that is γ_G is the unfavourable value in both equations. However, if this principal is not used then the formulae for R_d should be re-written by replacing γ_G with $\gamma_{G;fav}$ in the equation, as follows (the equation for E_d remains the same):

$$R_d = \tau_{Rd} = \frac{(c'_k/\gamma_c) + (1 - r_u)\gamma_{G,fav} \sigma_{vk} \cos^2 \beta(\tan \phi_k/\gamma_\phi)}{\gamma_{Re}}$$

The partial factor on resistance, γ_{Re}, takes the value of 1.0 in Design Approaches 1 and 3 and so, since Design Approach 2 has not been adopted by most countries for carrying out slope design (and is not considered further in this paper), the formula for design resistance can be simplified to (where '|' denotes alternative factors):

$$R_d = \tau_{Rd} = \frac{c'_k}{\gamma_c} + (1 - r_u)[\gamma_G | \gamma_{G;fav}]\sigma_{vk} \cos^2 \beta \frac{\tan \phi_k}{\gamma_\phi}$$

In Design Approach 3 (DA3) partial factors on permanent actions are taken as 1.0 coupled with the same partial factors on material properties as for Design Approach 1 Combination 2 (DA1-2). Thus within this paper discussions will centre on Design Approach 1 (DA1) with any comments on DA1-2 being equally relevant to DA3.

The implications of the two ways of applying partial factors to actions in the above equations for the design action and the design resistance are that, if the single source principle is adopted, design approaches that factor unfavourable actions and not material properties are unsafe except where the resistance due to effective cohesion is large compared with that due to friction. Thus for DA1, Combination 1 (DA1-1) is unsafe for normal angles of friction but would be overruled by DA1-2 which factors material properties and not permanent actions. The resulting equivalent minimum factor of safety normal angles of friction would be 1.25, that is the same as the partial factors on effective stress shear parameters, γ_c and γ_ϕ or 1.4 if undrained strength is used.

However, if the single source principal is not invoked DA1-1 is critical for effective stress calculations with the resulting equivalent minimum factor of safety of 1.35, that is the same as the partial factor on unfavourable permanent actions, γ_G. These comments assume that there are no variable actions.

Bond & Harris (2008) discuss this fully and present design charts for both translational slides and circular failures.

Whether to adopt the single source principle or not is a matter of debate. In the case of a slope where both the actions and the resistances are a function of the weight density of the soil, it would seem illogical to adopt different partial factors for the components of force parallel and perpendicular to the sliding surface.

Identifying favourable and unfavourable components of the equations becomes even more difficult when analysing circular and non-circular slip surfaces, where it is not possible to satisfy all the equations of equilibrium without making assumptions. For example, consider the standard formulation of Bishop's Simplified Method (Bishop 1955) for circular slips:

$$F = \frac{\sum_0^n [\{c'_i b_i + (W_i + Q_i - u_i b_i) \tan \varphi'_i\}(\sec \alpha_i / 1 + (1/F) \tan \alpha_i \tan \varphi'_i)]}{\sum_0^n (W_i + Q_i) \sin \alpha_i}.$$

In this formula the terms are calculated for each slice (i) and the equation solved iteratively to find the factor of safety, F, which appears on both sides of the equation. Both the stabilizing effects (above the line) and the de-stabilizing effects (below the line) involve the same factors: W, Q and α. It is

Fig. 5. Design approach adopted for slopes in different countries (after Bond pers. comm. 2012); see Table 2 for explanation of design approaches.

therefore difficult to identify clearly whether the effect of these factors is favourable or unfavourable for a particular slice.

Existing computer programs are not designed to treat favourable and unfavourable actions differently and it is debatable whether they should as the actions come from the same source. It is therefore recommended that DA1-1 is not used for slope design. When applying the single source principle all permanent actions are treated as unfavourable and for this case DA1-2 governs for all reasonable values of the angle of friction. This approach has been adopted in the revision of BS6031 (2009).

Approaches to applying partial factors in Europe

A survey of the Nationally Determined Parameters (NDPs) that have been adopted by each National Standards Body in Europe has been carried out by Bond (pers. comm. 2010, 2012). A total of 24 of the 31 members of CEN responded to the survey and Figure 5 records the choice of design approach (1, 2, or 3) made by each country. A majority (17 countries) have allowed DA3 to be used for the design of slopes; 9 countries have allowed DA1; and 5 countries DA2 (although each of these three countries also allows one or more of the other DAs to be used). Since the application of DA1 and DA3 to slopes results in almost identical outcomes, it appears that Europe is effectively united on the way to design slopes.

However, the survey has highlighted an important variation between the countries' interpretation of Eurocode 7. Three countries (Austria, Denmark and the Netherlands) have varied the values of the partial factors γ_{φ}, γ_c, and γ_{cu} according to the risk involved in the design (Austria has linked the factors to classes of consequence and the other

Table 3. *Partial factors adopted for slope design across Europe (after Bond pers. comm. 2010)*

Partial factor	*Country*	*Single value*	*Class of consequence (CC) or reliability class (RC)*		
			CC1/RC1	*CC2/RC2*	*CC3/RC3*
γ_{φ}	EN 1997-1	1.25	Not used		
	Austria	Based on CC	1.1	1.15	1.3
	Denmark	Based on RC	1.2	1.2	1.32
	Netherlands	Based on RC	1.2	1.25	1.3
γ_{cu}	EN 1997-1	1.4	Not used		
	Austria	Based on CC	1.2	1.25	1.4
	Denmark	Based on RC	1.8	1.8	1.98
	Netherlands	Based on RC	1.5	1.75	2.0

two countries to reliability classes, both defined in EN 1990 and summarized in Table 3).

Is a partial factor of 1.25 sufficient?

There has long been a debate on what is an acceptable factor of safety for a slope. Table 4 summarizes some examples of the available published advice on selecting a suitable value for factor of safety.

The actual operating factor of safety is a function of several factors:

- the selection and variability of the soil shear strength along the sliding surface and within the body of the slide;
- the operating pore water pressures;
- the weight density of the soil and its variations; and
- the ability of the method of analysis to model the real behaviour of the slope.

In a traditional analysis the selection of a desired level of safety would be based on an assessment of the quality/conservatism of the data being input into the analysis coupled with an assessment of the risks to life and property. For designs to EC7, the flexibility of varying the required factor of safety seems to have been removed with a specified partial factor of 1.25 on effective stress parameters (equivalent to an overall factor of safety of 1.25) and 1.4 for total stress analysis. However, as discussed above, Bond's 2010 survey of the approaches being adopted by European Union countries suggests that several countries are adopting a more risk-based approach to partial factors.

For some designers these values appear to be low but they should be taken in the context of other requirements of EC7 in relation to the selection of characteristic values for parameters and design groundwater conditions coupled with the definition of an appropriate set of design situations. For soil properties EC7 requires that a cautious estimate of the value affecting the limit state should be adopted (EN 1997-1 §2.4.5.2(2)P). The standard also requires that conservative groundwater levels be adopted: '... design values shall represent the most unfavourable values that could occur during the design lifetime of the structure ...' (EN 1997-1 §2.4.6.1(6)P). This statement applies to limit states

Table 4. *Published values for factors of safety for slopes*

Condition	*Factor of safety*		*Source*
	Short term (undrained)	*Long term (drained)*	
Cutting first time slide		1.3–1.4	BS6031 (1983)
Cutting slide on pre-existing surface		1.2	BS6031 (1983)
End of construction		1.3	Whitlow (1995)
Steady seepage condition		1.25	Whitlow (1995)
After sudden drawdown		1.2	Whitlow (1995)
Natural slope of long standing		1.1–1.2	Whitlow (1995)
All slopes	1.5	1.5	Berry & Reid (1987); Das (2000); Coduto (1999)
Negligible risk to life or property		>1.0	G C O (1983)
Low risk		1.2	G C O (1983)
High risk		1.4	G C O (1983)
"my view is that calculated safety factors are meaningless (except for those that imply failure)"			Bromhead (1992)

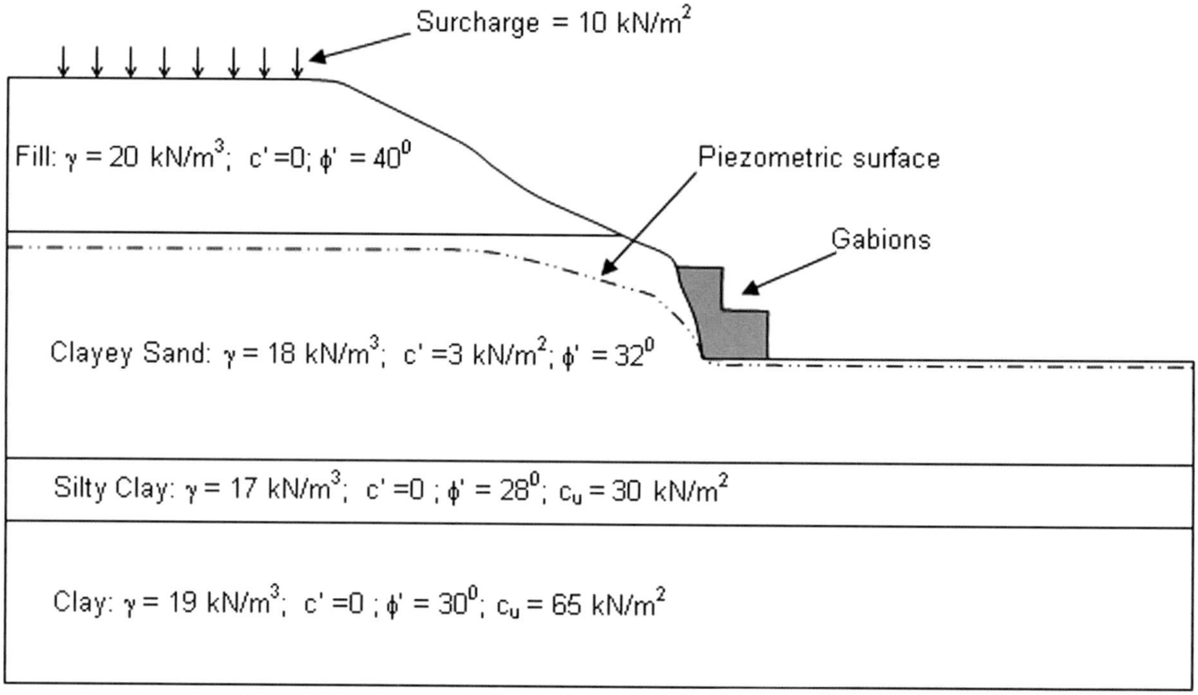

Fig. 6. Gabion wall remediation scheme, based on original section courtesy Beadman (2009), Byrne Looby & Partners.

with severe consequences, that is ultimate limit states. Thus provided the requirements in the selection of characteristic values for parameters and design groundwater levels/pore pressures are adopted it could be argued that an equivalent factor of safety of 1.25 is acceptable.

The design requirements and partial factors specified in EC7 relate to normal structures (Geotechnical Category 2). Where there are greater risks, for example seismic design, major civil engineering structures, unusual ground conditions, these requirements would need to be modified. Further, EN1990 allows the designer to modify the partial factors where risks are judged to be higher than for 'normal' structures (EN 1990 Annex B). Thus EC7 provides a level of flexibility and discretion to the designer to make sound engineering judgements to ensure that designs have an adequate level of reliability. Further, although deformations are not usually critical to the performance of embankments or cuttings, EC7 requires that serviceability limit state (SLS) conditions are satisfied and this may govern the design particularly for highly brittle soils.

Slopes comprising drained and undrained materials

EC7 provides different partial factors for undrained cohesion (c_u), effective cohesion (c') and internal friction (ϕ'). The partial factor for c_u is $\gamma_{cu} = 1.4$ and for c' is $\gamma_c = 1.25$. Similarly the partial factor for ϕ' is $\gamma_\phi = 1.25$ (the partial factor is applied to tan ϕ'). Thus where a slope comprising a layer or layers of a clay soil coupled with layers of granular materials is to be analysed in the short term an EC7 design to DA1-2 or DA3 would apply different partial factors to the strengths of the clay layers and to the granular layers. This may mean that the critical mode of failure is different for an EC7 analysis when compared with a traditional analysis.

Beadman (2009) presented an example for an embankment remediation scheme using a gabion retaining wall as shown in Figure 6 below. Traditional circular slip and non-circular slip analyses were carried out coupled with an EC7 analysis and a finite element study using PLAXIS. The critical slip surfaces are shown in Figure 7 with the factors of safety given in Table 5.

Beadman (2009) modelled the EC7 condition in PLAXIS by factoring (reducing) the undrained strength by γ_c/γ_{cu} that is dividing c_u by 1.12 and stipulating that the target factor of safety should be greater than 1.25. There is debate as to the best methods for applying EC7 principles within numerical modelling; this paper does not attempt to comment on this difficult area.

For an analysis to be satisfactory according to EC7 the design resistance must be greater than the design action which, for a limit equilibrium slope stability analysis, may be taken as satisfied if the overall or lumped factor of

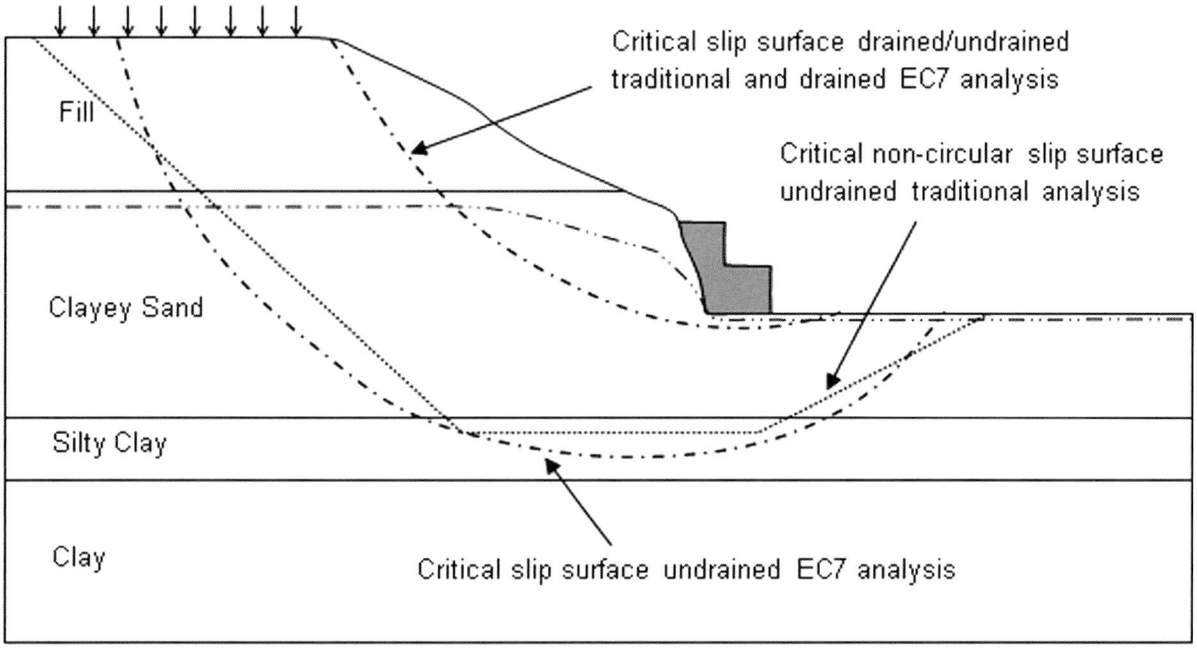

Fig. 7. Critical failure surfaces, original data courtesy Beadman (2009), Byrne Looby & Partners.

safety is greater than 1.0. Thus the EC7 limit equilibrium analyses suggest that the proposed gabion design is satisfactory with the undrained analysis being critical (lowest factor of safety). This analysis indicates a different (deeper) critical slip surface than that suggested by the traditional analysis which could potentially pose greater risks to any infrastructure running along the top of the embankment.

The finite element analysis suggests that the factor of safety in the undrained condition is lower than that derived from the standard limit equilibrium analysis, 1.27 cf. 1.35. For the EC7 case the finite element model suggests that the requirements of EC7 are not satisfied as the calculated factor of safety is less than 1.25. Unlike the limit equilibrium analysis the critical failure is still relatively shallow but there is evidence that a secondary deeper surface through the soft clay is developing.

The results of a limit equilibrium analysis of a granular embankment overlying a variable thickness of soft clay are presented in Figure 8. In the analysis, the embankment has been taken as 5 m high with the following parameters:

$$\gamma_k = 20\ \text{kN/m}^3;\ c' = 0;\ \phi' = 35^{\circ}.$$

Table 5. *Summary of stability analysis, courtesy Beadman (2009), Byrne Looby & Partners*

Program	*Description*	*FOS*
SLOPE W	Existing conditions	0.98
	Gabions drained – shallow	1.35
	Gabions undrained – shallow	1.35
	Gabions undrained – non-circular deep	1.40
	Gabions drained Eurocode 7 shallow	1.06*
	Gabions undrained Eurocode 7 deep	1.03*
PLAXIS	Gabions undrained shallow	1.27
	Gabions undrained EC7 shallow	1.22

*Factor of safety calculated using Design Approach 1 Combination 2.

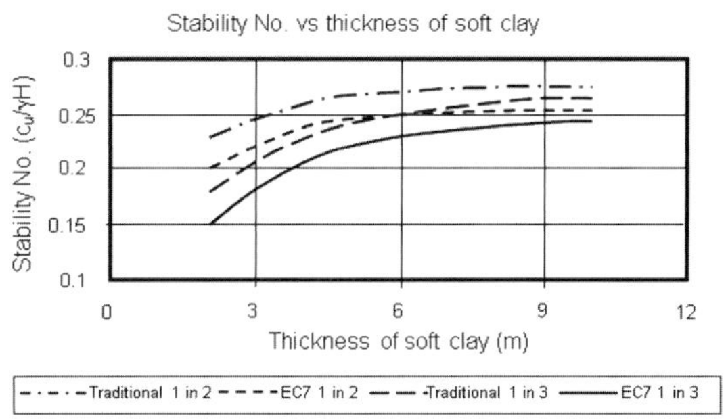

Fig. 8. Summary of stability analyses of granular embankment overlying soft clay.

The results show the required characteristic undrained strength stability number for the clay to provide a factor of safety of 1.5 for a traditional analysis and 1.0 for an EC7 analysis. For a 1-in-2 embankment slope the minimum factor of safety in a traditional analysis is for a shallow slip within the embankment material as $(\tan \phi')/(\tan \alpha)$) is 1.4 which is less than the required minimum of 1.5. However, for such slip surfaces a lower factor of safety would normally be acceptable so Figure 8 provides the results for a traditional analysis ignoring shallow failures in the embankment. For all analyses the critical surface is one that is tangential to the base of the soft clay up to the thickness of layer considered which is 10 m (2 × H).

For both a 1-in-3 and 1-in-2 embankment slope, EC7 provides a less conservative result than a more traditional analysis which might require a minimum factor of safety of 1.5.

For the 1-in-2 embankment a traditional analysis would suggest that the factor of safety is governed by the embankment material with the critical surfaces falling within the embankment. Failures of this type are relatively insignificant and it would be more appropriate to consider more deep seated failures. The EC7 analysis overcomes this problem by providing different partial factors for drained and undrained materials which leads to a deeper critical slip surface. In this regard the EC7 analysis is more useful as it identifies a failure mode which has greater consequences for the structure.

Conclusions

The introduction of EC7 has prompted further debate on the most appropriate methods of analysis to achieve a satisfactory level of reliability (safety) for embankments and slopes.

Current methods of analysis which include design approaches that factor unfavourable and favourable permanent actions differently will be difficult to apply. Design approaches that utilize partial factors on material properties are therefore recommended. Evidence from across Europe suggests that most countries are adopting this stance. By applying partial factors to material properties EC7 makes little practical change to analytical methods.

The UK National Annex has made no changes to the partial factors for slopes and there is some debate as to whether the current values provide an adequate level of reliability. Other European countries are adopting a more risk-based approach where the values of the partial factors vary depending on Consequence or Reliability Class. This would appear to be an approach that is worth further consideration within the UK geotechnical community.

Adopting different partial factors for drained and undrained parameters may lead to predicted failure modes which differ from those derived when using a single lumped factor of safety. The design situations where this occurs are limited but nevertheless may result in more significant volumes of soil involved and thus an increase in the potential consequences of the predicted failure mechanism.

Appendix: notation

Actions and effects

F_k = characteristic action
F_{rep} = representative action
F_d = design action
E_d = design effect of actions

Geometrical parameters

a_{nom} = nominal dimension
a_d = design dimension

Material properties and resistance

X_k = characteristic material property
X_d = design material property
R_d = design resistance

Combination and partial factors

γ_F = partial factor on an action (generic)
γ_G = partial factor on a permanent action (unfavourable)
$\gamma_{G,fav}$ = partial factor on a permanent action (favourable)
γ_Q = partial factor on a variable action
γ_E = partial factor on an action effect (generic)
Δa = safety margin on a dimension
γ_M = partial factor on a material property (generic)
γ_φ = partial factor on (coefficient of) shearing resistance of soil
γ_c = partial factor on effective cohesion of soil
γ_{cu} = partial factor on undrained strength of soil
γ_R = partial factor on resistance (generic)
γ_{Rv} = partial factor on vertical (bearing) resistance
γ_{Rh} = partial factor on horizontal resistance
γ_{Re} = partial factor on earth resistance
ψ = combination factors
ψ_i = combination factor on the *i*th action

General

GEO = ultimate limit state GEO as defined in BS EN1997-1
STR = ultimate limit state STR as defined in BS EN1997-1
SLOPE/W = limit equilibrium slope stability analysis program developed by GEO-SLOPE International, Canada
PLAXIS = finite element computer program for geotechnical analysis developed by Plaxis bv, Netherlands

References

BS 1377. 1990. *Methods of Test for Soils for Civil Engineering Purposes*. British Standards Institution, London.

BS 5930. 1999. *Code of Practice for Site Investigations*. British Standards Institution, London.

BS 6031. 1981. *Code of Practice for Earthworks*. British Standards Institution, London.

BS 6031. 1993. *Code of Practice for Earthworks*. British Standards Institution, London.

BS 6031. 2009. *Code of Practice for Earthworks*. British Standards Institution, London.

BS 8002. 1994. *Code of Practice for Earth Retaining Structures*. British Standards Institution, with Amendment 2 (dated May 2001).

BS 8004. 1986. *Code of Practice for Foundations*. British Standards Institution, London.

BS 8006-1. 2010. *Code of Practice for Strengthened/Reinforced Soils*. British Standards Institution, London.

BS 8081. 1989. *Code of Practice for Ground Anchorages*. British Standards Institution, London.

BS EN 1990. 2002. *Eurocode – Basis of Structural Design*. British Standards Institution, London.

BS EN 1997-1. 2004. *Eurocode 7 – Geotechnical Design, Part 1 – General rules*. British Standards Institution, London, 999.

BS EN 1997-2. 2006. *Eurocode 7 – Geotechnical Design, Part 2 – Ground Investigation and Testing*. British Standards Institution, London, 999.

Beadman, D. 2009. Case studies: Feedback on Designing to Eurocode 7. Slope Engineering 09, Ground Engineering, London, Nov 2009.

Berry, P. L. & Reid, D. 1987. *An Introduction to Soil Mechanics*. McGraw-Hill, Maidenhead, Berkshire.

Bishop, A. W. 1955. The use of the slip circle in the stability analysis of slopes. *Géotechnique*, **5**, 7–17.

Bond, A. J. & Harris, A. J. 2008. *Decoding Eurocode 7*. Taylor & Francis, Abingdon.

Bromhead, E. N. 1992. *The Stability of Slopes*. 2nd edn. Blackie Academic & Professional, Glasgow, 411.

Coduto, D. P. 1999. *Geotechnical Engineering, Principles and Practices*. Prentice-Hall, New Jersey.

Das, B. M. 2000. *Fundamentals of Geotechnical Engineering*. Brooks/Cole, Pacific Grove, USA.

GEOTECHNICAL CONTROL OFFICE. 1983. Geotechnical Manual for Slopes. Engineering Development Department, Hong Kong.

Whitlow, R. 1995. *Basic Soil Mechanics*. 3rd edn. Longman Scientific & Technical, Harlow, Essex.

Geotechnical risk and performance: with a focus on earthworks compaction and data management

John Perry

Mott MacDonald Ltd, Mott MacDonald House, 8–10 Sydenham Road, Croydon CR02EE, UK

Abstract: This paper discusses the various influences controlling risk and performance from new build to the in-service aspects of the transport infrastructure geotechnical asset of which earthworks are the major part. First it focuses on the risks associated with earthworks during construction. Compaction specifications, their origins and past supporting research results are described as are the advantages and disadvantages of the end-product. The method specification gives an insight into the development of earthworks design, construction and contracts. Recommendations are given based on modern practice and current contractual arrangements. Monitoring performance is critical to the management of geotechnical assets and the establishment of IT based methods in the UK is discussed to complete a 'cradle to grave' review of risk and performance aspects of earthworks.

Risk

The risk to earthworks during new-build has been well documented in conferences such as '*Failures in earthworks*' (Institution of Civil Engineers 1985) and the risks to performance when in use are comprehensively covered in the two CIRIA guides (Perry *et al.* 2003*a*, *b*).

The main influences on risk can be broadly classified at a high level as:

(1) engineering properties (Fig. 1) including climate (external) changes, ageing and unforeseeable events;
(2) contractual arrangements; and
(3) construction methodology including quality control.

In order to manage the risk of item 1, it is useful to consider the strategic approach advocated by Burland (1987) and illustrated in Figure 2. This triangle is useful in highlighting how geotechnical science and engineering still relies heavily on practical evidence and experience applied to data and theory. The need to show that geotechnical engineering reduces the negative impact on projects can be difficult to sell, when many other areas of engineering have visually enhancing positive aspects such as tall buildings and bridges.

Perhaps the biggest impact on engineering properties is the increased demand of earthworks with increased static and dynamic loadings and the changes in climate leading to different groundwater and pore water conditions. The profession is increasing its knowledge in these areas, for example '*Rehabilitation of urban railway embankments research, analysis and stabilisation*' (O'Brien 2007), and '*Influence of climate and vegetation on railway embankments*' (Scott *et al.* 2007) but there is still much work to do.

Contractual arrangements in the United Kingdom and the rest of the world are well known and now considerably diverse. The general policy is to try to place risk with the body most able to manage it. However, in the commercial world, the policy of most organizations is to carry as little risk as possible whether best able to manage it or not. Hence the risk and responsibilities are now well argued and need to be clear on commencement of the contract. If not implemented, the hazard and its probability may not be properly managed and the risk to the project will increase. This applies to earthworks as well as other things, and in earthworks the cost of a failure can be enormous. Modern geotechnical procedures, if properly followed with adequate funding and skills, allow this risk to be managed and provide a very positive contribution to the success of projects.

How an earthwork is built will be reflected in its longer term performance. Earthworks can be important structures with considerable volumes of natural (unmanufactured) materials. Therefore, sampling and testing to meet the design criteria are important areas for quality control.

Performance

The assurance of good performance in earthworks begins with construction and monitoring of construction methods. The major construction method that ensures good performance in service is compaction. A considerable amount of

From: Radford T. A. (ed.) 2012. *Earthworks in Europe*. Geological Society, London, Engineering Geology Special Publications, **26**, 21–31, http://dx.doi.org/10.1144/EGSP26.4

Fig. 1. Trafficability is one of the major risk assessments required during construction.

work has been undertaken in this field and some is described below. Monitoring performance in service is primarily controlled by reaction to managed data as a result of factors such as condition deterioration.

The results of the monitoring provide a measure of the 'resilience' of an earthwork and a network of earthworks.

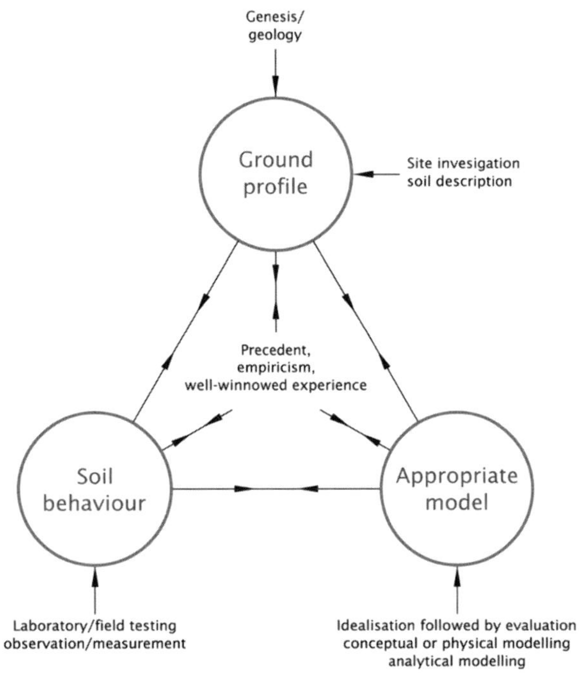

Fig. 2. Burland's triangle illustrating the main areas for geotechnical design risk management.

Monitoring new build with an emphasis on shallow compaction and performance of fills

Compaction is one of the main methods of ensuring performance of an earthwork. It increases shear strength and stiffness and reduces permeability. Achieving these improvements for a particular grading of soil and rock requires a reduction in air voids by compactive effort (mass per unit width/area and layer thickness) and suitable moisture content. There are two types of compaction specification: end-product, where the final density or air void content is measured; and method, where the method of construction is specified. Both have merits and disadvantages.

It is useful to look at the development of compaction as this relates to current and future performance. Until 1969 (with the publication of the Highways Specification for Road and Bridge Works (Ministry of Transport 1969)) the specification for highways was an end-product type, an approach which is again becoming popular today. The specification for bulk earthworks required 10% air voids to be achieved, with 5% air voids in the upper 600 mm, in 9 out of any 10 consecutive tests. This specification was in place for over twenty years and still remains in place in other countries today. However, in the UK this all changed as the research at the then Road Research Laboratory (RRL) showed problems with the end-product approach and funding was available to exploit the method approach. During the late 1950s and early 1960s RRL visited six construction sites (Table 1). All of these sites involved considerable earthworks.

In terms of measuring compaction, it is generally agreed that 'air voids' is a suitable method for monitoring

Table 1. *Sites and soil types (original terms) tested by RRL to see the degree the end-product compaction specification was achieved*

Site no.	*Site*	*Predominant soils tested*
1	Maidenhead by-pass (M4) (Berkshire section)	Gravel/hoggin/ stony clay
2	Maidenhead by-pass (M4) (Buckinghamshire section)	Hoggin
3	Slough by-pass (M4)	Hoggin/sandy clay/ silty clay
4	Oxford southern and western by-pass (A34)	Sandy gravel
5	Biggleswade by-pass (A1)	Fly ash Stony sand/sandy gravel Silty clay
6	South Mimms by-pass (A6)	Heavy clay (sometimes stony)

compaction for fine grained soils and overcomes some of the effects of variation in soils. It is not suitable for granular soils where relative compaction is preferred.

A statistical approach was adopted by RRL to see if the testing frequency of 9 out of 10 tests at 10% air voids or better was adequate at these six sites. The basic statistical data are given in Table 2. Table 3 below shows that only seven soil type/site combinations achieved compliance to 10% air voids in 9 or more out of 10 tests. Only on the Maidenhead by-pass (Buckinghamshire section) and South Mimms by-pass was full compliance of all soil types achieved. These actual results need to be compared with the statistical estimate of mean air contents required to achieve 9 out of 10 tests conformance (Table 4).

Very low mean air voids are required to achieve the specification and most of the sites and soil types failed to meet the specification. The number of tests recommended for an accurate determination of fill air content is shown in Table 5. Lewis & Parsons (1963) concluded that:

- A higher rate of testing than was being applied in practice was required to be statistically significant.
- Errors in the sand-replacement test in wet non-cohesive materials made compaction control difficult.
- The requirement that 9 out of 10 results should have an air content of 10% or less represented a very high state of compaction and an even higher state of compaction needed to be achieved in order to ensure compliance with the specification with a reasonable degree of confidence.

However, no failures due to inadequate compaction had been recorded in over 25 years (Parsons 1992). Parsons (1992)

Table 2. *Mean and standard deviation of soil air content at each site by soil type*

Soil type	*Site no.*	*No. of results*	*Air content (per cent)*	
			Mean	*Std dev*
Clean granular soils ≤ 10% passing 63 μm	1	67	5.4	±4.9
	2	51	2.3	±2.9
	4	178	5.5	±4.5
	5	23	1.8	±3.0
Granular soils with fines >10% passing 63 μm	1	88	6.4	±4.4
	2	92	1.5	±2.1
	3	24	6.1	±3.7
	5	75	3.8	±4.2
Intermediate and low plasticity clays; with or without stones	1	54	5.5	±3.2
	3	80	6.2	±3.7
	5	26	5.0	±3.6
	6	7	3.1	±2.7
High plasticity clays; with or without stones	5	9	6.1	±5.2
	6	85	2.9	±3.7
Uniformly graded materials	3	21	10.2	±4.9
	5	58	13.9	±8.9

confirms that the conclusions of the research of Lewis & Parsons (1963) explained above, led to the research, development and implementation of a method specification. The method specification was perceived to have the following advantages:

- there was less reliance on insufficiently accurate field testing methods;
- the rate of testing using conventional techniques was kept low and would therefore result in less interference of earthmoving operations; and
- it would lower costs.

The result of this initial research and the decision to move to a method specification lead to an extensive and thorough programme of research ultimately leading to Clause 609 and Table 6/4 in the Ministry of Transport *Specification for Highway Works* which was first published in 1969. The method specification has been used in the UK and in many other countries ever since, and on many forms of new build infrastructure. It is not unique to the UK; similar approaches have been adopted in Europe. An end-product approach was never excluded in the specification and both have been available since 1969. This approach is embraced in the *Design Manual for Roads and Bridges*, Advice Note HA 70 (Department of Transport 1994).

Times have changed. Does the method specification still stand up to today's construction industry culture? Certainly Eurocode 7 (1995) states that statistical approaches may be helpful, but warns a cautious approach. There is now a need to balance productivity, accuracy, assurance and variability of materials and plant. We have increased testing technology, a better understanding of testing and its accuracy, there is a drive toward end-product reflecting modern risk allocation and contracts and we have a quality assurance culture.

Both end-product and end-performance depend on measures of what is finally produced and this has been enforced by changes in risk responsibility and contract. There is now a structure where the contractor is responsible for the risk management and some clients no longer employ supervisors to ensure that the work is carried out to a standard. Many clients prefer to employ reviewers of the final product that is the end-product rather than monitor the construction in progress. This relies on compliance testing and is not designed to be statistically rigorous but to show that values are met.

Generally, earthworks are built to the method specification and then, because of the need to fulfil Quality Assurance requirements, compliance testing is carried out to provide evidence that the workmanship level has been achieved (an end-product approach). It is rare for a specific general fill material to require end-product only.

Currently, in the UK, the method specification need not be rigorously monitored and a compliance testing regime that has fewer tests than are statistically needed is used to prove an end-product. Is improvement required and what would be the effect on costs and programme? Would it be possible for the earthworks inspectors' to complete a

Table 3. *Proportion of test results with an air content of 10% or less**

Site	*Soil type*	*Number of tests*	*Percentage of results having an air content of 10% or less*
Maidenhead by-pass (Berkshire section)	Gravel	37	73
	Hoggin	88	77
	Hoggin and stony clay	31	87
	Stony clay	23	83
	All soils	188	78
Maidenhead by-pass (Buckinghamshire section)	Hoggin	126	97
	Hoggin and silty clay	16	100
	All soils	143	97
Slough by-pass	Hoggin	14	79
	Clayey hoggin	24	75
	Stony silty sand	21	33
	Silty sand	18	56
	Stony sandy clay	31	97
	Sandy clay	11	82
	All soils	134	71
	All well-graded soils	81	85
Oxford southern and western by-pass	Sandy gravel	156	81
	Sandy gravel and stony sandy clay	16	88
	All soils	176	81
Biggleswade by-pass	Sandy gravel	23	96
	Stony sand	75	92
	Stony silty clay	26	88
	Fly ash	28	0
	Fly ash and stony sand	28	64
	All soils	189	74
	All well-graded soils	133	91
South Mimms By-pass	Heavy clay	54	96
	Stony heavy clay	31	90
	All soils	89	94

*Only soils in which at least 10 tests were made included under individual soils types

reporting system as part of a QA system that confirms the method specification that is number of passes, depth of layer and type of plant and material? This is not popular as the concept of bands of inspectors on site is considered wasteful and it would be difficult to police. Contractors would prefer to work with soil tests that are familiar and considered to be much more accurate than they were 40 years ago (subject to correlations with trusted tests) and where a procedure can be easily demonstrated, a number produced and employs the contractors and designer's staff who are closely involved in the works. This process of testing is still vulnerable to the question of the frequency of testing. Lewis & Parsons

Table 4. *Estimate of mean air content required to achieve compliance with end product*

Soil type	*Mean air content (%) required to ensure compliance at 90% confidence level*	
	10 tests	*20 tests*
Clean granular soils ≤ 10% passing 63 μm	2.6	3.1
Granular soils with fines > 10% passing 63 μm	3.3	3.8
Intermediate and low plasticity clays; with or without stones	3.8	4.3
High plasticity clays; with or without stones	2.5	3.0
Uniformly graded materials	10% air void criterion not applicable	

Table 5. *Accuracies with which the mean air content of each successive 1000 cu.yards of fill could be determined for varying numbers of tests*

Site		*Actual number of tests*	*Air content per cent*		*Estimated number of tests for limits of error, (90% confidence), of –*	
			Mean	*Standard deviation*	*±1.0%*	*±2.0%*
Maidenhead by-pass (Berkshire section)	1st 1000 cu.yards	26	6.9	±3.9	43	12
	2nd 1000 cu.yards	39	6.7	±3.7	39	11
	3rd 1000 cu.yards	36	11.4	±4.4	54	15
	4th 1000 cu.yards	91	5.8	±3.5	35	10
Maidenhead by-pass (Buckinghamshire section)	1st 1000 cu.yards	41	1.4	±2.3	16	5
	2nd 1000 cu.yards	26	2.4	±2.7	22	7
	3rd 1000 cu.yards	16	3.7	±3.5	35	10
	4th 1000 cu.yards	19	1.8	±2.4	17	6
	5th 1000 cu.yards	17	2.7	±3.7	39	11
Slough by-pass	1st 1000 cu.yards	32	6.8	±3.3	32	9
	2nd 1000 cu.yards	39	7.0	±2.9	25	7
Oxford southern and western by-pass	1st 1000 cu.yards	54	7.6	±3.6	37	11
	2nd 1000 cu.yards	44	6.8	±5.1	73	19
	3rd 1000 cu.yards	38	5.9	±5.0	70	19
Biggleswade by-pass–/–	1st 1000 cu.yards	31	3.3	±4.0	45	13
	2nd 1000 cu.yards	31	4.6	±4.0	45	13
	3rd 1000 cu.yards	48	6.5	±4.4	54	15
South Mimms by-pass	1st 1000 cu.yards	14	2.7	±2.5	19	6
	2nd 1000 cu.yards	24	5.5	±5.4	81	22
	3rd 1000 cu.yards	35	2.1	±2.4	17	6

(1963) showed scientifically that the number of tests required was impractical if they are to be statistically rigorous.

Taking the A3 at Hindhead as a typical example, the first earthworks season in 2007 relied totally on method specification and few dry density tests were undertaken. The second season employed an 'end-product' approach as the contractor felt that this gave greater flexibility in the material he could use. In the 2008 season 195 000 m^3 was

Table 6. *Typical modern quantity of testing to achieve an end-product specification*

Total fill placed during 2008: 195 000 m^3		
Material class		
Class 1A	No. of dry density determinations	313
	Average degree of compaction	100.60%
Class 1B	No. of dry density determinations	346
	Average degree of compaction	99.80%
Class 2A1 (departure obtained for non-plastic Class 2A changing method compaction to be same as Class 1A)	No. of dry density determinations	24
	Average degree of compaction	98.20%
Class 2C	No. of dry density determinations	173
	Average degree of compaction	98.10%
Class 6I/6J	No. of dry density determinations	89
	Average degree of compaction	97.80%
Class 7J (departure obtained for non-plastic Class 7C/7D materials to simplify classification and on-site handling)	No. of dry density determinations	27
	Average degree of compaction	98.10%
Total dry density determinations	972	
Tests per 1000 m^3	4.98	

Table 7. *Sample sizes from statistical approach on A3 Hindhead*

Site	*Material*	*Actual number of tests per 1000 m^3*	*Air voids percent (Target 10%)*		*Estimated number of tests for limits of error (90% confidence) of:*	
			Mean	*Standard deviation*	*±1%*	*±2%*
A3 Hindhead	Class 1	4.7	7.57	3.05	25	6
	Class 2	4.5	5.97	2.34	15	4

Site	*Material*	*Actual number of tests per 1000 m^3*	*Percentage of maximum dry density (2.5 kg method) (Target 90%)*		*Estimated number of tests for limits of error (90% confidence) of:*	
			Mean	*Standard deviation*	*±1%*	*±2%*
A3 Hindhead	Class 1	4.7	100.14	3.08	26	6
	Class 2	4.5	97.73	2.25	14	3

emplaced and an average of 5 tests undertaken per 1000 m^3 (see Table 6). This is in agreement with the estimate made by Lewis & Parsons (1963) that 5–10 tests is a practical number. However, the statistical approach (Table 7) requires considerably more for 1% limit of error.

This is not a unique situation. The amount of testing required for statistical analysis is still considered to be impractical and has not been adopted by industry when using an end-product specification; the use of method specification is preferred.

In general, the end-product approach is used in circumstances where:

- there is a growing culture of end-performance by clients as it 'measures' what you get at the end of the works;
- QA requirements can be evidenced more easily;
- the method specification is considered to be too restrictive on the use of materials; and
- *in-situ* testing is quick and, with correlations, reasonably accurate.

It's a question of pros and cons; either method may be appropriate depending on the site and the contractor's requirements and methodology. The overriding requirement especially in varying materials is that the method chosen can allow materials to remain in place if outside the design moisture band; the approach needs to give an almost immediate answer and show adequate compaction. For more consistent clay materials a method approach may be more appropriate with Moisture Condition Value (MCV) used but even in these conditions the contractor may prefer to put in more compactive effort (i.e. compaction plant passes, weight and thinner layers) drier of the design dry moisture content limit. For wet fills, trafficability (i.e. the ability of construction plant to move over the fill material) in low height fills is an obvious observed criterion but for high fills then minimum air voids are no longer a good parameter for compaction and shear strength, and hence relative compaction should be adopted.

Good compaction does not completely ensure good performance at and near the top of embankments. In particular cohesive fills at formation level need to be assessed for long term degradation which involves predicting stiffness based on plasticity. This should not be tested *in situ* but relies on research results relating Plasticity Index (PI) to predicted/equilibrium California Bearing Ratio (CBR). It is very common to accept CBR tests at the time of construction but this does not ensure performance as degradation can occur. Capping layers designed to ensure performance should be placed to the thickness given in the design.

Figure 3 outlines a procedure which has satisfied the quality requirements on a number of schemes. It highlights the flexibility that can be achieved with both method and end-product specifications, the flexibility being governed by the amount of acceptable material and risk management. By adopting such an approach the method specification is taken as the default, because of its advantage of speed of construction, and the end-product used to include further materials where they are scarcer and the economic model changes as loss of time may be less cost than import of materials.

To estimate the amount of testing, the advice in the *Design Manual for Roads and Bridges*, HA70 (Department of Transport 1994) is applicable using Table 8. A limit of error appropriate to the works is required to use this table. Given past performance, 1% limit of error is too high and 2% is too low. The exact value of limit of error needs to be decided by those who are contractually required to do so and who are best placed to know the ground conditions and risks. This provides a rationale which meets both

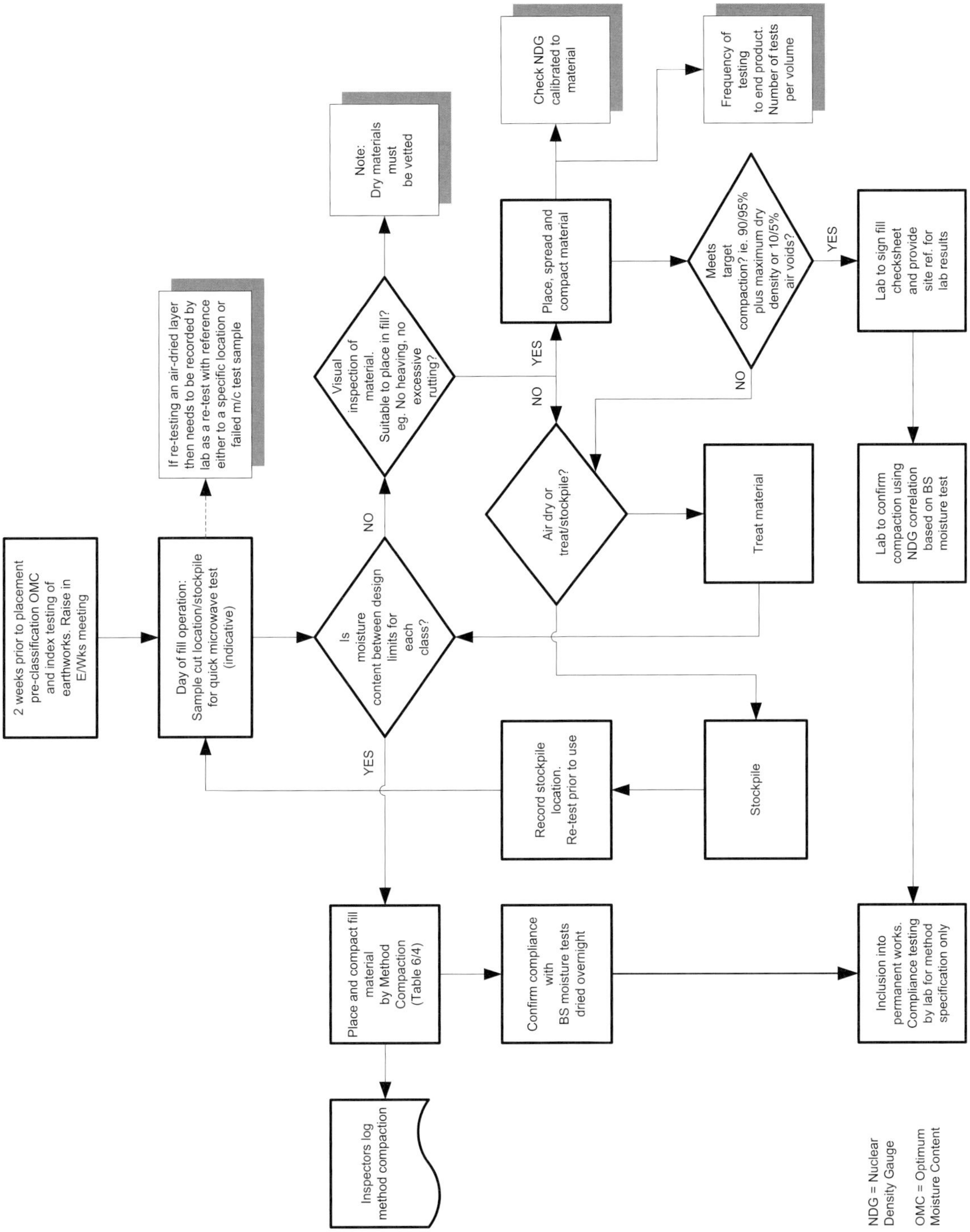

Fig. 3. A modern materials handling procedure.

Table 8. *Numbers of tests required to produce results having mean values of air voids within various limits of error with a probability of 90% (from HA70/94)*

Value of standard deviation of air voids %	*Number of test for limit of error*	
	± 1% air voids	*± 2% air voids*
±1	5	3
±1.5	8	4
±2	13	5
±2.5	19	7
±3	26	8
±3.5	35	10
±4	45	13
±4.5	57	16
±5	70	19
±5.5	84	23
±6	99	27
±6.5	116	31
±7	135	35

method and end product objectives, in a rigorous manner and, most importantly, is practical and cost effective.

Monitoring existing earthworks for performance

Monitoring earthworks as part of asset management, a discipline that has developed considerably over the last 5 years, mainly takes the form of measurements of condition, measurements of particular engineering parameters (such as movement) and prediction of future performance. The overriding outputs are efficient use of expenditure, customer focus, health and safety and cost. These are some of the information systems that are used to ensure performance. They are the core of performance monitoring. They provide sufficient data to make decisions, for example for whole life costing and resilience.

One of the main methods of geotechnical data management (to facilitate asset management and hence monitor and predict performance) is through GIS internet-based

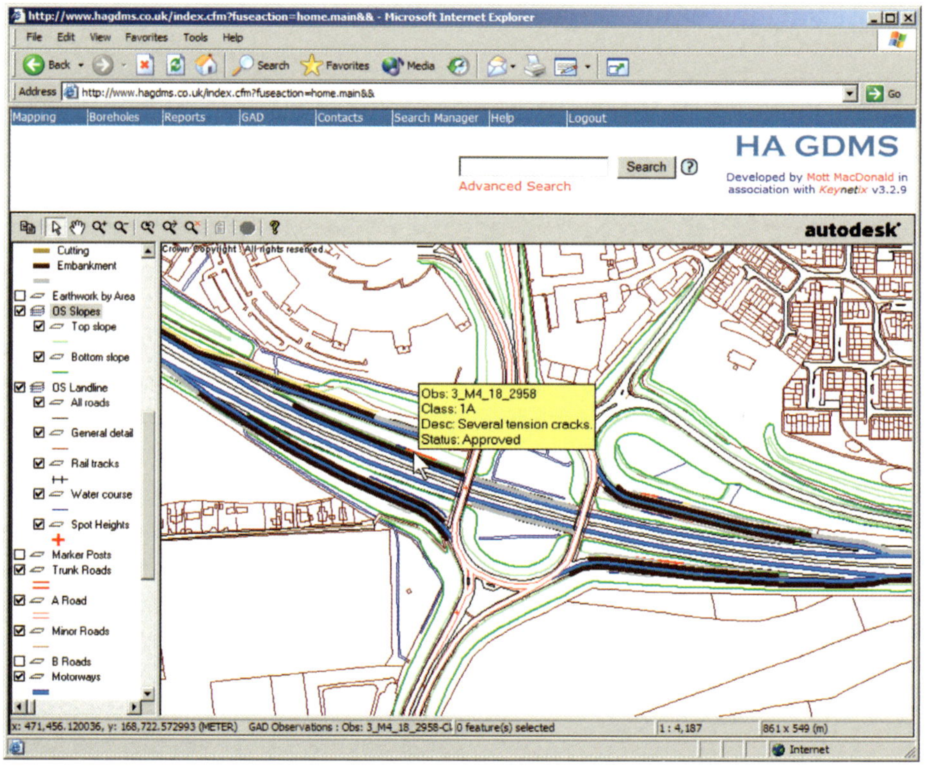

Fig. 4. Screen shot from HAGDMS showing general layout and some of the layers possible.

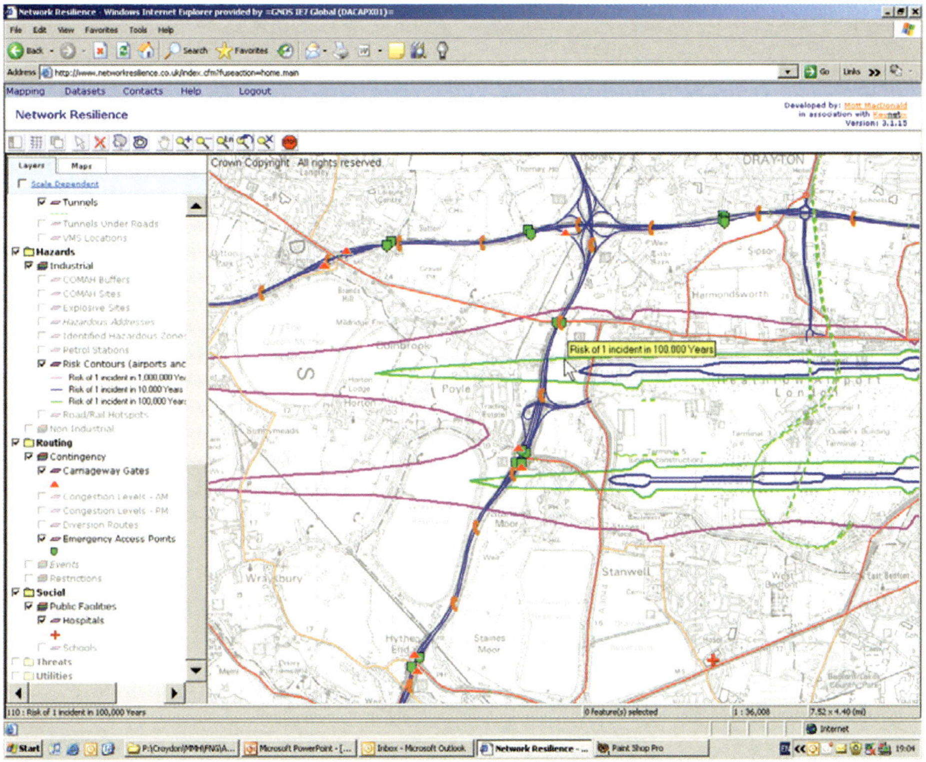

Fig. 5. Screen shot from HAGDMS showing overlay of effect of aircraft overrun on highway network.

systems where data are entered and managed on a regional or national level or at a scale sufficient to allow decisions to be made across a network. Such a system has been provided by the Highways Agency (HA) for its geotechnical assets. The Highways Agency Geotechnical Data Management System (HAGDMS) has been fully operational since July 2002 and provides the following:

- access over the internet via a standard web browser;
- a national database of all highways-related geotechnical reports and boreholes held by the HA and its agents;
- a repository for long-term electronic records complying with HA document management and Public Records Office requirements;
- a national database of all geotechnical inspection and maintenance records;
- GPS enabled PocketPC software for collection and transfer of earthwork inspection data, including support for Design, Build, Finance and Operate (DBFO) roads;
- maps showing identified geotechnical hazards to allow proactive and strategic asset management.

Since such data have been captured over a significant period and entered on a system that allows easy interrogation, it is possible to record where the earthworks and their problems are, and to overlay other information (Fig. 4). As data from different assets become available, performance can be influenced across disciplines. For example, highway earthworks performance in carrying traffic can be affected by hazards such as fires, plane overruns (Fig. 5) and flooding. The effects of these hazards can be estimated and overlain on the highway network to show the effect on earthworks and traffic. This allows contingency planning to ensure safe and efficient passage.

The Pitt Review (Pitt 2008) has emphasized the need for flood contingency planning to minimize disruption to critical infrastructure as a result of extreme rainfall events. Predictions of flooding levels are shown in Figure 6 with the sections of the M4 indicated that would be closed by flooding. The diversion route through Slough shows two localized areas of flooding that would close it, and helps to identify pinch points that need priority contingency plans or engineering solutions.

Since data are regularly stored on such systems, it becomes possible to interrogate the data and look for trends. This is an extremely powerful tool and leads to

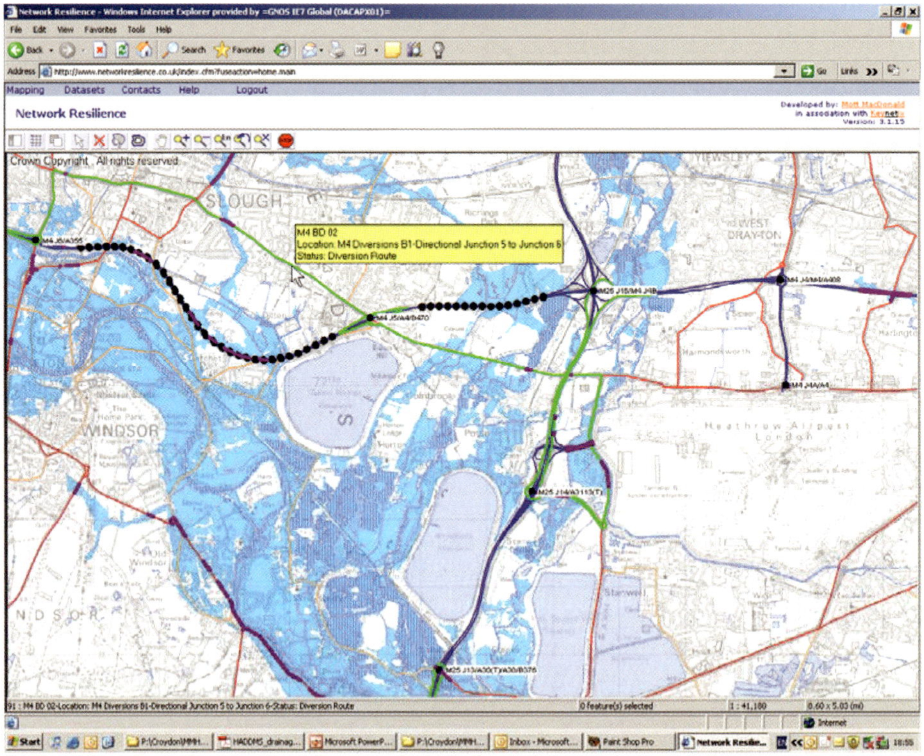

Fig. 6. Screen shot from HAGDMS showing effects of flooding around the M4/M25 motorway junction related to earthworks.

prediction. Such management of the geotechnical asset was not possible until these systems became available and because these systems have been engineering led, decisions on the engineering assets can be made more efficiently. Figures 7, 8 and 9 show typical examples of simple presentations which inform and provide evidence of performance and justification for funding. This type of strategic information is essential in showing the scale of a problem, identifying where action needs to be taken and assists in decision making of asset managers. With an established

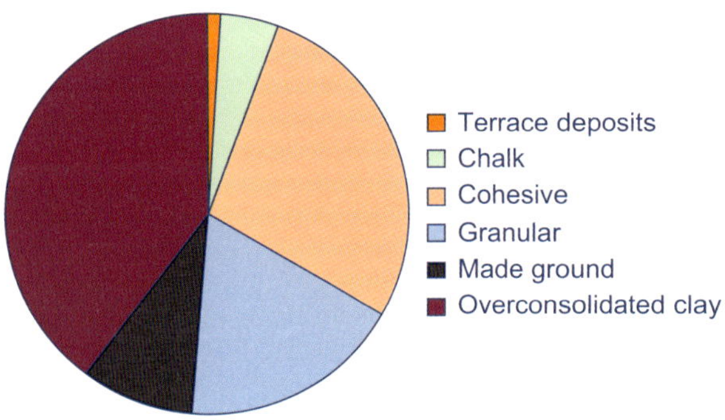

Fig. 7. Proportion of failed embankment slopes in different deposits.

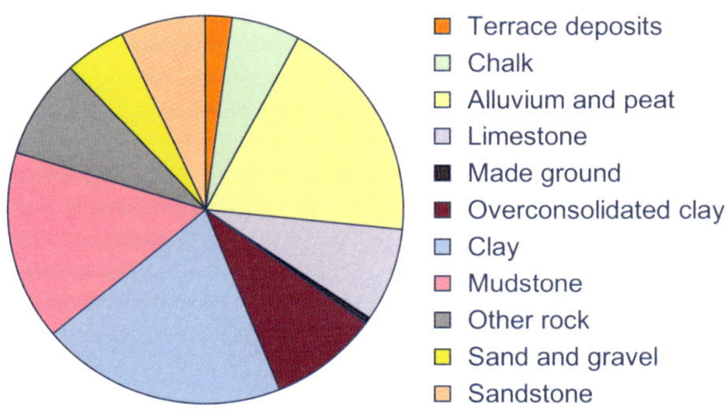

Fig. 8. Proportion of failed cutting slopes in different deposits.

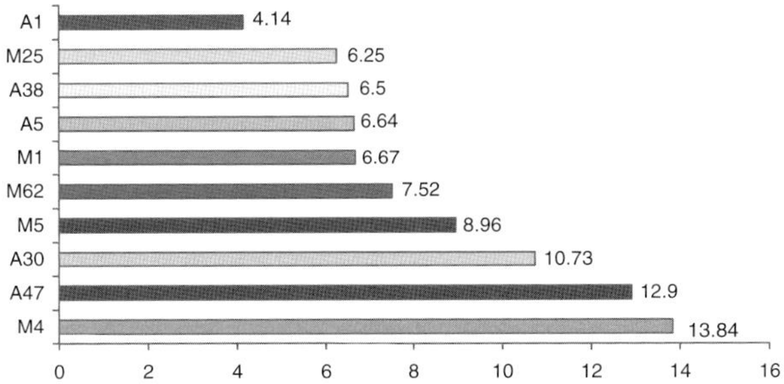

Fig. 9. The highest lengths in kilometres of failed slopes on different highways.

system this information is readily available, but without it funds can be difficult to justify.

Conclusions

The following conclusions can be drawn from this paper.

- Performance is governed by construction methods and quality as well as ageing factors.
- The rigorous statistical approach can be adopted for compaction provided a reasonable level of Limit of Error is agreed, and forms an auditable method of sizing the amount of sampling and testing required.
- The method specification and end-product specification can work together to provide an assured approach to earthworks compaction.
- Systems are an essential part of monitoring performances should be engineering led and adopted more widely.

The author would like to thank the Highways Agency and Balfour Beatty Major Projects for their permission to publish this paper. The paper presents the author's views which are not necessarily those of Mott MacDonald Ltd. The assistance of many colleagues at Mott MacDonald is much appreciated, especially that of Anthony Drake who assisted collecting much of the data presented in this paper.

References

BURLAND, J. B. 1987. The teaching of soil mechanics: a personal view. *Proceedings of the 9th European Conference on Soil Mechanics and Foundation Engineering*, 31 August–2 September 1987, Dublin. A. A. Balkema, Rotterdam, **3**, 1427–1441.

DEPARTMENT OF TRANSPORT. 1994. *Design Manual for Roads and Bridges*. Volume 4 Geotechnics and Drainage, Section 1 Earthworks, Part 5 HA 70/94 'Construction of Highway Earthworks'. Highways Agency. http://www.dft.gov.uk/ha/standards/dmrb/vol4/section1/ha7094.pdf

EUROCODE 7. 1995. *Geotechnical Design – Part 1: general rules, as contained in the UK National Application Document DD ENV 1997-1*. London, BSI. http://www.eurocodes.co.uk/PartDetail.aspx?EurocodePartID=46

INSTITUTION OF CIVIL ENGINEERS. 1985. Failure in earthworks. *Proceedings of the Symposium on Failures in Earthworks*, 6–7 March 1985, Institution of Civil Engineers, London. Thomas Telford.

LEWIS, W. A. & PARSONS, A. W. 1963. *An analysis of the status of compaction measured during the construction of embankments on six major road schemes*. Laboratory Note LN/399/WAL.AWP. Road Research Laboratory.

MINISTRY OF TRANSPORT. 1969. *Specification for Road and Bridge Works*. 4th edn. HMSO, London.

O'BRIEN, A. 2007. Rehabilitation of Urban Railway Embankments: Research, Analysis and Stabilisation. *In*: CUÉLLAR, V., DAPENA, E. ET AL. (eds) Geotechnical Engineering in Urban Environments. *Proceedings of the 14th European Conference on Soil Mechanics and Geotechnical Engineering*. 24–27 September 2001, Madrid, Spain. Mill Press, Rotterdam, **1**, 125–143.

PARSONS, A. W. 1992. *Compaction of soils and granular materials: a review of research performed at the Transport Research Laboratory*. HMSO, London.

PERRY, J., PEDLEY, M. & BRADY, K. 2003*a*. *Infrastructure Cuttings: Condition, Appraisal and Remedial Measures*. CIRIA Report C591, Construction Industry Research and Information Association, London.

PERRY, J., PEDLEY, M. & REID, M. 2003*b*. *Infrastructure Embankments: Condition Appraisal and Remedial Treatment*, 2nd edn. CIRIA Report C592, Construction Industry Research and Information Association, London.

PITT, M. 2008. *The Pitt Review: Lessons learned from the 2007 floods*. Cabinet Office, London.

SCOTT, J. M., LOVERIDGE, F. & O'BRIEN, A. 2007. Influence of climate and vegetation on railway embankments. *In*: CUÉLLAR, V., DAPENA, E. ET AL. (eds) Geotechnical Engineering in Urban Environments. *Proceedings of the 14th European Conference on Soil Mechanics and Geotechnical Engineering*, 24–27 September 2001, Madrid. Mill Press, Rotterdam, 659–664.

Geotechnical asset management for the UK Highways Agency

C. M. Power[1]*, D. A. Patterson[2], D. M. Rudrum[3] & D. J. Wright[4]

[1] Mott MacDonald, 111 St Marys Road, Sheffield S2 4AP, UK
[2] Highways Agency, 2/15E Temple Quay House, Bristol BS1 6HA, UK
[3] Arup, 13 Fitzroy Street, London W1 T 4BQ, UK
[4] Atkins, 500 Aztec West, Almondsbury, Bristol BS32 4RZ

**Corresponding author (e-mail: christopher.power@mottmac.com)*

Abstract: This paper outlines the Geotechnical Asset Management Strategy of the Highways Agency (HA). The strategy operates in a series of key stages, from setting standards and advice, through data collection and analysis through to the ultimate aim of providing and maintaining an asset that meets the service level that it is required to attain. These key stages are outlined, and illustrated with examples where appropriate. The HA, and its managing agents, have progressed a significant way along the 'roadmap', but further work is still ongoing. A key tool for the delivery of the HA Geotechnical Asset Management Strategy is the Geotechnical Data Management System (HA GDMS). This is discussed throughout the paper, and examples of the functionality of the system are given.

The Highways Agency (HA) is responsible for the operation, maintenance and improvement of the 7050 km of motorways and trunk roads (major 'A' roads) in England, estimated to be valued at over £85 billion (Highways Agency 2009). The key aims of the HA are to provide 'Safe roads, reliable journeys, informed travellers'. Operation and maintenance of the HA road network is undertaken by a series of 13 managing agents (MA), each responsible for defined geographical areas, and undertaking a wide range of tasks on behalf of the HA.

The HA has a diverse range of assets that requires management and maintenance; one of the key assets being the geotechnical asset, which comprises both earthworks (cuttings, embankments and acoustic/aesthetic bunds) and sections considered to be at-grade. These assets vary considerably in age (refer to Fig. 1), geological conditions, design and construction method. Following peaks in construction of the HA network in the 1970s and 1990s, the road infrastructure is largely complete, and the key activities now undertaken by the HA are centred on network improvement, optimization and maintenance.

As a key part of the HA network, the geotechnical asset presents a considerable maintenance challenge for the HA and its managing agents, which is addressed through the strategic management methodology outlined in this paper.

Asset management strategy

Development of the geotechnical asset management strategy of the HA has been undertaken in accordance with a development 'roadmap' (see Fig. 2). This roadmap sets the framework within which a series of research and development, consultancy and implementation tasks have been undertaken. The aim is to move from an early position of being able to produce specific outputs (e.g. inventory information about the geotechnical assets) through to achieving desired outcomes for the HA business; the key outcome being the ability to provide an asset that delivers the required service level for the road user.

Standards and advice

The geotechnical asset management strategy of the HA is underpinned by a series of documents that form part of the *Design Manual for Roads and Bridges*. The key document is the HD41/03 standard, *Maintenance of Highways Geotechnical Assets* (Highways Agency 2003). This document outlines the key aspects of the asset management strategy, and was released to coincide with the national implementation of the asset management system described in the next section. In particular, HD41/03 deals with:

- required competencies for key geotechnical staff contributing to the maintenance process (the key role being the managing agent's Geotechnical Liaison Engineer);
- asset inspection methodologies and frequencies (the key being the detailed Principal Inspections);
- the risk assessment framework for asset condition; and
- workflow and certification procedures for geotechnical maintenance works.

From: Radford T. A. (ed.) 2012. *Earthworks in Europe*. Geological Society, London, Engineering Geology Special Publications, **26**, 33–39, http://dx.doi.org/10.1144/EGSP26.5

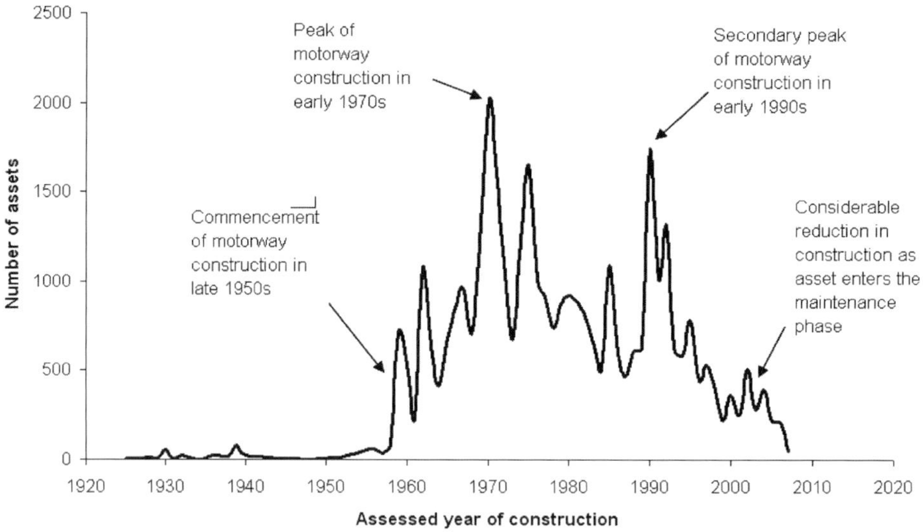

Fig. 1. Assessed age of the geotechnical assets of the HA road network.

The other applicable standard is HD22/08, *Managing Geotechnical Risk* (Highways Agency 2008). This document gives a more generic framework for the management of geotechnical risk on all applicable HA geotechnical projects, which includes major projects outside of the maintenance arena. It also outlines reporting and certification procedures

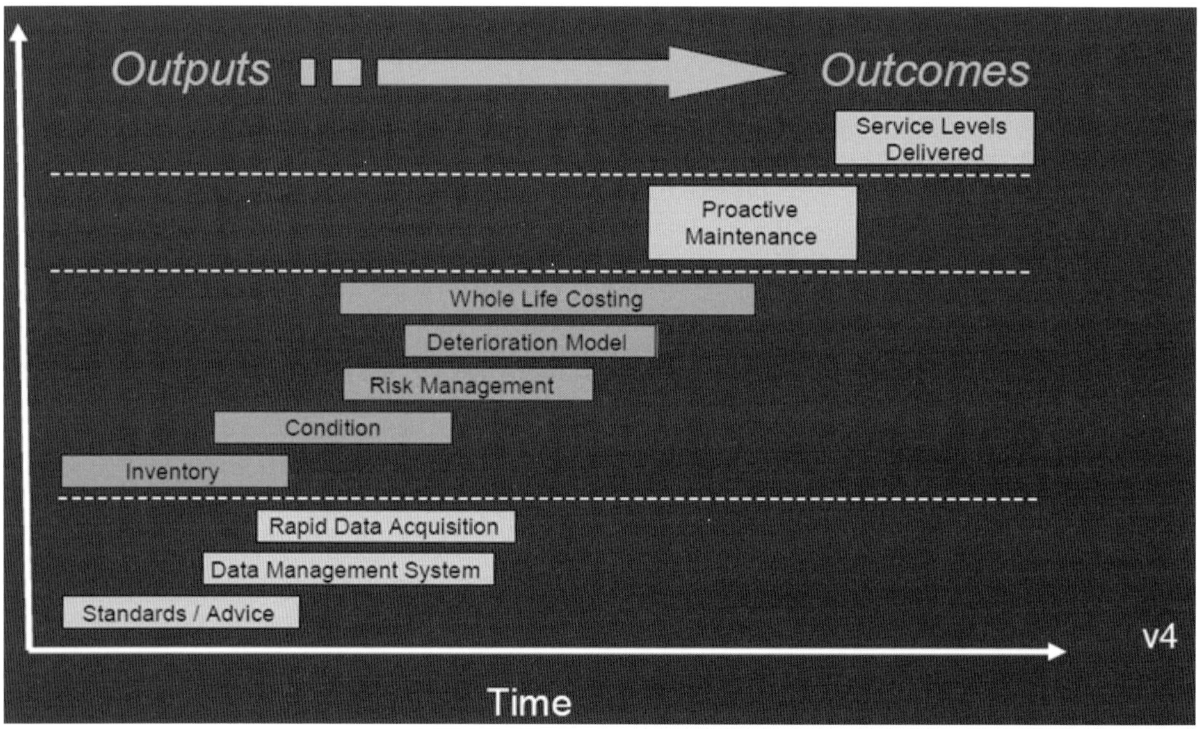

Fig. 2. Geotechnical asset management strategic roadmap.

for such projects, and as such imposes a minimum required level of quality assurance on reports submitted to the HA.

In addition to the formal documentation contained within the *Design Manual for Roads and Bridges*, more *ad hoc* advice and training has also been provided to those involved with geotechnical asset management for the HA. Most of this advice has been in the form of guidance documentation relating to survey methodologies and to the use of the asset management system (in the form of user guides). A series of formal office and field-based training courses has also been provided over many years to support the asset management processes. A support team has been in place since 2003 to provide advice and training to managing agents, and others taking part in maintenance or wider design activities.

The HA Geotechnical Data Management System (HA GDMS)

Key to progressing the geotechnical asset management strategy of the HA has been the development and use of the HA Geotechnical Data Management System (HA GDMS), an online Geographical Information System (see Fig. 3).

HA GDMS was rolled out across England in 2002. It is an easy-to-use, centralized system that can be accessed by all those requiring information on the HA geotechnical asset, whether from within the HA, as a managing agent, design agent or other party to a geotechnical project. HA GDMS comprises a mapping interface with a series of mapping layers, combined with databases of key geotechnical information, including:

- A contacts database of geotechnical staff (also the user details database). This acts as a register of authorized system users, and of key organizations that contribute to the management of the geotechnical asset. It also records the rights that users have to functionality within the system, such as the ability to approve data or provide technical approval for remediation works. There are approximately 1000 registered users of the HA GDMS, from within the HA and the organizations working for (or in collaboration with) the HA.
- A technical archive of over 13 000 geotechnical reports, including nearly 9000 scanned copies of the reports available to view or download. These reports are stored with key descriptive metadata, allowing intuitive searching, and are linked to the map to allow discovery of reports that are applicable to certain roads or sections of roads.
- An archive of exploratory hole data held by the HA, and the ability to upload scanned exploratory hole logs or Association of Geotechnical and Geoenvironmental Specialists (AGS) format electronic ground investigation data.
- Links to the borehole archive of the British Geological Survey, 1:625 000 scale geological solid and drift maps and a reference mapping layer of BGS district geologists.
- The HA Geotechnical Asset Database (GAD), containing detailed inventory and condition information for nearly 45 000 individual geotechnical assets and over 250 000 observations.

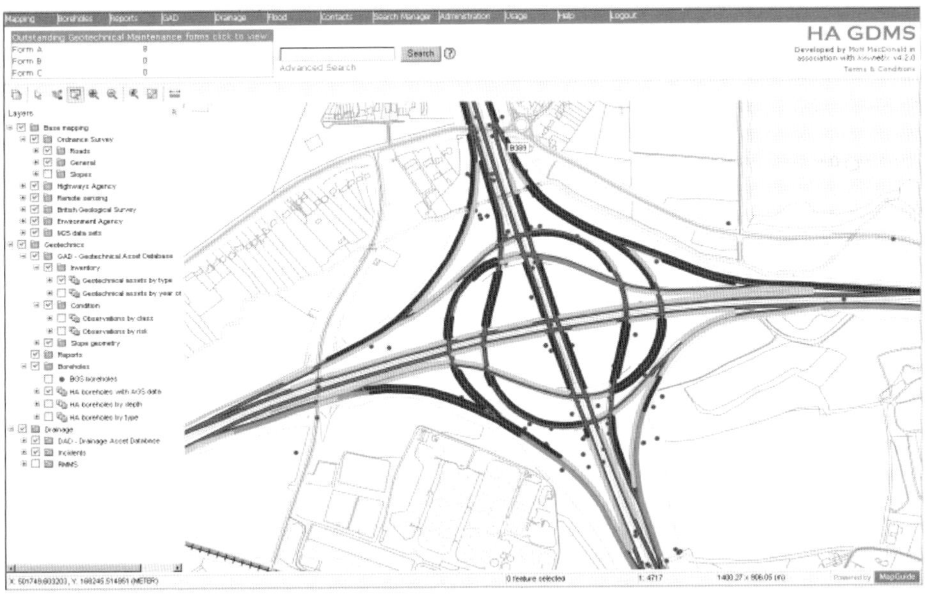

Fig. 3. Screenshot of the HA Geotechnical Data Management System.

- Links to the HA Drainage Data Management System (HA DDMS), the sister system that provides the complete asset management system for the HA drainage asset.

The HA GDMS Geotechnical Asset Database has associated field data capture software (PocketGAD) that operates on small hand-held computers, with links to real time GPS positioning. This software allows surveying data to be captured in the field, with in-built error capture logic, which can then be uploaded to HA GDMS for instant population of the Geotechnical Asset Database. PocketGAD includes detailed mapping and aerial photography, which can be viewed in the field, allowing instant verification of the asset information being collected, or being verified and updated by re-survey. Visualization tools (GADGET, the Geotechnical Asset Database Graphical Earthwork Tool), available in both PocketGAD and HA GDMS (see Fig. 4), allow the assets to be visualized for ease of understanding.

Since its introduction in 2002, principal inspection surveying by managing agents ensured that the HA GDMS inventory of assets, with associated condition information, was complete by 2008. Full details of a total of over 10 000 km linear length of assets are held by the GAD, and ongoing surveying ensures that the condition information remains relevant and accurate. Data quality is maximized as far as possible through use of online tools (such as QuASaR, the Quality Assurance: Surveys and Reporting tool) and an approval process that allows self-certification of data by managing agents.

Critically, the HA GDMS has been developed by an engineering team, with experience as HA managing agents, rather than being driven by the information technology that facilitates the tool. Throughout its development, user feedback, through the use of 'wish lists' has ensured that enhancements to the system have met the key needs of the HA business and the operational needs of the managing agents that make up most of its users.

Inventory and condition data acquisition

On an annual basis, the managing agents are required by the HD41/03 standard to submit a Geotechnical Asset Management Plan (GAMP), which is the starting point of the maintenance cycle. The GAMP includes a summary of work undertaken in the previous year, and an outline of the planned maintenance work to be carried out in the coming 5-year period.

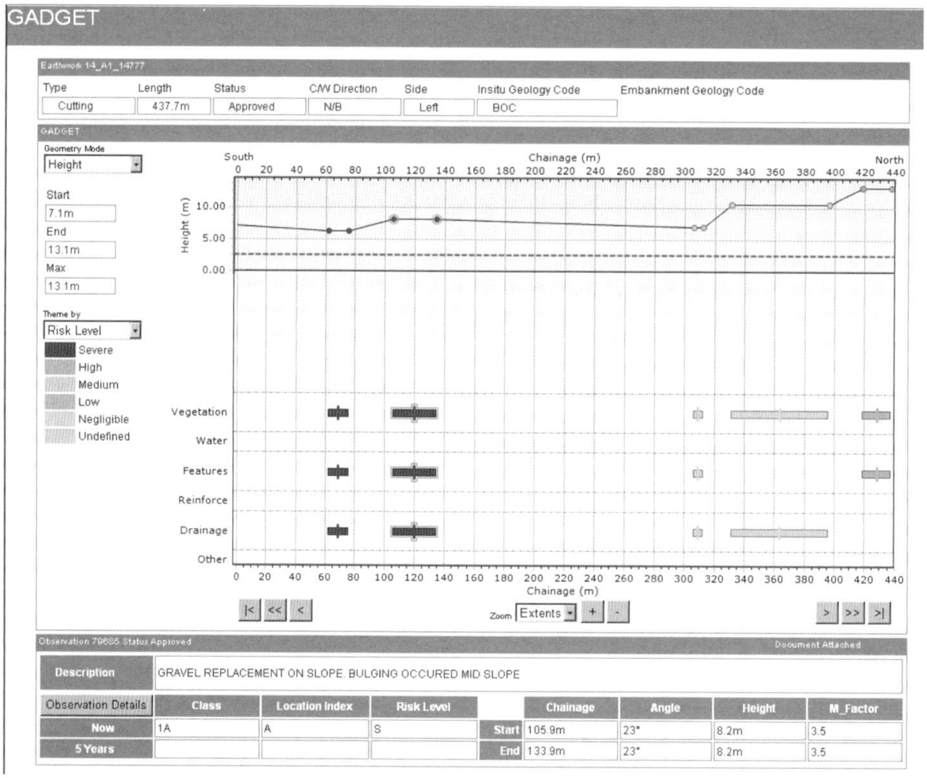

Fig. 4. Screenshot of GADGET, the visualization tool for the Geotechnical Asset Database.

As part of the GAMP, geotechnical asset inspections as planned for the coming year are described. Whilst annual inspections of all assets are required to identify any defects, the key surveys undertaken are the Principal Inspections, which are typically undertaken on 20% of the network in an MA area per year, such that the survey cycle is completed in a five-year period. The Principal Inspections collect information that can be broadly described as:

- inventory information that can essentially be considered to be static. This includes information such as the location of the asset, geological conditions and slope geometry; and
- condition information, which is time limited and considered to be live, or transient data. This would include observations of a wide range of geotechnical features that may be observed (e.g. tension cracks, subsidence, rock falls) but also other observations that affect the asset, such as water observations (e.g. ponding on slope, marshy ground).

The collected condition information can be very diverse on a geotechnical asset, and can include features which might be considered to present a risk to the safety or operation of the network, and others that present a lower, or no risk. This variation in the risks presented by the observations is addressed through the risk assessment framework set out in the HD41/03 standard.

Risk assessment and management

On discovery of an observation of asset condition during a survey, the surveyors are tasked to determine the risk that the observation poses to the performance of the asset. Whilst this risk assessment can be reviewed and approved back in the office, the key decisions are best made in the field, and for this reason surveyors are expected to be qualified and trained to recognize and assess geotechnical problems.

The risk assessment is carried out in a series of stages, which determine two key variables:

- An observation class is assigned, which is either within:
  - Class 1: defects;
  - Class 2: areas considered to be at risk;
  - Class 3: areas of repair, strengthening or preventative works;
- The location of the observation is assessed, relating to the proximity of the feature to the carriageway, or other highway infrastructure. A location index is applied, ranging from A (e.g. affecting a carriageway running lane) to D (e.g. remote from the carriageway).

This assessment is carried out based on the feature as observed now, but a secondary assessment is made of the class and location index that is expected (based on expert knowledge) in five years' time. This expert forecasting is included to allow for an assessment of features that are currently only minor (e.g. tension cracks), but can reasonably be expected to worsen in the coming five years (e.g. the tension cracks representing a precursor to a slope failure).

Table 1. *HD41/03 risk assessment matrix for observations of class and location index now*

Location index	Class				
	1A/1B/1C	*1D*	*2A/2B*	*3A/3B*	*3C*
A	S	H	M	N	N
B	S	M	M	N	N
C	H	M	L	N	N
D	M	L	N	N	N

Table 2. *HD41/03 risk assessment matrix for observations of class and location index in five years time*

Location index	Class				
	1A/1B/1C	*1D*	*2A/2B*	*3A/3B*	*3C*
A	H	M	N	N	N
B	H	L	N	N	N
C	M	L	N	N	N
D	L	N	N	N	N

Based on the assessed classes and location indices (now and five years), risk levels are assigned (Tables 1 and 2). By way of the example above, the tension cracks may be considered a minor (Class 1D) defect away from the carriageway (location index C) now, but in five years may be assessed as a potential failure (Class 1A) that affects the hard shoulder (location index B). This would result in a 'risk level now' of moderate (Table 1), but of high in five years' time (Table 2). This higher assessed risk in five years allows for proactive, preventative works to be undertaken. HD41/03 provides recommended actions related to risk levels (Table 3).

Survey data approval and notification of defects

Whilst surveying and classification of asset condition is generally undertaken by qualified, but generally more junior staff, the approval of data collected is undertaken by an approved geotechnical liaison engineer within each of the MAs, a role that requires approval by the Highways Agency. The HA GDMS Geotechnical Asset Database has in-built user rights that allow certain tasks to be available only to qualified users, which includes the approval of geotechnical asset data. This approval process is designed to ensure that the quality of data held in HA GDMS is maintained at a high level.

Table 3. *Recommended actions relating to risk levels*

Risk level	*Description*	*Recommended geotechnical action*
S	Severe	Remedial action must be undertaken with highest priority and H&S/traffic management requirements considered and kept under constant review. Consider value for money (VFM) of preventative works on adjacent Class 2 assets and potential impact on other routine or capital maintenance activities.
H	High	Remedial action required, timescale to be determined by the Overseeing Organisation Geotechnical Advisor and Overseeing Organisation Area Manager, but within 5 years. Interim monitoring/inspection may be called for and H&S/traffic management requirements considered. Consider VFM of preventative works on adjacent Class 2 assets and potential impact on other routine or capital maintenance activities.
M	Medium	Remedial action may not be required but preventative action advisable within 5 years. Review inspection and/or monitoring regime and potential impact on other routine or capital maintenance activities.
L	Low	No immediate action required. Review inspection and/or monitoring regime. Consider VFM of preventative works. Review potential impact on other routine or capital maintenance activities.
N	Negligible	No immediate action required. Re-inspect in five years.

Following the completion of principal inspections and the subsequent approval of data, any observed defects are reported to the Highways Agency using the reporting procedure outlined in the HD41/03 standard. Functionality built into the HA GDMS allows automated e-mail notification of defects submitted through the system, to inform key staff rapidly. Follow-on reporting ensures that a record of key, risk-based decisions is kept for any notified defects, from identification, through investigation and remediation, to feedback reporting.

Geotechnical Asset Database analysis and reporting

With the development of the HA GDMS, a number of facilities have been implemented that allow data held to be analysed for tactical and strategic purposes (largely by the MAs) and reported on to provide an overview for the HA. These tools form a crucial part of the geotechnical asset management strategy of the HA, and are used at all stages of the maintenance process.

The ability to search for all key asset-based knowledge, either through a text-based query (such as finding all reports in the database containing the name of a particular scheme) or through interaction with the map (for example finding all the geotechnical assets on a given stretch of the network) is inherent in the functionality of the system. Output from such searches can be exported to a spreadsheet for further offline analysis, or if the search has a spatial output it can be plotted onto the HA GDMS map screen for simple spatial analysis.

Overview reporting of the geotechnical asset is achieved through a series of automated reporting tools in-built to HA GDMS. These are generated on a monthly basis, and report on:

- geotechnical asset lengths by asset types (see Fig. 5);
- total lengths of asset resurveying planned by the MA for the coming year, and the amount undertaken to that point in time (for each MA Area);
- total lengths of classified observations of asset condition, by HD41/03 class and risk level;
- total lengths of classified observations of asset condition by risk level as a percentage of the length of the network surveyed;
- the latest assessment of asset data coverage and quality; and
- a map showing survey completion against MA areas and HA network operations regions.

In addition to the monthly reporting, an annual *State of the network report* is prepared that provides high level statistics and key performance indicators on the state of the HAs geotechnical asset. This report gives summary information in a similar manner to the monthly report described above, but also includes wider information on the amount of information held in the HA GDMS databases (i.e. the reports and boreholes databases).

Whole life cost assessment of the geotechnical asset

In order to fully understand the cost to maintain an asset, it is necessary to consider how the asset will degrade, require repair and perform over its serviceable lifetime. Such a consideration of the whole life cost of an asset, measured against a defined performance or spending scenario, is the key to asset management.

Whereas the risk assessment, work prioritization, value management and repair/prevention regime outlined so far in this paper goes some way towards consideration of asset whole life cost, it currently works within a relatively *ad*

Area	Total Asset Length (km)	May 2009 / April 2010 Reinspection targets (km)	May 2009 / April 2010 Reinspection achieved to date (km)	May 2009 / April 2010 Reinspection achieved to date (%)
1	552	110	0	0
2	1048	282	5	1.8
3	1020	258	13	5
4	864	120	0	0
5	861	155	110	71
6	1164	303	19	6.3
7*	793	158	0	0
8	1090	180	0	0
9	911	83	0	0
10	938	246	66	26.8
11	811	285	186	65.3
12	1055	200	0	0
13*	635	127	0	0
14	516	131	0	0
Sum	12258	2638	399	15.1

Geotechnical Asset re-inspection targets and actual re-surveying completed

Fig. 5. Geotechnical asset re-inspection summary information from the monthly HA GDMS report.

hoc framework, where budgets are allocated annually on an area-by-area basis, based on best available information and experience both within the HA and the MAs. To gain the best performance from the geotechnical asset within the budgetary constraints set for its maintenance, a more holistic methodology for whole life cost is required.

The HA has undertaken a proof-of-concept pilot project to investigate methods of whole life cost assessment for their key assets, including the geotechnical asset. To date, the key tasks that have been undertaken have included:

- understanding how the principles of whole life cost can be applied to the geotechnical asset (through use of a small 'test bed' model;
- determining the age of geotechnical assets from available information;
- reducing the volume of geotechnical asset data available from the HA GDMS into a series of groupings based on asset location (within the HA areas), age, geological conditions, height and angle;
- determining the current condition of the geotechnical asset using the available data from HA GDMS;
- determination of preliminary deterioration models to describe the degradation of the asset groupings over time; and
- preliminary investigation of asset management strategies based on a series of asset performance and budget scenarios.

Key to the investigation of whole-life cost undertaken so far has been the data available from the HA GDMS Geotechnical Asset Database. Ongoing development in this area is highly dependent on the asset condition data being kept up-to-date, and the quality of the data held being continuously assessed and improved. It is likely that further activities and enhancements undertaken to the HA GDMS will be centred on improving the data that feeds into improved whole-life cost tools and the associated processes of the risk assessment approach and holistic appraisal of the geotechnical asset.

References

HIGHWAYS AGENCY. 2003. HD41/03: *Maintenance of Highway Geotechnical Assets. Design Manual for Roads and Bridges*, **4**, Part 3.

HIGHWAYS AGENCY. 2008. HD22/08: *Managing Geotechnical Risk. Maintenance of Highway Geotechnical Assets. Design Manual for Roads and Bridges*, **4**, Part 2.

HIGHWAYS AGENCY. 2009. Highways Agency Website. http://www.highways.gov.uk/knowledge/16956.aspx

The identification and management of cuttings at risk from deep-seated, delayed failure on the Highways Agency network

L. N. Parry

Mott MacDonald, Mott MacDonald House, 8-10 Sydenham Road, Croydon CRO 2EE, UK
(e-mail: lee.parry@mottmac.com)

Abstract: The UK trunk road network includes a considerable number of cuttings constructed through high plasticity, overconsolidated clays. These materials are prone to the process of delayed, deep-seated failure whereby instability can occur some considerable time after construction. The timescale between excavation and failure can be anywhere between 20 and 120 years. Many of the recorded deep-seated failures in the literature are from railway cuttings in the UK which are generally much older than cuttings on major highways. The bulk of trunk road construction in the UK occurred during the 1960s and 1970s and many cuttings are at, or approaching, an age when delayed, deep-seated failures might be expected to occur.

The Highways Agency (HA) has been aware of this issue and has sought to understand the causes and associated risks of these types of failure. The first phase of a study (concluded in 2007) provided a framework for assessing the likelihood of failure. The selection criteria were based on a combination of cutting height and slope angle. Mott MacDonald was commissioned to undertake the second phase of the study to focus on the application of the Phase 1 recommendations to the HA asset base. This paper describes the work undertaken during this study and conclusions reached.

The UK trunk road network includes a considerable number of cuttings constructed through high plasticity, overconsolidated clays. Theses materials are prone to the process of delayed, deep-seated failure whereby instability can occur some considerable time after construction. The timescale between excavation and failure can be anywhere between 20 and 120 years.

The term deep-seated failure is defined as those failures extending to more than 2 m depth below the slope surface and these may extend beyond the toe or crest of the slope (Perry *et al.* 2003). Shallow failures are less than 2 m deep and occur entirely within the cutting side slope. The results of a review of the occurrence and distribution of shallow failures on the Highways Agency (HA) network is contained within the Transport Research Laboratory report RR199 (Perry 1989). At the time of preparing RR199, shallow failures were by far the most common.

This study concentrates on first time slides only, that is, it is assumed that there were no pre-existing shear surfaces present prior to construction. Cutting slopes that contain pre-existing shear surfaces will behave in a different manner and have been excluded from this study. Many of the recorded deep-seated failures in the literature are from railway cuttings which are generally much older than cuttings on major highways. The bulk of trunk road construction in the UK occurred during the 1960s and 1970s and many cuttings are at, or approaching, an age when delayed, deep-seated failures might be expected to occur.

The HA has been aware of this issue and through seminars and research studies has sought to understand the mechanisms and associated risks of delayed, deep-seated failures as they apply to the HA network. The Geotechnical Consulting Group (GCG) was commissioned in November 2005 to conduct a study entitled '*Numerical Techniques and Appropriate Factors of Safety including Models for Delayed Failure of Clay Slopes*'; this was reported in February 2007 (Geotechnical Consulting Group 2007). This study provided a framework for assessing the likelihood of a particular cutting being prone to delayed, deep-seated failure. The selection criteria were based on a combination of cutting height and slope angle and three categories of stable, marginal and unstable were proposed.

In late June 2008, Mott MacDonald was commissioned to undertake Phase 2 of the study which was to focus on the application of the Phase 1 recommendations to the HA asset base. A final report was issued in July 2009 (Mott MacDonald 2009). This paper describes the work undertaken and the conclusions reached.

Failure mechanism

Excavations within stiff plastic clays generate negative pore pressures due to stress relief. These materials are generally

From: Radford T. A. (ed.) 2012. *Earthworks in Europe*. Geological Society, London, Engineering Geology Special Publications, **26**, 41–54. http://dx.doi.org/10.1144/EGSP26.6

of low permeability; equilibration of pore pressures to long term values can take a considerable period of time. Field measurements have demonstrated that it can take between 20 and 60 years or sometimes longer for this process to complete. Stiff, plastic clays also behave in a brittle manner such that their strength is strain dependent and drops significantly from the peak value with increasing deformation.

Following excavation and coupled with pore pressure equilibration, the material forming the cutting undergoes swelling due to stress relief which in turn causes strains to develop particularly at the toe where unloading has been the greatest. The strains within the slope cause the shear strength to reduce such that the average mobilized shear strength along a potential slip surface is part way between peak and residual. Figure 1 illustrates the development of this mechanism, simulated by the FLAC numerical modelling of progressive failure (Ellis & O'Brien 2007). Thus the two, time dependent, features of delayed, deep-seated failure are a long-term increase in pore pressures coupled with a reduction in available strength. The combination of these two factors can eventually lead to failure.

It has long been recognized that mobilized shear strengths calculated from back analysed, first time failures in stiff plastic clays are transitional between peak and residual; fully softened values were often proposed as representing an appropriate operational value of shear strength. However, without a theoretical basis the correlation might be coincidental.

With the development of numerical modelling techniques, it has been possible to develop a much better understanding of how internal strains develop within cutting slopes and thereby how available strengths reduce. It has been demonstrated in various technical publications (Ellis & O'Brien 2007) and in the modelling conducted for Phase 1 of this study that strains initiate at the toe and subsequently progress into the slope along a near horizontal surface. Strain levels increase with time until there is a zone at the base of the cutting where the soil has reached residual strength. The material in the remainder of the slope remains close to peak strength but the zone of residual strength material grows until the overall mobilized strength reduces to a point where failure occurs.

Thus the mechanism for delayed, deep-seated failure appears to be based on a distinct two-part strength model (peak and residual) operating along potential slip surfaces rather than a uniform, 'averaged' value of strength. A similar result was found for shallow failures (Chandler & Skempton 1974; Perry 1991).

Geological formations considered

Delayed, deep-seated failure occurs mainly in overconsolidated clays and particularly in those of high or very high plasticity due to their reduction in post peak strength with

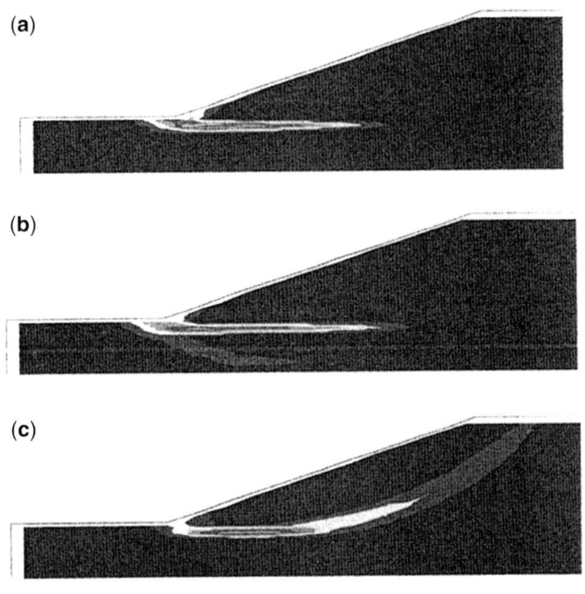

Fig. 1. Development of plastic shear strain with time (Ellis & O'Brien (2007) fig. 7).

Table 1. *Geological formations containing material with a potential for delayed, deep seated failure.*

Geological formation	*HA designation*
Atherfield Clay Formation	AC
Ampthill Clay Formation and Kimmeridge Clay Formation (undifferentiated)	AMKC
Charmouth Mudstone Formation (replacement name for Lower Lias Clay)	CHAM
Gault Formation	GLT
Gault Formation and Upper Greensand Formation (undifferentiated)	GUGS
Kimmeridge Clay Formation	KC
Kellaways Formation and Oxford Clay Formation (undifferentiated)	KLOX
London Clay Formation	LC
Lias Group	LI
Lower Lias Clay Formation	LLIC
Oxford Clay Formation	OXC
Reading Formation	RB
Upper Lias Clay Formation	ULC
Weald Clay Formation	WC
Wadhurst Clay Formation	WDC
Whitby Mudstone Formation (replacement name for Upper Lias Clay)	WHM
Woolwich and Reading Beds Formation	WRB

increasing strain. Table 1 lists the geological formations that were considered as potentially prone to delayed, deep-seated failure. The selection was chosen to align with other HA studies into over consolidated clays supplemented with other materials thought to be prone to delayed failure. Also shown are the geological codes adopted within HA Geotechnical Data Management System (GDMS).

Cuttings within the Lias formation are collated in one of three designations within HA GDMS namely Lias Group, Lower Lias Clay and Upper Lias Clay. More recently the Lower Lias Clay has been termed Charmouth Mudstone Formation and the Upper Lias Clay the Whitby Mudstone Formation. The Upper Lias Clay has a number of recorded delayed, deep-seated failures (Mott MacDonald (2009)) and may represent the highest risk of the three groupings. However, certain areas of the Upper Lias outcrop have very high sand contents (Green 1992) and so not all cuttings will be at risk. Similarly the Lower Lias is known in part to contain limestone bands that would increase mass strength but these may not occur throughout.

At the strategic level of this study all of the Lias groupings were assumed to be potentially prone to delayed, deep-seated failures but further review of site specific ground conditions could significantly reduce the number of cuttings with the potential for this type of failure to develop.

Identification of at-risk cuttings

Analytical basis

Stability criteria, based on a combination of limit equilibrium and finite element analyses, were developed during the Phase 1 studies presented in Geotechnical Consulting Group (2007); these limits are summarized in Figure 2. The limit equilibrium envelope was based on calculated safety factors of unity for various soil strength and groundwater

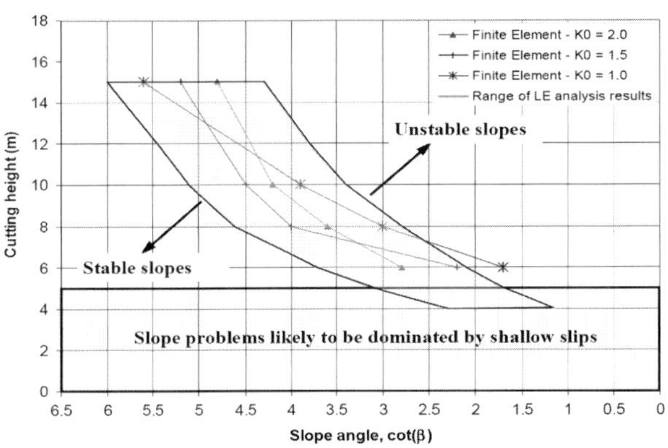

Fig. 2. Phase 1: delayed, deep-seated failure stability limits (Geotechnical Consulting Group (2007) fig. 41).

assumptions. Limiting conditions from finite element analyses are based on London Clay parameters for varying K_o conditions (Geotechnical Consulting Group 2007). The adopted limits apply to cuttings of between 5 and 15 m in depth. It is believed that instability in cuttings less than 5 m deep will be dominated by shallow failures whereas cuttings in excess of 15 m depth are likely to comprise significant proportions of fresher material and will not develop delayed, deep-seated failures.

The criteria presented in Figure 2 were used as a starting point in identifying cuttings at risk but the following points need to be considered when applying the criteria.

- The stability limits have been based mainly on parameters applicable to weathered London Clay. Care needs to be exercised when applying the findings to other geologies.
- The stability limits have been developed assuming that single sets of strength parameters are applicable irrespective of cutting depth. Weathering, and thereby strength, is depth dependent and deeper cuttings will have stronger material present at the toe than shallower cuttings. In an attempt to address this it has been assumed that cuttings in excess of 15 m depth will not be subject to delayed, deep-seated failure but Chandler (Mott MacDonald 2009) has shown that the depth of weathering in London Clay is unlikely to exceed some 7–10 m and may be considerably less. Thus for London Clay cuttings it is likely that the stability criteria are progressively conservative as cutting depths exceed 7 m. Similar comments may apply to other geologies. Depth of weathering as a control on the depth of deep-seated failure has been clearly shown in the deep-seated failures at New Cross (Skempton 1977) and Northolt (Chandler & Skempton 1974) railway cuttings where the failure surface passed along, not through, the top of the unweathered London Clay (the interface was above the cutting toe).
- The stability limits assume that deep-seated failures will only occur in cuttings greater than 5 m high. This is likely to be the case for most of the HA cuttings but there may be individual sites where the mechanism may affect cuttings that are only 4 m high. Recent numerical modelling studies (Potts *et al.* 1997) have shown that mobilized strength is a function of cutting height (Fig. 3). The failures analysed by Chandler & Skempton (1974) had an average height of about 7 m, and the back analysed strength is consistent with those derived from the numerical modelling. Therefore, it should be anticipated that mobilized strengths would increase and the risk of failure decrease as cutting height reduces, especially for heights less than 5 m.
- The stability limits are based on a hydrostatic groundwater table at between 1– 2 m depth which may not apply to all cuttings. Stability is very sensitive to groundwater pressures. This is clearly shown by the numerical modelling studies carried out by Ellis & O'Brien (2007). For example, for a 1:3 (V:H) slope in weathered

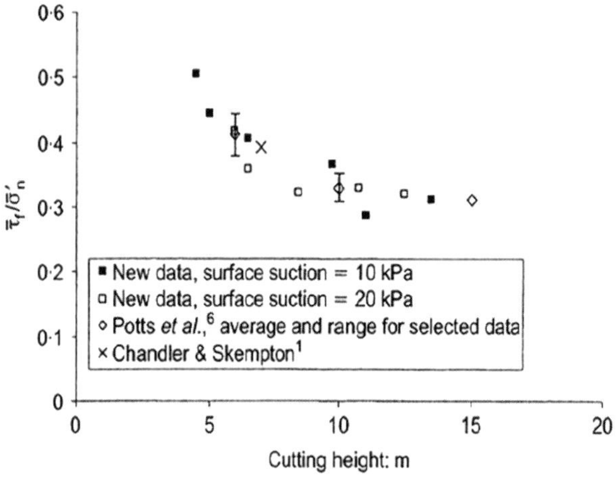

Fig. 3. Variation of stress ratio with cutting height (Ellis & O'Brien (2007) fig. 10).

London Clay the stable height increases from about 6.5 m to 8.5 m if the water table level drops from 1.0 m to 2.0 m below ground level.

Application to HA asset base

The Highways Agency Geotechnical Data Management System is an on-line repository for geotechnical data collected on behalf of the Highways Agency. The system can display geotechnical asset data as well as a large number of datasets including aerial photography, borehole locations, reports and Master Map details. The ability to query different datasets simultaneously makes the HA GDMS site extremely valuable for conducting desk studies and for planning site investigations. Given the amount of data captured, it was deemed that the information held on HA GDMS was suitable for use in investigating the risk of delayed, deep-seated failures in overconsolidated clay cuttings.

The stability limits proposed in the Phase 1 study were based on slope height and angle. This information is amongst the data collected during HA inspections and contained within HA GDMS for each cutting. Individual cuttings were compared using the 'morphology factor' which is derived from a combination of slope height and slope angle (Fig. 4).

Morphology factor contours were compared with the stability limits of Figure 2 (Figure 5). The morphology factor contours have a similar orientation to the derived stability limits confirming that the former is a suitable parameter for assessing stability. A morphology factor of 3.0 is a close approximation to the edge of the unstable zone and a value of 1.6 to the edge of the stable zone. Although these values are not an exact match to the stability limits they are a close enough fit for assessing the HA asset base at a strategic level. To

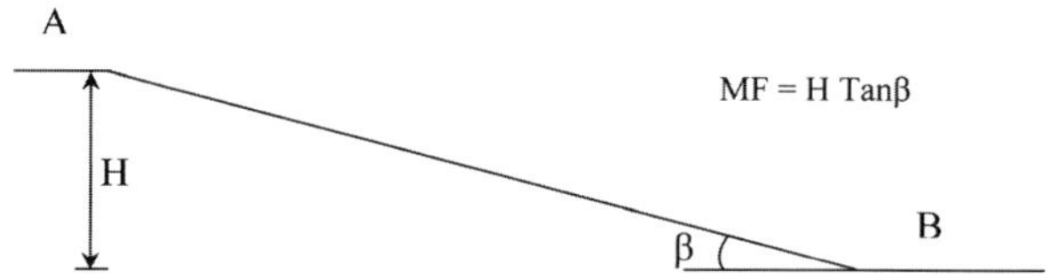

Fig. 4. Derivation of morphology factor.

summarize, Figure 6 illustrates how cuttings with a potential for delayed, deep-seated failure were identified from the data held in HA GDMS.

Age effects

General

As discussed earlier, delayed, deep-seated failure is a time-dependent process. The Phase 1 study did not address this issue but proposed a classification system based solely on cutting height and slope. In order to manage risk effectively and identify those cuttings most at risk of failure an understanding of the additional factors controlling the time to failure is required. These factors include:

- soil permeability
- soil stiffness
- pore water pressure regime
- geology/hydrogeology (especially the presence of high permeability material above, within or below the cutting slope)
- climate (mainly precipitation rate)
- vegetation type, size and distribution
- slope drainage (if present)
- presence of retaining structures.

It is beyond the scope of the present study to undertake the detailed modelling that would be required to quantify age effects fully. However, in view of the significance of this aspect a review of delayed, deep-seated failures reported in the literature for London and Upper Lias Clay was undertaken to identify the likely range of timescales to failure.

London Clay cuttings

Skempton (1977) reviewed the case records for a number of delayed, deep-seated failures in London Clay cuttings. He concluded from back analyses that average ru values (water pressure/total vertical stress) within the cuttings are initially negative and increase with time to reach a steady state value of about 0.3 some 50 years after excavation. The age of the individual cuttings at the time of failure ranged between 5 and 65 years with all but two of the failures occurring within a 20–50 year time span after excavation.

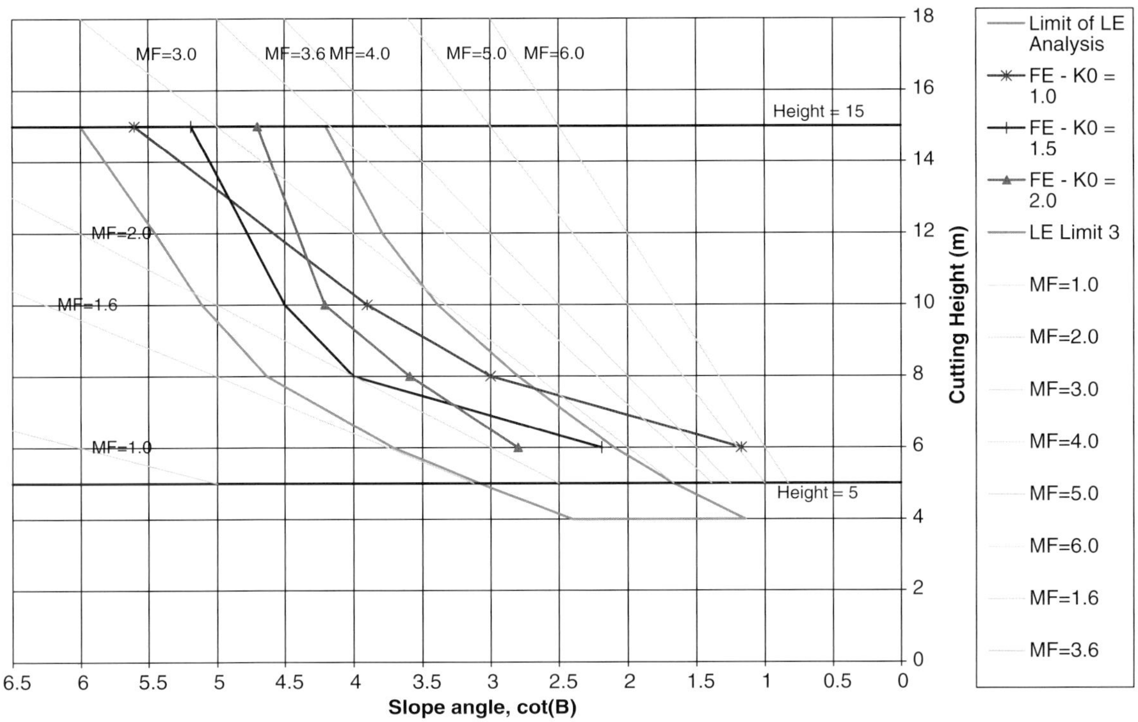

Fig. 5. Stability criteria and morphology factors.

The data presented by Skempton (1977) has been re-plotted as time to failure against slope angle by Leonards (1979), Tavenas (1981) and Chandler (1984). The results indicate a general increase in time to failure with a reduction in slope angle (Fig. 7). It would be expected that the time to failure would increase with cutting height for a given slope angle (due to the increase in drainage path length). The dominant parameter is likely to be the permeability of the stiff clay, which for near surface, weathered clays could be variable (Leonards 1979).

Upper Lias Clay cuttings

Chandler (1984) made a study of twelve failures within cuttings excavated though the Upper Lias Clay and his findings are also plotted on Figure 7. Two early failures that occurred some 5–10 years after construction have been ignored as these were initiated by over steepened excavation during construction with the failure then continuing after the cuttings were re-graded to their design profile. The remaining failures occurred over a period ranging from 40–130 years after construction with most failing 50 or more years after excavation. In contrast to the London Clay, there appears to be no correlation between time to failure and slope angle.

Review of HA asset base

Methodology

An extraction of the HA GDMS was obtained on the 23 August 2008 and filtered so that only relevant assets and asset information were considered. The data were sorted by component geology tagged to cutting height and slope angle. A recent check by the HA of data within HA GDMS has revealed that the database includes a small amount of erroneous data. Although it is believed that the information extracted for this study is adequate for the identification of overall trends, data intended for site specific usage should be independently verified. Actions are planned by the HA to address these issues.

A total of 73 cuttings were recorded in HA GDMS as containing reinforcement. This might include anything from soil nails to geogrids and might be localized at the toe or spread across the slope. Further refinement would only be possible by reviewing inspection reports for individual cuttings which was beyond the scope of the current study. The HA also has a Structures Database (SMIS) that includes information relating to retaining walls but again asset specific information would need to be reviewed to assess the level of additional support being provided. To avoid skewing

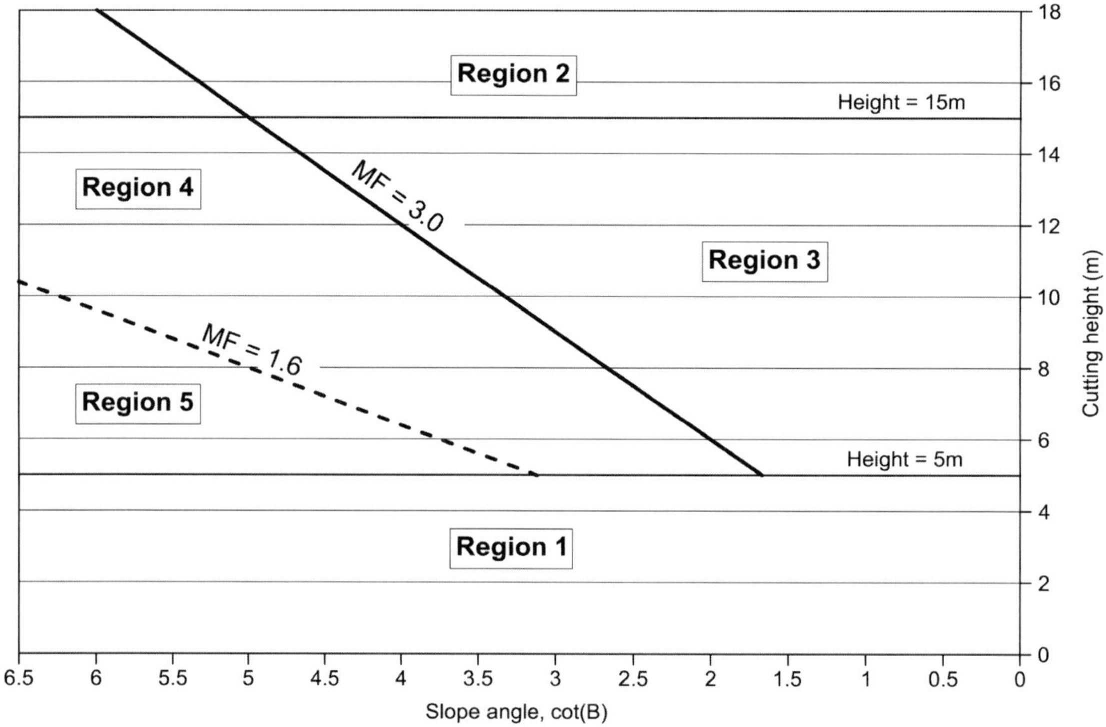

Fig. 6. Final stability model adopted showing risk regions. Region 1, shallow cuttings unlikely to fail in a deep seated manner; Region 2, high cuttings, possibly in unweathered material or other geology groups which are likely to have a low frequency of failure; Region 3, potentially unstable cuttings; Region 4, marginal cuttings; Region 5, stable cuttings.

the data if an asset has been reported as containing either reinforcement or a retaining structure, it was placed within a separate subset from the conventional earth structures and considered separately.

Review by geology

General

Using the HA GDMS data extract, individual plots based on Figure 6 were produced for all cuttings within each geological formation. Examples of the data obtained for London Clay and Upper Lias Clay are shown on Figures 8 and 9 which also contain previously reported failures on railway cuttings (Chandler & Skempton 1974). The previous failures in the Upper Lias Clay plot almost entirely within the unstable region whereas those in the London Clay plot in both the marginal and unstable regions.

The data obtained for all of the subject geologies are summarized in Table 2. They illustrate the distribution of overconsolidated clay cuttings within the HA asset base by geology and risk region as per Figure 6. The following points should be noted regarding the information extracted from HA GDMS.

- Lias Group: The guidance within HA GDMS states that Lower Lias Clay (LLIC) is now an obsolete geology code and Charmouth Mudstone Formation (CHAM) should be used instead. A check of the dataset extracted from HA GDMS showed that there were a small number of cuttings with a CHAM code. To avoid mixing data the latter have been collated separately.
- Upper Lias Clay Formation: The guidance within HA GDMS states that Upper Lias Clay (ULC) is now an obsolete geology code and Whitby Mudstone Formation (WHM) should be used instead. However, a check of the dataset extracted from HA GDMS showed that there were no cuttings within this formation.
- Woolwich and Reading Beds (Lambeth Group): There are data within HA GDMS for both ‘Woolwich and Reading Beds’ and ‘Reading Beds’. To avoid mixing data information relating to the two groups has been collated separately.

Although the data in Table 2 are based on numbers of cuttings rather than cutting length it is a start to understanding the scale of the issue and demonstrates the distribution of geologies across the various risk regions. For example the

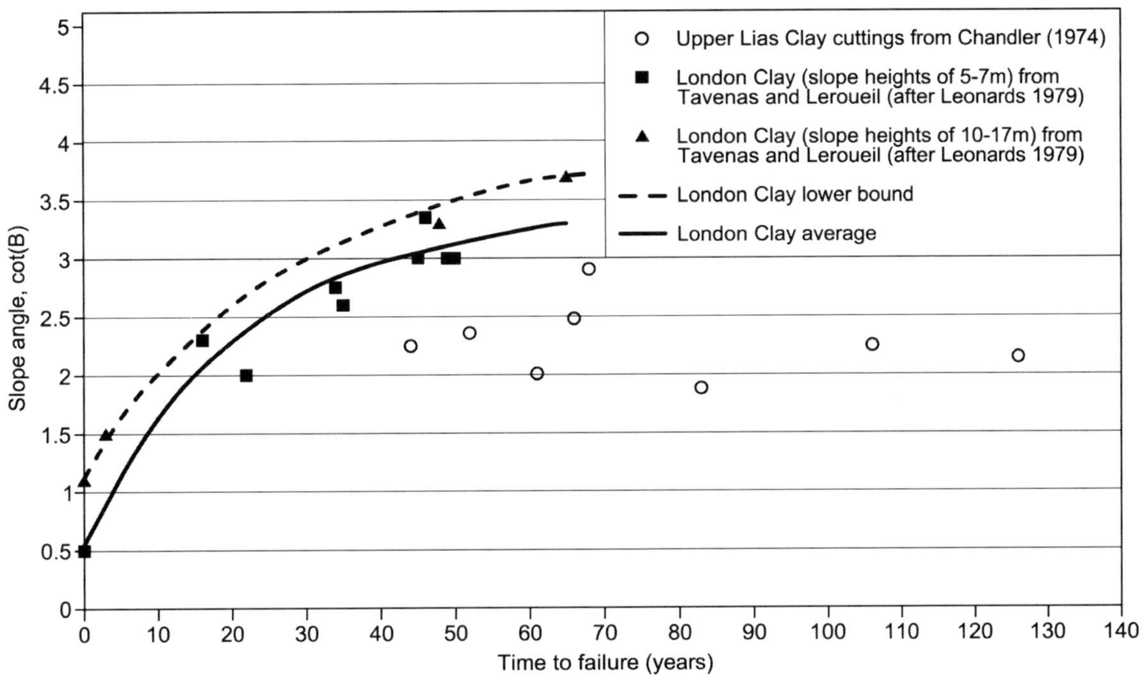

Fig. 7. Time to failure relative to slope angle.

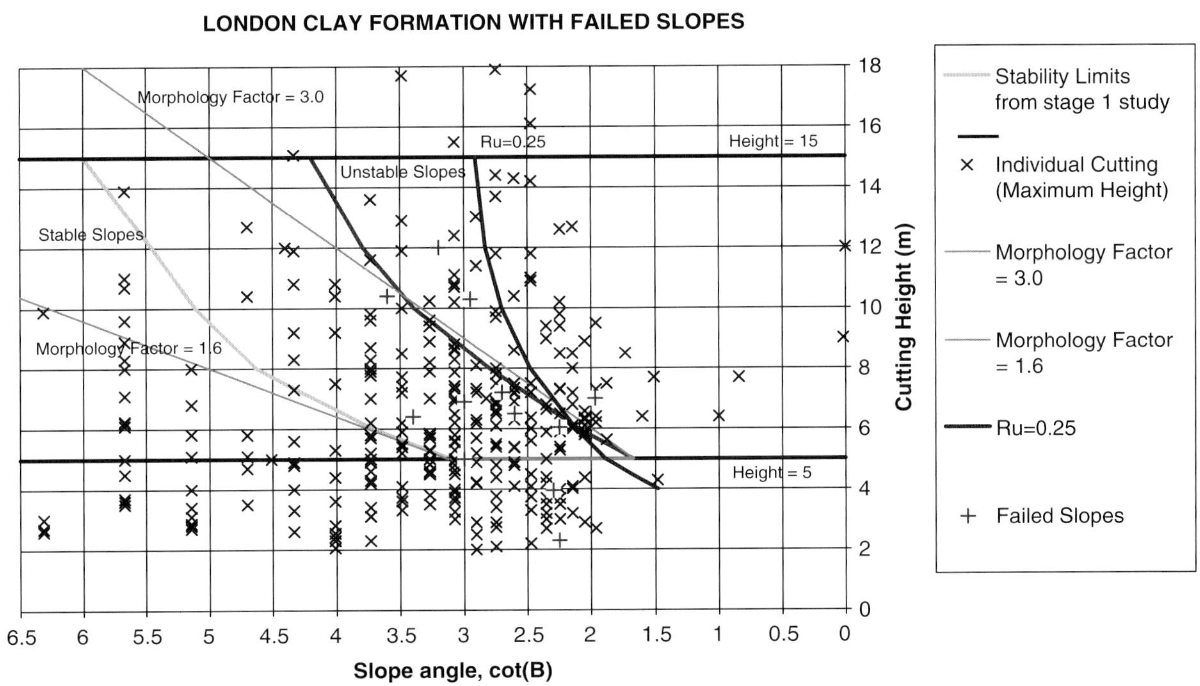

Fig. 8. London Clay data with previous failures and ru = 0.25 line.

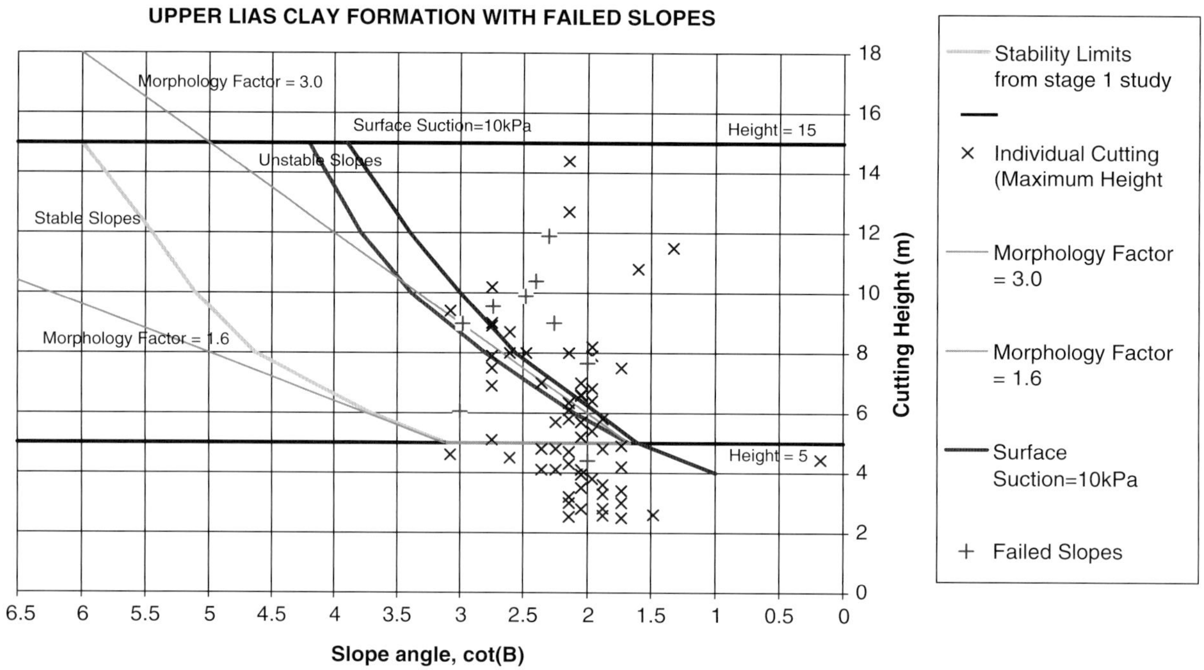

Fig. 9. Lias Clay data with previously recorded failures shown.

largest grouping of overconsolidated clay cuttings within the HA asset base are through London Clay which also represents the greatest individual numbers of potentially marginal or unstable cuttings. If all Lias data are considered together (Upper, Lower, Lias Group and Charmouth) it would represent a larger number of potentially unstable cuttings, 85 compared with 61 for the London Clay. As most of the Lias cuttings are in SW England, where rainfall is higher

Table 2. *Numbers of cuttings without reinforcement or retaining walls within the various risk regions for particular geologies*

Geology	*Cuttings below 5 m high (Region 1)*	*Cuttings above 15 m high (Region 2)*	*Unstable (Region 3)*	*Marginal stability (Region 4)*	*Stable (Region 5)*	*Total*
AMKC	2	0	0	0	0	2
CHAM	3	0	1	9	1	14
GLT	64	5	19	44	23	155
GUGS	3	0	2	9	1	15
KC	20	0	0	8	0	28
KLOX	1	0	5	1	0	7
LC	113	8	61	122	32	336
LI	27	6	17	22	1	73
LLIC	51	2	45	53	7	158
OXC	46	2	21	12	2	83
RB	9	0	11	9	2	31
ULC	30	0	22	12	0	64
WC	33	1	11	36	8	89
WDC	22	100	505	709	100	36
WHM	0					0
WRB	14					28
Total	438	25	225	353	78	**1119**

For key to geology refer to Table 1

than the SE, pore water pressures would also be anticipated to be higher in Lias slopes.

The data in Table 2 are in contrast to the findings of RR199 (Perry 1989) which found that the greatest percentage of cutting failures on the HA network had occurred in Gault Clay with virtually none in the London Clay. RR199 addressed shallow failures which would have been the dominant failure mechanism given the age of the network at the time the data was collected (1980–1988). This may indicate that as the HA network ages, the geologies giving rise to most of the failures may vary as failure mechanisms change.

Table 2 also shows that there are a total of 578 cuttings falling within the marginal/unstable region. This is a large number corresponding to almost 50% of all of the overconsolidated clay cuttings within the HA asset base. Clearly a ranking system is required in order to enable the effective management of risk.

Review by age

Delayed, deep-seated, failure is a time dependent process and cutting age is, therefore, an important factor in assessing risk. A further exercise was undertaken to review the data extracted from HA GDMS by age as well as geology and morphology factor. To understand the scale better, data was collated by cutting length rather than numbers of cuttings. The data collated from HA GDMS is based on morphology factor measured at the deepest point of each cutting. This means that the lengths quoted represent the total lengths of cuttings that contain an element of marginal or unstable slope rather than indicating that the total length is wholly marginal or unstable.

The collated data is summarized in Figures 10 and 11. These show the total length of cuttings containing a marginal or unstable component by geology and cutting age range. The actual total length of marginal or unstable cutting slopes will be somewhat less than the figures shown, but the data provides a starting point to understanding the distribution of potentially unstable cuttings.

The following conclusions were drawn;

- There is a clear skew to the data whereby the bulk of marginal and unstable cuttings tend to fall into the older age groupings prior to 1980. This is probably a combination of the fact that this period coincided with the zenith of road construction and that as understanding of the

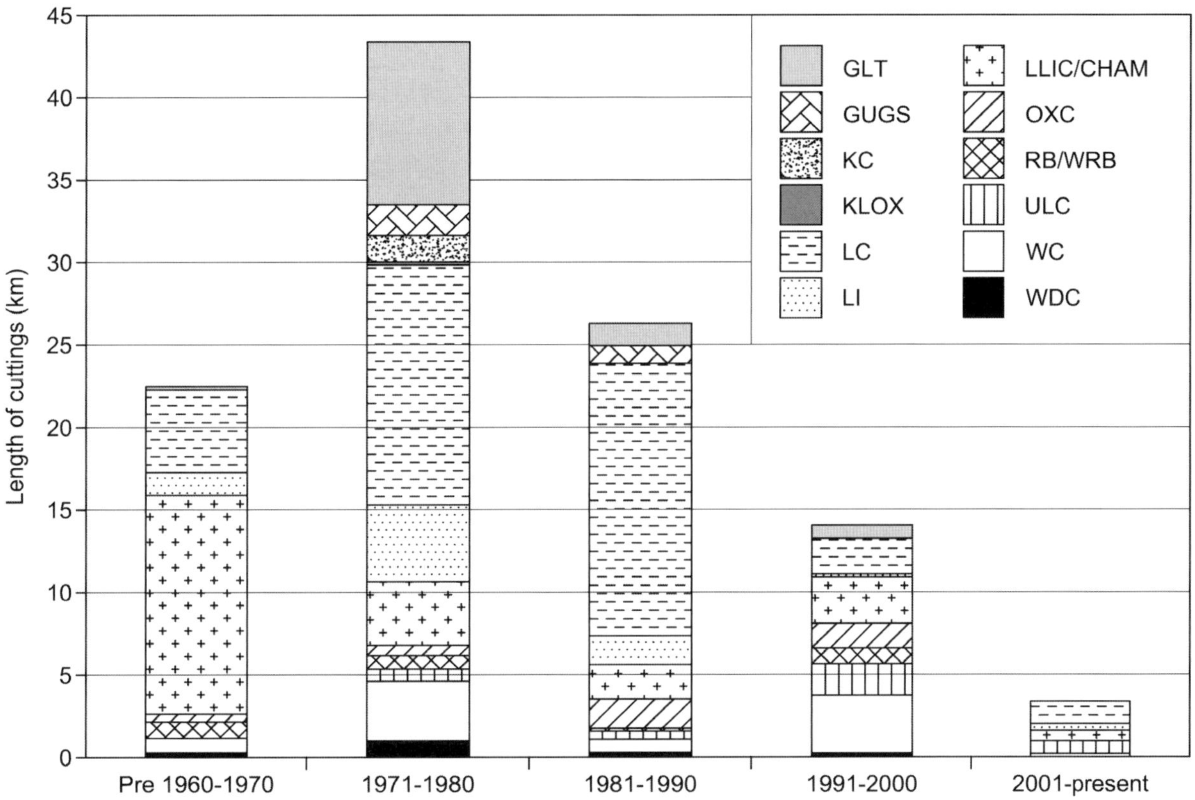

Fig. 10. Cutting length by age bands for marginal cuttings (Region 4). For key see Table 1.

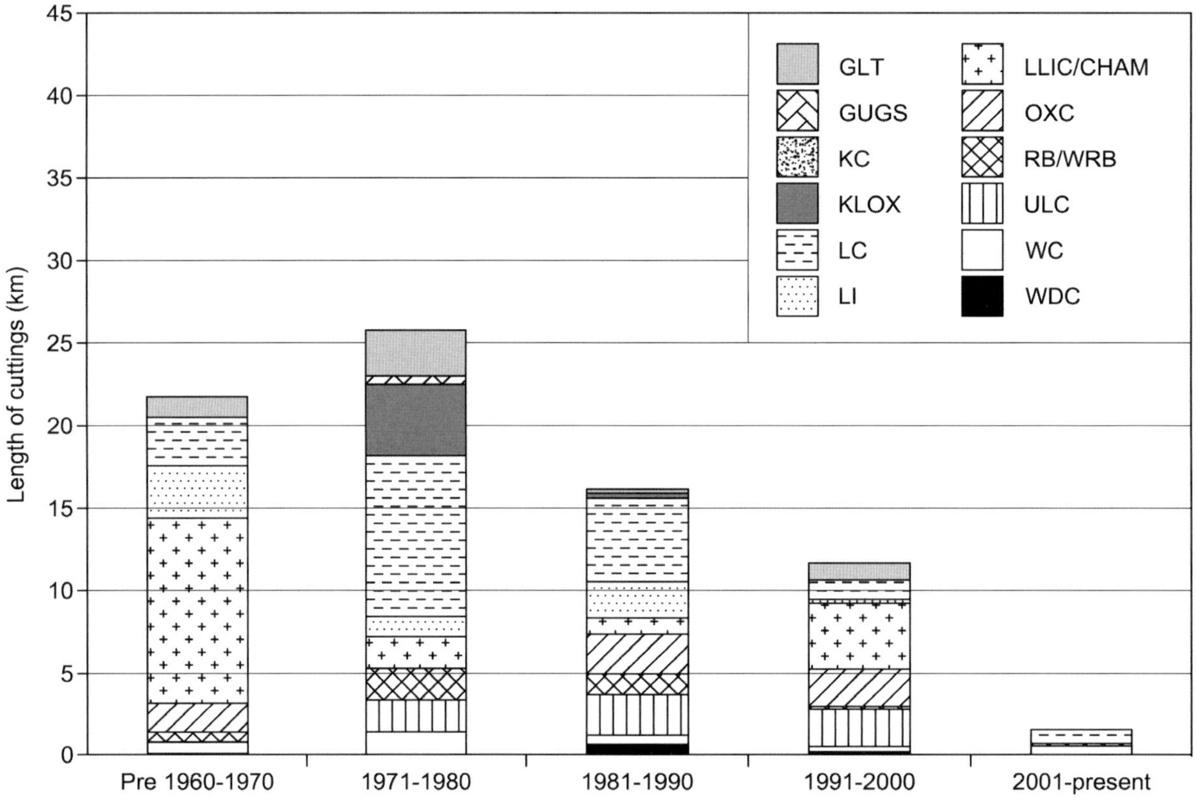

Fig. 11. Cutting length by age band for unstable cuttings (Region 3). For key see Table 1.

mechanisms controlling delayed, deep-seated failure have become more clearly understood, cuttings have been designed with flatter side slopes. In addition, the findings of RR199 led to a general flattening of side slopes to tackle shallow failures.

- The Lower Lias Clay Formation (LLIC) and Lias Group (LI) magnify the skew of the data towards the older age bands. This is particularly true for the Lower Lias Clay Formation (LLIC) which contains a significant length of pre-1970 cuttings within the unstable category.
- Excluding the Lower Lias Clay Formation (LLIC) and Lias Group (LI) the greatest length of cuttings within the marginal or unstable category are within London Clay. The bulk of the unstable cuttings are within the 1971–1980 age range and the bulk of marginal cuttings within the 1971–1990 age range. The Gault Clay and undifferentiated Kellaways and Oxford Clay Formations vie for second place for cutting length in the marginal or unstable category with the bulk of these cuttings dating from 1971 to 1980.

To summarize, the bulk of marginal or unstable overconsolidated clay cuttings within the HA asset base are some 30–50 years old which, based on UK rail experience, is the age range at which delayed, deep-seated failures are likely to become apparent. In addition, cuttings through the Lower Lias Clay Formation (LLIC) may significantly affect the total lengths of marginal or unstable cuttings and the median age of the latter. Therefore, it is important to establish the interrelationship of the three Lias groupings within HA GDMS and in particular if materials in the Lower Lias Clay Formation (LLIC) and Lias Group (LI) are likely to be subject to delayed, deep-seated failure.

Identification of high-risk HA cuttings

The review of the data within HA GDMS against the failure criteria proposed in the Phase 1 study has revealed that a considerable number of overconsolidated clay cuttings fall into the marginal and unstable category. The total length of cuttings within the marginal category is 75.8 km and in the unstable category 49.6 km. These figures increase to 102 km and 74.4 km, respectively, if the Lower Lias Clay Formation (LLIC) and Lias Group (LI) are included. These figures represent a considerable length of earthwork to designate as at risk of failure and further rationalization of risk is required to enable the effective management of these elements of the HA asset base.

Cutting age is an important factor in the process of delayed failure and therefore, not all cuttings classified as potentially unstable from the Phase 1 criteria pose the same risk at a particular point in time. An investigation was undertaken to consider how models could be developed to identify the most at risk cuttings within the London and Upper Lias clays.

London Clay cuttings

The limit equilibrium analyses undertaken in the Phase 1 study generally assumed a 10 kPa surface suction equating to a groundwater level 1 m below the slope surface. This represents a very high groundwater condition equivalent to a ru value of about 0.45 on the critical failure surface. Modelling and field evidence indicates that it can take some considerable time to develop such high pore pressures (see Fig. 12). The data in Figure 12 demonstrate that an overall ru value in excess of 0.25 is unlikely to occur in cuttings less than 35 years of age and therefore, the stability of such cuttings should be considered using lower pore pressures.

Pore pressures are not the only factor to be considered in identifying high risk London Clay cuttings. Weathering is also important and cuttings in excess of 9 m depth are more likely to contain less weathered (stronger) material at the toe of the cutting. Records of past failures also indicate that the failure time of London Clay cuttings appears to be related to slope angle, with flatter slopes taking longer to

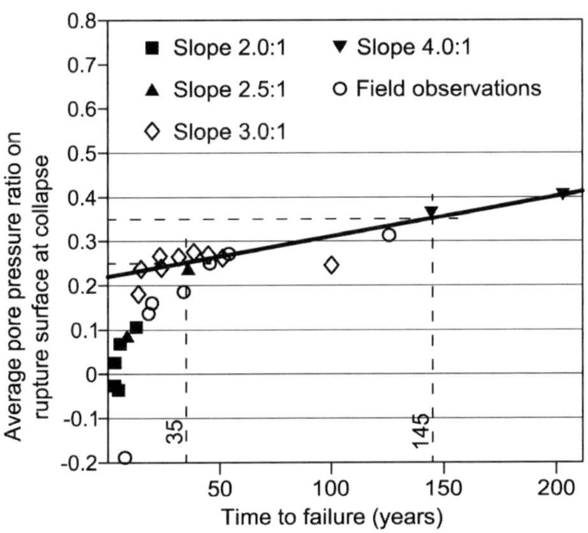

Fig. 12. Development of pore pressures in clay cuttings model *v.* observations (Geotechnical Consulting Group (2007) fig. 30).

Table 3. *London Clay cuttings: failure times relative to slope angle*

Slope angle (cot beta)	*Time to failure (years)*	*Time to failure/ cot beta*
2	15	7.5
2.5	25	10
3	42	14

fail (see Fig. 7). The general relationship presented in Figure 7 can be approximated as shown in Table 3.

Table 3 demonstrates that to select one particular age of cutting as representing the onset of critical conditions would be an oversimplification. Normalizing time to failure against slope angle produces a much smaller variation. A normalized value of 10 (time to failure/cot beta), for the range of slope angles considered, would seem a reasonable first approximation as a threshold to identify high risk London Clay cuttings.

The factors described above have been used to derive an envelope for the identification of high-risk London Clay cuttings (Fig. 13). This envelope combines elements of both marginal and unstable cuttings as defined by the overall stability criteria for this study. It is confirmed by the recorded failures in London Clay (see Fig. 8) which indicate that failures occur in both stability groups.

Upper Lias Clay cuttings

As discussed earlier (Fig. 7) the youngest recorded age of a delayed, deep-seated failure of an Upper Lias Clay cutting was 40 years with the remainder of failures exceeding 50 years of age. All but one of the previous cutting failures exceeding 5 m in height plot within the unstable region of the Phase 1 failure criteria (see Fig. 9). Therefore, for the Upper Lias Clay cuttings the highest risk examples can be taken as those plotting within the unstable region of the stability criteria and with an age exceeding 50 years.

Other cutting geologies

Within the confines of the current study it has not been possible to establish the failure-time profile for the other subject geologies and whether they are more likely to behave like the London Clay or Upper Lias Clays. The conservative view is to assume behaviour more akin to the London Clay, that is, failures are likely to commence after 20 years (rather than the 50 years in the Upper Lias) and failures are also likely to occur within the marginal stability zone as well as the unstable zone. In order to strike a balance, cuttings constructed prior to 1971 that is, about 40 years old or more were taken as the high risk group for the remaining geologies. This assumption should be reviewed as data is gathered on the performance of cuttings through other geologies.

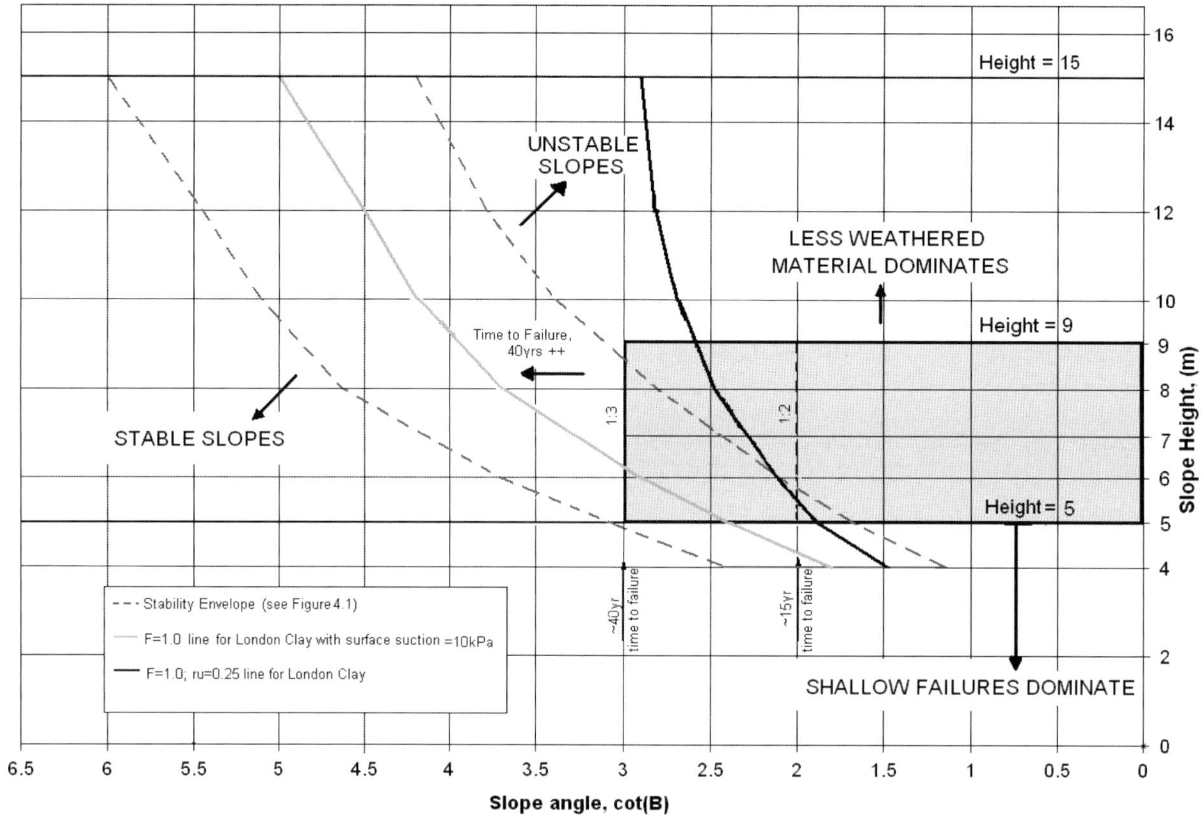

Fig. 13. Envelope for high risk London Clay cuttings.

Proposed risk ranking

A risk ranking was developed to identify those cuttings in the HA network most at risk of delayed, deep-seated failure (Table 4). The four risk groups identified in Table 4 provide a basis for developing an effective management system for these earthworks. Management options range from desk study, assessment and monitoring, to enhanced inspection, to standard HA inspection procedures as the risk group reduces.

The HA already has a process in place for the identification of at-risk earthworks based on the past performance and maintenance needs of earthworks within the same geology on the same road. The process is outlined in *Maintenance Standard HD41* (Highways Agency 2003). HD41 deems an earthwork to be 'at risk' if it has a morphology factor equal to or in excess of an earthwork (within the same geology on the same road) with a significant defect or previously subject to remedial works. It has not been possible to integrate the findings of the present study with the approach outlined in HD41 but this is an area that should be studied in more detail.

Conclusions

It has been demonstrated by numerical modelling, field observation and back analysis of previous failures on the railway network that delayed, deep-seated failure in plastic, overconsolidated clay is a real phenomenon with several examples reported in the literature. Using modern soil mechanics theory it is possible to identify those cuttings potentially at risk on the HA network. Adopting this process this Phase 2 study has shown that there are 187 km of cuttings that potentially contain a marginal/unstable element. This is just over half the length of all HA cuttings passing through the subject geologies considered in this study (341 km). During the study the possibility of using a GIS mapping layer to identify cuttings at risk was investigated and a prototype developed. It is believed that, once fully validated data is available, a GIS layer could be added to HA GDMS to assist with the identification and management of critical assets.

The review of the data contained within HA GDMS has shown that there is sufficient information available to enable cuttings at risk from delayed, deep-seated failure to

Table 4. *Risk ranking strategy*

Risk Group	*Geology / stability zone*		
	A	*B*	*C*
1 (105 sites; 36.52 km)	'Marginal' or 'unstable' London Clay cuttings constructed before 1991 with a slope steeper than 1:3, depth < 9 m and age/cot ß > 10	'Unstable' Upper Lias Clay cuttings constructed before 1961	'Unstable' AMKC, CHAM, GLT, GUGS, KC, KLOX, LC, LI, LLIC, OXC, RB, ULC, WC, WDC, WHM, WRB constructed before 1971
	58 sites 17.69 km	None (oldest cutting constructed 1974)	47 sites 18.83 km
2 (157 sites; 51.77 km)	All other 'unstable' London Clay cuttings	'Unstable' Upper Lias Clay cuttings constructed in 1961 or later	'Unstable' AMKC, CHAM, GLT, GUGS, KC, KLOX, LC, LI, LLIC, OXC, RB, ULC, WC, WDC, WHM, WRB constructed 1971 or later
	40 sites 13.36 km	22 sites 6.73km	95 sites 31.68 km
3 (316 sites; 98.39 km)	All other 'marginal' London Clay cuttings	'Marginal' Upper Lias Clay	'Marginal' AMKC, CHAM, GLT, GUGS, KC, KLOX, LC, LI, LLIC, OXC, RB, ULC, WC, WDC, WHM, WRB
	85 sites 28.45 km	12 sites 3.87 km	219 sites 66.07 km
4a (78 sites; 26.16 km)	'Stable' cuttings		
4b (438 sites; 115.61 km)	Cuttings with depth < 5 m		
4c (25 sites; 12.30 km)	Cuttings with depth > 15 m		

See Table 1 for key to abbreviations.

be identified. Coupled with case studies relating to the manner and timescale for these failures a preliminary risk ranking has been developed to identify the most critical groups of cuttings. This approach has demonstrated that some 37 km of the 187 km fall into the highest risk group of delayed, deep-seated failure. A staged response is recommended with the highest risk cuttings selected for desk study and more detailed assessment whereas enhanced inspection is recommended for less critical assets with normal HA management procedures continuing for the lowest risk assets.

A number of assumptions have had to be made in deciding the relative risk between cutting types. There may also be a small amount of erroneous data within HA GDMS and there is a need to test the conclusions drawn from the present study against the processes outlined in HD41 for the identification of at risk earthworks. Therefore, it would be beneficial to undertake a pilot study within one of the HAs management areas to test and benchmark the conclusions drawn from the present study. This could take the form of a series of desk studies undertaken on the apparently highest risk cuttings, plus a few other check sites from lower risk groups to allow comparison with risk categories derived from HD41. The various groups of the Lias Formation contain a significant proportion of the potentially at-risk earthworks but there is some uncertainty if all of the materials present in the formation would be susceptible to the development of delayed, deep-seated failure. Further research of the Lias Formation is required to establish if certain elements can be excluded as not containing materials likely to be subject to delayed, deep-seated failure.

Excluding the Lias, the greatest length of potentially unstable cuttings is through London Clay. This is in contrast to the findings of RR 199 which, based on actual failures, indicated that Gault Clay was the most problematic material and that no failures were recorded in the London Clay. This may reflect the fact that the surveys for RR 199 were carried out in the mid 1980s when the asset base would have been less than 25 years old and failures would have been shallow rather than deep-seated. It is possible that the risk profile of the asset base will change over time as the dominant failure mechanism moves from shallow to deep. RR199 provided a very useful snapshot of asset condition at 25 years age and it would be of benefit to undertake further research to align the current study with RR199 to address whole life risk profiles and assist with asset management.

Although to date there has been a lack of delayed, deep-seated failures on the HA network, the asset base and particularly those cuttings most at risk, is approaching an age when delayed, deep-seated failures might be expected. This is because of the large number of highways built in the 1960s, 1970s and 1980s and the fact that younger cuttings tended to be built with flatter slopes as the understanding of delayed, deep-seated failure mechanisms improved. Therefore, there is a real possibility of disruption to the highway network occurring in the short to medium term.

Despite the findings of this study the fact remains that based on UK railway experience, where the asset base is much older, a number of deep-seated failures on the HA network would have been expected to have occurred by now. The key question is whether these deep-seated failures will ever occur (the analytical analysis needs refinement) or if the timescales to failure are in excess of UK rail experience. There are a number of factors that could explain the latter including improved initial quality of construction on the HA network, better drainage and different vegetation management practices. In order to resolve this aspect it is recommended that a number of key sites be subject to detailed assessment and monitoring. The sites could be identified from the pilot study recommended above to confirm the risk ranking model.

This paper reports on work that has been carried out by Mott MacDonald under a commission from the Highways Agency and the author wishes to thank D. Patterson of HA for reviewing this paper and giving permission to publish. Thanks are also expressed to T. Messafer and R. Lung from HA who managed the work. Special thanks to colleagues at Mott MacDonald for their input into the project and in particular to J. Perry, A. S. O'Brien and T. Spink for guidance and support.

References

Chandler, R. J. 1984. Delayed failure and observed strengths of first-time slides in stiff clay, a review. *Proceedings of the 4th International Symposium, Landslides*, Toronto.

Chandler, R. J. & Skempton, A. W. 1974. The design of permanent slopes in stiff fissured clays. *Geotechnique*, **24**, 457–464.

Ellis, E. A. & O'Brien, A. S. 2007. Effect of height on delayed collapse of cuttings in stiff clay. *Proceedings of the ICE Geotechnical Engineering*, **160**, 73–84.

GEOTECHNICAL CONSULTING GROUP. 2007. *Numerical Techniques and Appropriate Factors of Safety (Including Models for Delayed Failure of Clay Slopes).* GCG Phase 1 Report for the Highways Agency, Issue 02 Feb 2007.

Green, G. W. 1992. *Regional Geology Guide – Bristol and Gloucester Region*, 3rd edn. HMSO, London.

HIGHWAYS AGENCY. 2003. Maintenance of Highway Geotechnical Assets (HD41/03). *In*: *Design Manual for Roads and Bridges, Geotechnics and Drainage, Earthworks*, **4** (1), Part 3 http://www.dft.gov.uk/ha/standards/dmrb/vol4/section1/hd4103.pdf

Leonards, G. A. 1979. Stability of slopes in soft clays. Special Lecture. *6th Pan American Conference on Soil Mechanics and Foundation Engineering*, Lima.

MOTT MACDONALD. 2009. *Delayed deep-seated failure of over consolidated clay cutting slopes*. Mott MacDonald Report Ref: 250353/01/B, July 2009 for the Highways Agency.

Perry, J. 1989. *A Survey of Slope Condition on Motorway Earthworks in England and Wales*. Transport Research Laboratory, Research Report RR199.

Perry, J. 1991. *The extent and an analysis of shallow failures on the slopes of highway earthworks*. PhD Thesis. University of Durham.

Perry, J., Pedley, M. & Brady, K. 2003. *Infrastructure Cuttings – Condition Appraisal and Remedial Treatment.* Publication C591, CIRIA, London.

Potts, D. M., Kovakevic, N. & Vaughan, P. R. 1997. Delayed collapse of cut slopes in stiff clay. *Géotechnique*, **47,** 5.

Skempton, A. W. 1977. Long term stability of cuttings in brown London Clay. *Proceedings of the 9th International Conference of Soil Mechanics Foundation Engineering*, Tokyo, **3**, 261–270.

Tavenas, F. & Leroueil, S. 1981. Creep and failure of slopes in clays. *Canadian Geotechnical Journal*, **18**, 1.

Spanish technical regulation for road construction

E. Dapena[1]* & A. Parrilla[2]

[1] Doctor Ingeniero de Caminos, Canales y Puertos, Laboratorio de Geotecnia, CEDEX, Alfonso XII, 3, 28014 Madrid Spain
[2] Ingeniero de Caminos, Canales y Puertos, Dirección Técnica de Carreteras, Ministerio de Fomento, Pº de la Castellana, 67, 28071 Madrid, Spain

*Corresponding author (e-mail: e.dapena@cedex.es)

Abstract: The new *Technical Regulation for Road Construction* (PG3) has applied in Spain since 2002 and governs execution of all the highways, expressways and motorways run by State Administration.

The most important aspects introduced in this regulation in terms of the construction of fills are (1) extending the characteristics of the fill material to use as large a quantity as possible of the excavated ground *in situ* and to keep the amount sent to dump sites to a minimum; (2) to set up standards to regulate finished fills; and (3) to define a new type of random or quarry-run fill consisting of materials with a grading between earth and rock fill that were not in the earlier PG3 regulation.

The new PG3 regulation refers to the construction of earth, rock and random fills, and pays particular attention to the placement, installation and the relevant control of the methods involved. In the case of earth fills, the control method involves fixing some preliminary characteristics for the soils to be used in fills, determined by laboratory tests, whereas for random and rock fills, the control consists of first creating test fills through which the previously established compaction procedure is accepted or modified. For random and rock fills, the control of the finished fill requires dividing the layers into set-sized sections and running tests on predetermined strips. The regulation also defines the criterion for acceptance or rejection of the zone being tested.

In 2002 a new *Technical Regulation for the Construction of Roads, Motorways and Masonry Bridges* (PG3 2002) was brought out in Spain to replace the previous 1975 regulation (PG3 1976).

In two articles, the 1975 regulation laid down the conditions for placing and controlling compaction of the materials used in the execution of earth and rock fills. The current regulation updates these articles on earth and rock fills and adds a third on the conditions for the placement on site and control of the compaction process of some intermediary materials between earth and rock fills known as random fills. This paper reports on the content of these articles, particularly as regards the compaction of earth fills, rock fills and random fills.

Fill zones

To define the construction procedure and the material to be used in a fill, the fill zone is divided into five different sections (Fig. 1).

Foundation The bottom layer, in contact with the natural soil, may be subjected to water action as a result of flooding. It may sometimes be necessary to build it with drainage materials. It must be at least 1 m thick.

Shoulders These are the zones of the body of the fill where the effect of external agents is felt. In general they should comprise stable materials, which are sometimes required to be impermeable. Their width is not defined as it will depend on atmospheric conditions and the type of material involved in their composition; however an average of 2 m could be taken.

Core The core is the internal zone of the fill and generally has the greatest volume. It is here that the largest volume of the materials extracted from the excavation can be placed.

Transition The transition is the top layer of the fill, only specified in rock and random fills as this layer supports the crown. It must be at least 1 m thick.

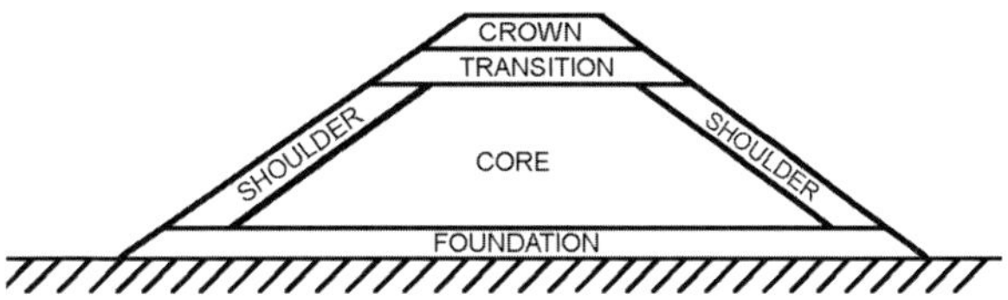

Fig. 1. Embankment sections.

From: RADFORD T. A. (ed.) 2012. *Earthworks in Europe*. Geological Society, London, Engineering Geology Special Publications, **26**, 55–60, http://dx.doi.org/10.1144/EGSP26.7

Crown This is the bottom layer of the pavement and is sized according to the amount and type of traffic that the carriageway is designed to withstand.

Classification of fills by grading

The type of material, classified by grading (Fig. 2), serves to distinguish the three types of fill: earth, rock and random. Rock fills are built of a material with a fines content of less than 10% and containing less than 30% of material under 20 mm in particle size. The maximum grain size is also required to be greater than 100 mm (Fig. 3). Earth fills have either a fines content greater than 35% or over 70% with a particle size less than 20 mm (Fig. 2). Random fills comprise all fill types built of a material that does not fulfil the conditions demanded for rock fills or earth fills, for example, a material with a fines content between 10 and 35% and where particles smaller than 20 mm in size account for between 30 and 70%.

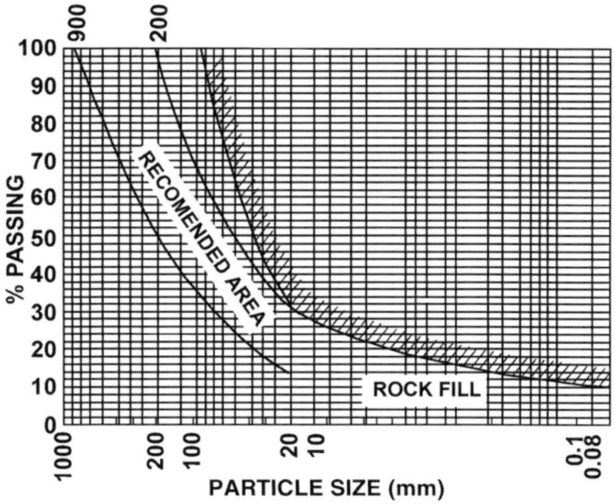

Fig. 3. Valid area for rock-fill gradings and recommended envelope for the coarser particles.

Earth fills

Use of marginal materials

Marginal materials are those which do not fully meet existing specifications. They can be used provided an adequate study of fill behaviour is carried out. Such so-called marginal materials include plastic soils with a liquid limit between 65% and 90% (Fig. 4); soils with a gypsum content greater than 20%; soils containing over 1% of other soluble salts such as magnesium sulphate; soils with over 2% organic matter; collapsible soils, that is which under an 0.2 N mm^{-2} load undergo over 1% settlement when flooded and soils with free swelling of 3–5%.

Marginal materials may only be used in fill cores, subsequent to a study identifying the properties making them marginal, analysing the effect of these properties on the behaviour of the fill and describing the appropriate measures to be taken for the fill behaviour to be classified as acceptable. The following soils cannot be used in earth fills: plastic soils with a liquid limit greater than 90%, soils with over 1% settlements caused by collapse, with over 5% free swelling or with over 20% gypsum content or over 2% organic matter.

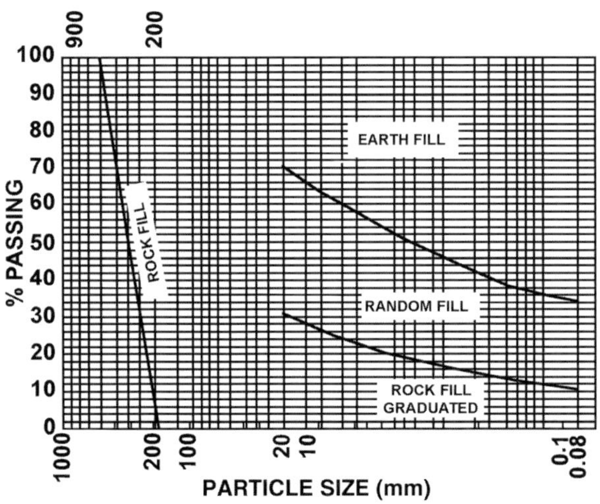

Fig. 2. Granulometric sections of different fills.

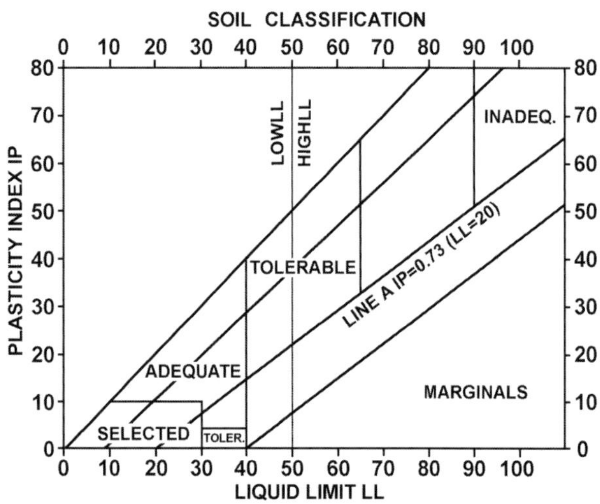

Fig. 4. Soil classification.

Layer thickness

Layers should generally be 30 cm thick and must be greater than 3/2 of the maximum particle size of the soil to be placed.

Compaction degree

In Spain, the tendency is to demand compaction that will provide densities higher than 95% of the density obtained in the Modified Proctor test. The optimum water content (W_{opt}) of the Proctor test defines $W_{opt}-2\%$ and $W_{opt}+1\%$ values (Fig. 5) to which SL_{-2} and SL_{+1} levels of saturation correspond. The compacted sample must have a level of saturation between SL_{-2} and SL_{+1}. For collapsible or expansive soils, the level of saturation of the compacted soil must fall within the interval defined by the $W_{opt}-1\%$ and $W_{opt}+3\%$ limits. The new regulation also requires that in the plate loadings test standard the vertical deformation (VD) modulus in the second VD_2 load cycle for earth fills is greater than or equal to 30 MPa and that the VD_2/VD_1 ratio is smaller than or equal to 2.2 (NLT 357 1998).

Compaction control

In earth fills, compaction control is carried out on the finished product. The section defined for control corresponds to a 500 m length of placed layer. This layer should cover the full width of the carriageway under construction and will be divided into three areas: two edges and the central strip (Fig. 6). The control is based on determining the density and water content at five points (one every 100 m) in the central area and a further five points in each of the edge strips. A plate-bearing test or a rutting test standard (NLT 256 1999) must also be carried out every 500 m section.

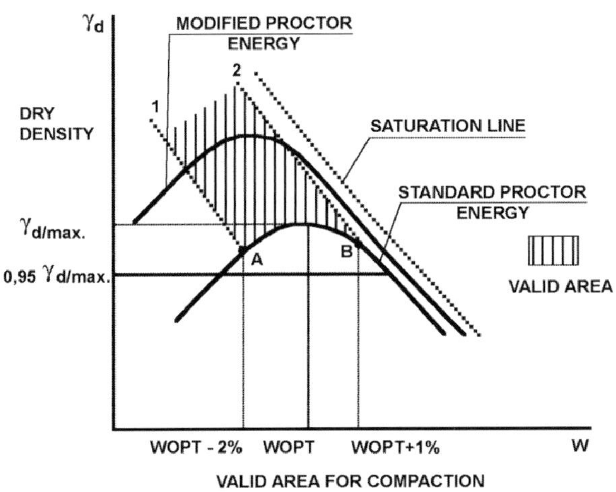

Fig. 5. Valid area for compaction in earth fills.

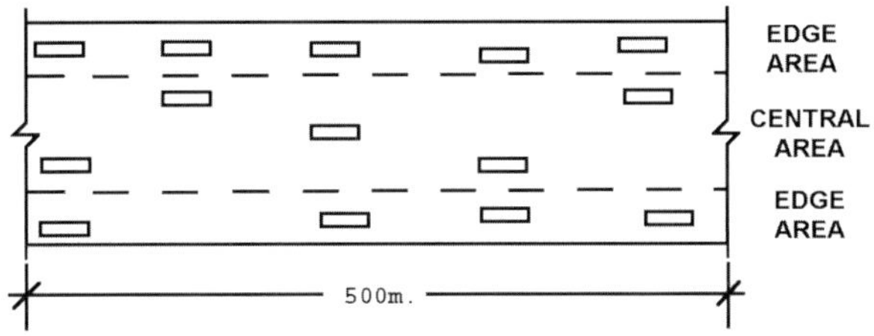

Fig. 6. Earth fills batch definition for compaction control.

Acceptance or rejection of the compacted earth fill

The following conditions must be met for the controlled layer to be accepted:

- The average value for density and water content must be in the zone defined as valid in the water content-density diagram (Fig. 5).
- In addition, 70% of the water content–density values obtained in the compaction control must be within the zone defined as valid.
- The modulus obtained in the plate loading test (NLT357 1998) must meet the required conditions of $VD_2 \geq 30$ MPa and VD_2/VD_1 less than or equal to 2.2.

Rock fills

Materials for rock fills

The new regulation requires materials for rock fills to be stable against water and chemical action. Water stability implies a less than 2% loss of weight of materials subjected to 25 wet/dry cycles. Chemical stability refers to the fact that they should not contain soluble salts, gypsums in particular, nor pyrites nor combustible minerals.

Layer thickness

Layers must be thick enough to obtain the desired compaction with the machinery available. In general this will be 60 cm, and never over 1.35 m nor three times the maximum particle size of the aggregate. It should also be greater than 3/2 of the maximum size.

Degree of compaction

The compaction of the rock fill should be such that:

- porosity is less than 30% ($n \leq 30\%$) (density $\gamma_d \geq 1.9$ g cm^{-3});
- the settlement of the rock fill after the last passage of the roller must be less than 1% of the layer thickness.

Compaction control

The control of the construction of a rock fill relates to the procedure applied. The compaction procedure to be used for each type of material, must be checked to ensure that the required degree of compaction is obtained. To this end, a test rock fill is built using the assigned material and following the proposed procedure. The procedure must specify layer thickness, the number of passes of the roller and the type of machinery to be used. In all cases, the roller must pass over more than four times.

The volume of the test rock fill must be greater than 3000 m^3, constructed in two or more layers with a width equal to or greater than 10 m. A topographic control test must be carried out on the compacted layer to ascertain the number of roller passes necessary to make the settlement produced in the last pass less than 1% of the layer thickness. The grading, density or porosity of the experimental rock fill must be measured by making three trial pits with a surface size of over 2 × 2 m and a depth equal to the layer thickness (the volume of these trial pits must be greater than 4 m^3).

The construction procedure will be accepted if these two conditions are met: that settlement after the last passage of the roller is less than 1% and that porosity is less than 30%. The procedure will only be valid so long as there is no change in the characteristics of the material.

Acceptance or rejection of the placed rock fill

The placed rock fill will be accepted if the compaction is carried out following the procedure used in the test rock fill so long as the material is the same as that used therein.

Random fills

Materials for random fills

The materials used in random fills must meet the relevant granulometric conditions for such a fill, that is between 10 and 35% of the material passing through a 0.063 mm sieve and between 30 and 70% of the material passing through a 20 mm sieve. Fills composed of rocks which are susceptible to weathering and degradation are also considered to be random fills, in other words, rocks that in 25 wet/dry cycles undergo a loss greater than 2%, rocks with soluble minerals and rocks with combustible minerals.

Rocks considered to be marginal, and therefore requiring a special study before their placement on site as random fills are rocks whose fraction passing a 20 Una Norma Española (UNE) sieve have characteristics of marginal or unsuitable soils, rocks containing oxidizable sulphurs, rocks with a gypsum contents higher than 20%, rocks with over 1% content of soluble salts other than gypsum and rocks with over 2% combustible mineral content.

Use of marginal rocks in random fills

Marginal rocks may be used in random fills provided that they are the subject of a special study covering the following aspects:

(a) appraisal of the properties causing the marginal nature;
(b) effect of these properties and their evolution on the behaviour of the fill;
(c) monitoring of the long-term deformation and stability of the fill comprising marginal rocks;
(d) handling care, construction arrangements and technical regulations to be adopted in each of the different uses of the marginal rocks included in the work.

Layer thickness

It is recommended that each layer carried out should be 40 cm thick. Thicknesses of over 60 cm are not permitted. This thickness must be greater than 3/2 the maximum size.

Degree of compaction

The degree of compaction of random fills should be such that:

- the density of the fill is over 95% of the maximum density obtained in the Modified Proctor test carried out on the material with a particle size less than 20 mm;
- the settlement undergone by the random fill after the last passage of the roller must be less than 1% of the initial layer thickness (there must be at least four passes of the roller);
- there must be less than 5 mm settlement in the rutting test (Fig. 7) (NLT 256 1999), except in the transition layer where settlement must be less than 3 mm.

Compaction control

The control of the construction of a random fill adopts a similar procedure to that for rock fills. For each material, the regulation proposes a construction procedure subject to experiments on a test fill. The test fill must be designed so that it has the degree of compaction defined for the random fill and that the procedure for this type of material is considered valid.

The test random fill must have a volume of over 3000 m^3, be constructed in three or more layers and have a width equal to or greater than 8 m. For the control process, the test random fill is divided into three strips, namely a central strip whose width is half the total fill width and two edge strips whose width is a quarter of the total fill width. Each of the three strips is subjected to a topographic control to ascertain the number of passes necessary so that the settlement obtained on the last passage is less than 1% of the layer thickness. Two rutting tests should be run on each of the strips to check that the settlement obtained is less than 5 mm. When these conditions have been met, five trial pits are to be made, two in the edge strips and one in

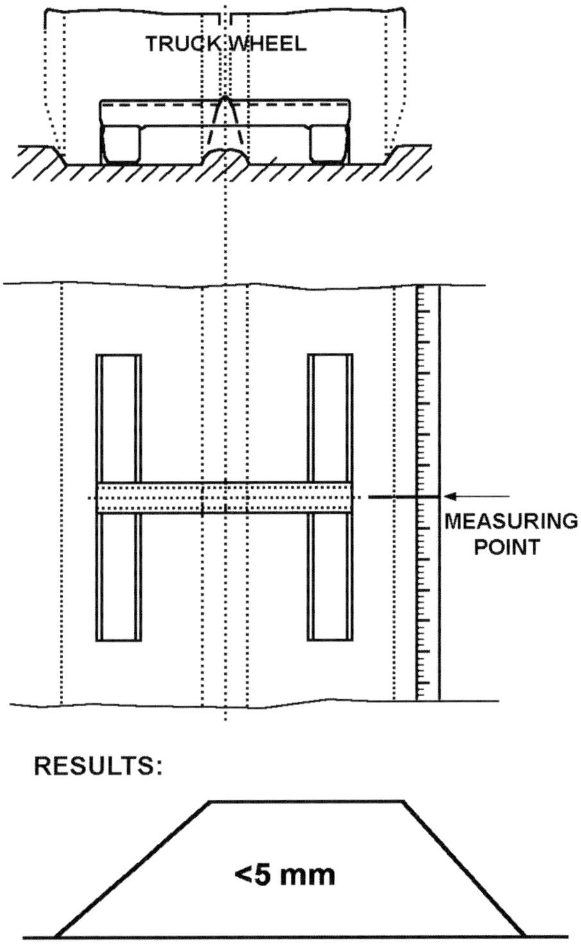

Fig. 7. Random fill rutting test.

the central strip, in order to determine the grading, water content and dry density of the compacted material. The trial pits must have a volume greater than 1 m^3, a ground plan surface area over 1 m^2 and sufficient depth to affect the entire layer thickness and only this layer thickness.

The construction procedure will be accepted if the following above-stated conditions are verified:

- the settlement provoked by the last passage of the roller in the topographic control carried out on the three strips (in all the determinations involved) is less than 1% of the layer thickness;
- there have been at least four roller passes;
- less than 5 mm settlement is obtained in the six rutting tests carried out on the three strips;
- the density determined in the five trial pits carried out is over 95% of the density obtained in the Modified Proctor test run on the material with a particle size under 20 mm.

Although the compaction control of the random fill is based on a procedure control, a control must be carried out on the finished fill using rutting tests and by determination of water content and density. To this end, the unit to be controlled is defined as a layer whose width is the same as the width of the carriageway under construction and that is 500 m long. Three strips are to be defined in this unit, namely an internal one half as wide as the total width of the layer and two edge strips a quarter as wide of the total layer width. A rutting test must be run on each of these strips (Fig. 7) and five determinations made of the water content and density of the fill, generally applying the sand replacement method (UNE 103503 1995) and a rutting test has to be run on one of these three strips.

Acceptance or rejection of the placed fill

The random fill already in place will be accepted if it is executed using the procedure established for this type of material and if the rutting test gives a settlement of under 5 mm (3 mm in the transition zone). In addition, the average value of the density test results and 60% of the density tests must be greater than the value required for the degree of compaction.

Conclusions

The PG3 2002 General Technical Regulation for the Construction of Roads and Bridges, published by the Ministry for Development in Spain, states the conditions that must be met in the construction of earth fills, rock fills and random, quarry-run fills in its Articles 330, 331 and 333.

An important idea that guided preparation of these articles was extension to the types of natural materials permitted for use in fills to minimize the quantity of material sent to dump sites. This idea includes the wording for Article 333 to be able to build random fills.

The three types of fill are defined by their grading, in terms of their fines content (material passing a 0.063 mm sieve). Materials are considered to be rock fills if the fines content is less than 10%, random fills if it is between 10 and 35% and earth fills if the fines content is over 35%. In addition, in rock fills the percentage of grains smaller than 20 mm must be under 30% whereas in random fills it should fall between 30 and 70%.

In terms of soils, the field of use of natural grounds has been extended, opening up the group of marginal soils that, with a particular study for placing them on site, are permitted for use in earth fills. This group of marginal soils includes gypsiferous materials with up to 20% gypsum content (extremely abundant in Spain) and which could cause environmental problems if sent to dump sites.

In random fills, the regulation permits the use of marginal rocks, such as those rocks which are susceptible to weathering and degradation with wet/dry cycles, rocks with oxidizable sulphurs, rocks with over 20% gypsum content, rocks with over 1% content of other soluble salts and rocks with combustible minerals in a greater proportion than 2%. Rocks with a fraction passing a 20 mm sieve considered to have characteristics of marginal soils are also considered to be marginal rocks.

The recommended layer thickness in each of the types covered is 30 cm in earth fills, 40 cm in random fills and 60 cm in rock fills. In any case, the layer thickness in the last two fill types is limited to 60 and 135 cm.

The compaction required in earth fills must enable a density to be obtained that is 95% of that corresponding to the reference Proctor test and an area of validity is defined for fixing the water content. For rock fills and random fills, the compaction procedure is fixed by building an experimental embankment at the start of the works. In any event, the number of roller passes must be at least four and the settlement in the last passage of the roller must be under 1% of the layer thickness. Compaction control is carried out on 500 m sections. In earth fills, the density, water content and modulus of deformation are determined in the second cycle of the plate-bearing test. In random fills, the density, water content and settlement are determined in the rutting test. In rock fills, only the compaction procedure has to be controlled, fixed by an experimental rock fill made with each type of material.

References

NLT 256. 1999. *Ensayo de Huella en Terrenos.* Normas NLT. CEDEX, Spain.

NLT 357. 1998. *Ensayo de Carga con Placa.* Normas NLT. CEDEX. Spain.

PG3. 1976. *Pliego de Prescripciones Técnicas Generales para Obras de Carreteras y Puentes.* Ministerio de Obras Públicas y Urbanismo. Spain.

PG3. 2002. *Pliego de Prescripciones Técnicas Generales para Obras de Carreteras y Puentes.* Ministerio de Fomento. Spain.

UNE 103503. 1995. *Determinación 'in situ' de la Densidad de un Suelo por el Método de la Arena.* Normas UNE. AENOR. Spain.

Czech standards and specification for earthworks

Vítězslav Herle

ARCADIS Geotechnika, a.s., Praha, Geologicka 4, 152 00 Praha 5, Czech Republic
(e-mail: herle@arcadisgt.cz; herle@geotechnika.cz)

Abstract: General requirements for earthworks in the Czech Republic are given in the national standard CSN 73 3050 '*Earthworks*' which has been in place since 1987. It provides information on excavation including the rather controversial seven excavatability classes, recommendation for slope and shoring of excavation, requirements for fill construction, control of compaction and tolerances of finished surfaces.

CSN 73 6133 '*Design and execution of earth structures on roads*' is currently being revised. Beside classical soil and rock fills it gives recommendations for the use of difficult or marginal materials such as uniform (wind drifted) sand, loess, colliery spoil, fly ash, slag.

Control of earthworks is described in CSN 72 1006 '*Compaction control of soils and other fills*' which is valid for both highway and railway infrastructure as well as for buildings, landfills etc.

Apart from national standards, the Ministry of Transport supports publications of technical specifications for design (TP) and construction control (TKP) focused on certain material or technology used in highway and railway infrastructure. TP coversfor example soil improvement, soil reinforcement, use of fly ash, colliery spoil or lightweight fill. TKP givesfor example detailed requirements for the site quality control of earthworks, reinforced fill, soil nailed slopes and gabions.

The Czech standardization system for earthworks on infrastructure projects is based on European design standards (e.g. EN 1997 Eurocode 7), European execution standards (e.g. EN 14475 *Reinforced fill*), Czech design and execution standard (e.g. CSN 73 6133 *Design and execution of earthworks on highways*; CSN 75 2310 *Earth dams*), technical design codes (e.g. TP 76 *Geotechnical investigation of highways*; TP 93 *Design and execution of earthworks on highways with use of fly ash*; TP 97 *Geosynthetics in earthworks on highways*) and standard specifications for quality control (e.g. TKP 4 *Earthworks*; TKP 16 *Piles and diaphragm walls*; TKP 30 *Special geotechnical constructions – reinforced fill, soil nailing, gabions*). A similar system of standards, codes and specifications is also used in Slovakia.

Czech standards for earthworks

The basic Czech standard for earthworks CSN 73 3050 *Earthworks* is almost obsolete and will soon be replaced by the thoroughly revised CSN 73 6133 *Design and execution of earthworks on highways*. It has 12 chapters and 5 appendices. The aim of this standard is to provide engineers both designing and on construction sites with advice on the suitability of soil and rock materials, approaches to the use of secondary by products such as fly ash, slag or colliery spoil as well as information on the use of lightweight materials or geosynthetics.

Earthwork categories

A key chapter of the new standard is the design of earth structures on highways (cuts, fills). Methods of design follow the principles given in Eurocode 7-1. Three geotechnical categories for earth structures are defined as follows:

Category 1
Cut not deeper than 3 m or fill not higher than 3 m. The earth structure is not in contact with running water and underground water has no influence on the structural behaviour. There is no peat, bog or other highly compressible soil, or collapsible soil (loess) in the subsoil. The area is stable (no landslide) and it is not influenced by subsidence due to mining activity in the past or at present. Design of the fill does not include lightweight materials, secondary by-products, soil improved by binders or geosynthetics reinforcement.

Category 1 earth structures can be designed without geotechnical investigation or with random test pits if necessary.

Category 2
Cut deeper than 3 m or fill higher than 3 m. Cuts not deeper than 3 m or fills not higher than 3 m may also belong to Category 2 if the conditions for Category 1 are not met or if there is a possiblility that the building may lead to slope instability or may cause excessive deformation of other constructions in the area. Design of earth structures in Category 2 must be based on geotechnical investigation with standard

From: Radford T. A. (ed.) 2012. *Earthworks in Europe*. Geological Society, London, Engineering Geology Special Publications, **26**, 61–65, http://dx.doi.org/10.1144/EGSP26.8

methods of field and laboratory testing. Special technical specification (TP 76) defines the minimum extent of the geotechnical investigation.

Category 3
Structures that do not belong into first two categories (extremely loaded retaining structures, unstable slopes and landslide areas, mining areas with risk of subsidence, use of by-products with unpredictable long term behaviour). Design of earth structures must be supported by detailed geotechnical study with standard and special tests, trial embankments or models. Structures in Category 3 must be monitored (settlement, horizontal movement, pore pressures, extension of geosynthetics reinforcement, etc.) over a long period until it is possible to declare that the behaviour is stable.

Spread footings

Due to the existence of new CEN and ISO standards, in particular Eurocode 7, the widely used Czech standard CSN 73 1001 *Subsoil under spread footings* (published in 1988) was withdrawn in April 2010. Many Czech geotechnical engineers had become used to its simple approach to the design of spread footings particularly in categories 1 and 2. The standard has a lot of tables giving for example 'allowable bearing capacity' of spread footing based on soil classification according to modified Unified Soil Classification System (USCS), ranges of undrained strength and compressibility in relation to soil classification and consistency or ranges of effective shear strength related also to soil class. This simple approach has been very popular, particularly with less experienced geologists who provided designers with recommendations for the design of footings based on few test pits or boreholes and based on visual classification or an estimate of soil classification recorded in the field.

A designer was able to use the table values of geotechnical data mentioned above and, commonly, due to lack of time did not consider whether these values were appropriate or, for most of the cases, overestimated. Over the past twenty years, this practice has lead to a lower quality of drilling and sampling (nobody required high quality samples), less laboratory testing particularly in mechanical tests (direct shear, triaxial tests, oedometer tests) and to very conservative or overestimated foundation design. In fact designers were content that geotechnical investigations were cheap and that they had all of the data they needed for the design. Nobody disputed the conservative design.

Experience with the Czech standard CSN 73 1001 led to the decision to eliminate table values from any subsequent standard that dealt with foundation design and to keep the positive parts that do not appear in other standards. It is hoped that both geologists and geotechnical engineers working on investigations will be encouraged to rely more on field data and laboratory testing of quality samples than on field estimate and standard table values. Some useful parts of CSN 73 1001 such as a modified USCS classification will be included in the revised CSN 73 6133 *Design and execution of earthworks on highways*. Design of spread footings will have to follow procedures in Eurocode 7.

Table 1. *Safety factor in relation to soil/rock and shear strength (CSN 73 6133)*

Cut		
Soil type	*Shear strength (effective)*	*Safety factor*
Fine grained	Peak	1.5
	Critical	1.15
	Residual	1.1
Coarse grained	Peak	1.2
	Critical	1.15
Rock	Peak	1.3
	Critical	1.15

Safety factor slope design

The safety factor approach to the stability of slopes has been retained in the informative annex to CSN 73 6133 to cover the period of the introduction of Eurocode 7 into design practice . In this standard the safety factor approach is refined and facilitates use of peak, critical and residual shear strength based on real situation in the field. The corresponding safety factor changes its values based on type of shear strength used in the calculations. Table 1 shows values of the safety factor for a cut in relation to soil/rock and type of shear strength.

Excavatability

The highly variable geological conditions in the Czech Republic cause problems of excavatability. The Czech standard CSN 73 3050 *Earthworks* specifies 7 classes of excavatability based on soil type, relative density, consistency, presence of coarse particles, cobbles, stones and rock blocks. As far as rock is concerned the standard gives a detailed rock specification from soft weathered rock (class 3, 4) which can be excavated with loaders to sound hard compacted rock (class 7) that needs blasting. Nevertheless, a long practice with this standard shows how easily this classification can be misused. In current practice, classification of excavatability is part of the geotechnical investigation report. This classification is usually prepared on the basis of randomly distributed small diameter boreholes; however; it serves for estimation of the building costs. When the rock mass changes in both vertical and horizontal directions or if the boreholes were poorly drilled or logging was carried out by a less experienced geologist then the situation in the actual excavation may be quite different from that described in the report.

The shifting of a cut excavation from a lower class to a higher one always results in extra costs and disputes with the client. In order to avoid or minimize these discussions

Table 2. *Classification of excavatability (CSN 73 6133)*

Class	*UC strength*	*Average density of discontinuities distance (mm)*		
CSN 73 6133	*MPa*	*< 150*	*150–250*	*> 250*
R 1	>150	II	III	III
R 2	50–150	II	III	III
R 3	15–50	II	III	III
R 4	5–15	I	II	II
R 5	1.5–5	I	I	I
R 6	<1.5	I	I	I
F 1–F 8				I
S 1–S 5				I
G 1–G 5				I
G a S with cobbles and boulders of 100–250 mm diameter in volume over 50% or with boulders over 250 mm up to 0.1 m^3 in 10–50% of total volume of excavated rock.				II

R, rock; F, fine-grained soil; S, sand; G, gravel; GaS, gravel and sand.

a new classification for excavatability was prepared for the revised CSN 73 6133. This new classification of excavatability was based on USCS classification for soil. Seven classes of excavatability in accordance with CSN 73 3050 were replaced with 3 classes described in CSN 73 6133. CSN 73 3050 was cancelled. Rock mass is classified according to unconfined compression (UC) strength as well as on the average spacing of the principle discontinuities. The number of classes was reduced from seven to three (Table 2):

Class I: Common excavation
All kinds of soil, decomposed rock with UC <5 MPa, secondary by products (slag, fly ash, colliery spoil). Shovel, dozer, loader or similar is used for excavation.

Class II: Rippable rock
It is necessary to use special disintegration tools such as rippers, pneumatic or hydraulic rock breakers. It is possible to use blasting for excavation if it is justifiable by speed of progress or economy of the work.

Class III: Rock for blasting
It is necessary to excavate the rock by blasting. Where seismic effects of the blasting might cause damage near buildings it is possible to use alternative methods of excavation (pneumatic or hydraulic breakers, wedges, etc.).

Special soils and secondary materials

As the Czech Republic is a highly industrialized nation, there are always secondary materials available in the vicinity, in particular siliceous and calcareous fly ash, high furnace slag, foundry sand and slag, and colliery spoil from deep and open cast mining . Further details on the specification for fly ash are given below as this material is frequently used.

Fly ash

The Czech Republic produces approximately 1 ton of fly ash per inhabitant per year or 10 millions tons in total. Some of this is used in industrial products, in particular cement production. However, most is deposited in landfills either in wet state (lagoons) or stockpiled dry. Contractors are encouraged to use it for fill. Some fly ash is rich in calcium (carbonate fly ash), commonly in the form of gypsum. Such products are not volumetrically stable and may cause heave in the upper parts of the fill. The standard CSN 73 6133 oulines a basic requirement (environmental, technical) for the use of fly ash and technical code TP 93 specifies how to design, construct and control embankments comprising fly ash. The current practice is to cover the external slope of the fly ash body with a soil layer which is compacted together with the fly ash. Figure 1 shows spreading of fly ash in the fill.

Loess

Loess is a low plastic, fine silt, wind deposit with high carbonate content that is typically very porous. The void content is usually over 40% which results in low density deposit. In the case of low moisture contents, this type of soil frequently demonstrates collapsible behaviour after wetting and is easily eroded. Consequently protection against erosion, in particular by rain water, is required if deep cuttings are to be formed in loess soil slopes.

If loess is present at the subgrade level it should be improved with lime. Loess reacts very well with hot lime to create a stable subgrade with low compressibility. If granular material is scarce in the project area then lime improved

Fig. 1. Spreading of fly ash in the road embankment.

soil is a good and environmentally friendly solution (there is no need to move the less suitable excavated soil out and bring granular material in). Details on design and application are in newly revised CSN 73 6133.

Soil improved with binders

Soil improvement is the preferred option in many projects. Lime is preferred for wet, fine grained soils as this reduces the plasticity and has a long term stabilization effect. It also works well in cold weather, even when temperatures are slightly below zero. Soil mixed with hot lime quickly increases in temperature, making spreading and compaction simpler. In Figure 2 the temperature of excavated clay soil is 3.5 °C. Figure 3 shows temperature of the same soil (22.9 °C) 30 min after mixing 2% of hot lime into the soil. In addition to hot lime, there are several special binder mixes (e.g. Dorosol) which usually contain fly ash, lime and dust from furnaces where lime is produced. They are cheaper and have a slower binding effect than lime.

Calcareous fly ash has also been used on several projects as a binder instead of lime. Although cheaper, a laboratory investigation is necessary to confirm that the mix of ash with soil will stay stable. Conditions for the use of improved soil are described in detail in the design code TP 94 '*Soil improvement on highway construction*'.

Fig. 2. Temperature of excavated soil.

Fig. 3. Temperature of soil after mixing with lime.

Lightweight fill

Economic comparison of various techniques of embankment construction on soft subsoil has sometimes led to the use of lightweight fill. The Czech Republic is a principal producer of lightweight aggregates that are made of clay which expands during process of burning in the furnace. This is one of the reasons why the technique of construction of lightweight fill has been included in the basic standard CSN 73 6133. In addition, a technical design code TP 198 *Lightweight fill on highways* was compiled in 2008. The current method of construction involves spreading a layer of expanded clay aggregates of 0.2 m up to 0.8 m in thickness and then covering it with *c*. 0.2 m of soil as an interlayer between thick layers of expanded clay aggregates (Fig. 4). Heavy vibrating rollers the compact the soil interlayer. This technique has been used successfully in many transition zones between bridge abutment supported by piles and a highway fill on soft ground. Due to its low weight (less than 25–30% of load of soil fill) settlement has been reduced accordingly. This has been confirmed by comparative numerical models and by settlement measurements on many highway fills.

The use of expanded polystyrene (EPS) for lightweight fill has not been popular due to its higher costs and an

Fig. 4. Spreading of lightweight expanded clay aggregate in the road fill.

uncertainty regarding its long term behaviour; expanded clay aggregates are substantially cheaper and behaves as any other soil fill in an embankment.

Reinforced fill

In the Czech Republic the design of reinforced fill is made in accordance with Eurocode 7 and CSN 73 6133. Design of the reinforcement is carried out in accordance with technical design code TP 97 *Geosynthetics in highway fill*. This manual gives detailed information on the use of geosynthetics for reinforcement, separation and filtration. There are also simple graphs for the design of reinforced fill which give guidance to investors, clients and less experienced designers who are less familiar with soil reinforcement.

Execution and quality control of reinforced fill is carried out in accordance with EN 14475 *Reinforced fill*, technical manual TP 97 *Geosynthetics in highway fill* and standard specification for control on site TKP 30 *Special geotechnical structures*which comprises three parts: reinforced fill, soil nailing and gabions.

The use of reinforced fill in highway construction is reasonably common; however, companies carrying out maintenance of highways request a maximum slope inclination of 1 V:1.5 H or 33.5° for slopes where grass has to be cut. For steeper slopes some form of erosion protection is required. A reinforced earth retaining wall is usually designed with soft facing (wrap around soil), gabions or concrete hollow blocks (see Fig. 5). The use of reinforcement at the base of embankments on soft subsoil (peat, bog, very soft clay) is particularly popular.

When building highways in mining areas with the risk of subsidence or in places where colliery spoil covers vast areas of great thickness and composed of loosely deposited clay lumps, reinforcement at the base of embankments (sometimes in combination with lightweight fill) can substantially reduce construction costs.

Fig. 5. Reinforced fill. Facing with hollow concrete blocs (lower), steel mesh (upper part).

Conclusion

This chapter provides a very brief overview of European and Czech standards, technical specifications for the design and standard specification for the quality control of earthworks used in the Czech Republic. These standards work well for infrastructure projects constructed under the auspices of the Ministry of Transport. Czech standards and specifications for other earthworks such as dams, canals, service lines (water, sewage, gas and electricity), buildings, landfills and others are not elaborated in such detail as the standard and specification for road infrastructure. Preparation of new specifications is commonly limited by finances and the willingness of experts to cooperate in their preparation.

References

CSN 72 1006. 1998. *Soil Compaction Control*. Cesky normalizacni institut, Praha.

CSN 73 1001. 1987. Foundation of Structures. Subsoil under Shallow Foundations. Urad pro normalizaci a mereni, Praha.

CSN 73 30 50. 1986. *Earthworks. General Requirements*. Urad pro normalizaci a mereni, Praha.

CSN 73 6133. 1998. *Road Earthworks – Design and Execution*. Cesky normalizacni institut, Praha.

CSN 75 2310. 2006. *Embankment Dams*. Cesky normalizacni institut, Praha.

EN 1997-1. 2004. *Eurocode 7: Geotechnical Design – Part 1: General Rules*. CEN Brussels.

EN 14475. 2006. *Execution of Special Geotechnical Works – Reinforced Fill*. CEN Brussels.

TP 76. 2009. *Technical Code for Geotechnical Investigation of Highways*. Ministry of Transport of the Czech Republic, Praha.

TP 93. 2003. *Technical Code for Use of Fly Ash in Design and Execution of Earthworks on Roads*. Ministry of Transport of the Czech Republic, Praha.

TP 97. 2008. *Technical Code for Use of Geosynthetics in Earthworks on Roads*. Ministry of Transport of the Czech Republic, Praha.

TP 198. 2008. *Technical Code for Use of Lightweight Fill on Roads*. Ministry of Transport of the Czech Republic, Praha.

TKP 4. 2009. *Standard Specification for Earthworks*. Ministry of Transport of the Czech Republic, Praha.

TKP 16. 2002. *Standard Specification for Piles and Diaphragm Walls*. Ministry of Transport of the Czech Republic, Praha.

TKP 30. 2009. *Standard Specification for Special Geotechnical Structures*. Ministry of Transport of the Czech Republic, Praha.

The work of ISSMGE TC3 (Geotechnics of pavements) and how it links to earthworks

A. Gomes Correia[1], A. Quibel[2] & M. G. Winter[3]*

[1] University of Minho, Campus de Azurém, 4800-058 Guimarães, Portugal
[2] CETE NC/CER. 10 Chemin de la Poudriére -BP245 76121 Grand Quevilly, France
[3] Swanston Steading, 109 Swanston Road, Edinburgh EH10 7DS, UK

*Corresponding author (e-mail: mwinter@trl.co.uk)

Abstract: The main goal of Technical Committee TC3 '*Geotechnics of Pavements*' of the International Society for Soil Mechanics and Geotechnical Engineering (ISSMGE) is to promote co-operation and exchange of information and knowledge about the geotechnical aspects in design, construction, maintenance and monitoring of pavements in transportation infrastructure (roads, railways, airports). Some of these aspects cover: earth structures in pavement and railway construction, compiling and homogenizing earthwork specifications, standardization of the continuous compaction control and promotion of the use of non-traditional by-products and recycled materials. This paper covers these main aspects and highlights the use of non-traditional materials in transportation earthworks.

The Technical Committee, TC3 '*Geotechnics of Pavements*', of the International Society for Soil Mechanics and Geotechnical Engineering (ISSMGE), started in 2001 under the proposal of the ISSMGE Board 2001–2005. It follows the previous activities of ETC11 (1997–2001). During all these activities important dissemination work related to earthworks was carried out:

- compaction of soils and granular materials (Gomes Correia & Quibel 2000);
- geotechnics for roads, rail tracks and earth structures (Gomes Correia & Brandl 2001);
- geotechnics in pavement and railway design and construction (Gomes Correia & Loizos 2004); and
- design and construction of pavements and rail tracks – geotechnical aspects and processed materials (Gomes Correia *et al.* 2007).

Furthermore, a workshop in connection with the 14th European Conference on Soil Mechanics and Geotechnical Engineering was held on the '*Management of Materials for Infrastructures in Urban Environments*' (Gomes Correia 2007). Themes such as: innovations in equipment, accelerated construction, flowable fills, micro-trenches and biomaterials for construction were discussed.

This paper addresses the specific work carried out related to intelligent compaction, mainly concerning harmonization and standardization, and the use of new, particularly recycled, materials in earthworks.

Intelligent compaction: a clearer understanding

In the framework of TC3 terms of reference a working group was set up to deal with management of compaction, including intelligent compaction (IC) and continuous compaction control. This involved promoting liaison between those dealing with road equipment and standardization. The main achievements of the work carried out under the leadership of Alain Quibel (secretary of French committee for road techniques, CFTR) are described here.

Implementation

From a European perspective, important work has been carried out to clarify the functionalities of the equipment proposed for intelligent compaction (Dussaugey 2009). The organizations involved in the project were:

- the ISSMGE TC3 committee, who identified the need for clarification;
- the 'road equipment committee' of the CFTR composed of manufacturers, contractors and administrative members, who identified the use of intelligent compaction in their area;
- CISMA, trade association of manufacturers and suppliers for the construction, infrastructure, steel, and material handling, who attend the CFTR road equipment committee meetings;

From: Radford T. A. (ed.) 2012. *Earthworks in Europe*. Geological Society, London, Engineering Geology Special Publications, **26**, 67–77, http://dx.doi.org/10.1144/EGSP26.9

- CECE, committee for European construction equipment. The secretariat of section I, for road equipment, is currently handled by the CISMA general secretary.

CISMA has played an important role in establishing a link between the ideas of TC3 reported in the CFTR road equipment committee, and the further discussions at CECE, making them practical. The methodology of the work implementation started with an enquiry to the main European manufacturers, prepared by CISMA, on the basis of a preliminary matrix describing the possible functions for intelligent compaction. The manufacturers were invited to present their own equipment for intelligent compaction in the matrix, and could also propose some improvements of the matrix itself.

This has resulted in an agreement between the main manufacturers about the way the equipment functions should be described and allowing each of them to present the possible uses of their equipment for intelligent compaction in a consistent way. Some details have yet to be finalized, but the present chapter outlines these descriptions, and gives some comments about their justification.

Description of the main functions and sub-functions in the IC rollers

For the time being six main functions (F) were agreed:

- *F1*: behaviour of the material, dealing with the dynamic response between the vibratory roller and the soil.
- *F2*: end of compaction, indication to the operator after comparison between successive passes.
- *F3*: positioning, traceability during process, satellite and associated methods; mapping of results.
- *F4*: operational information, showing the operator the parameters used that influence compaction.
- *F5*: control report, documentation, allowing the presentation of the end of compaction characteristics.
- *F6*: communication, etc., when the equipment is capable of transferring information.

The description of the available sub-functions helps to clarify how each main function is fulfilled. It is not an assessment of the 'performance' of the equipment for intelligent compaction, but a complete description. The comments are made to present the value of each sub-function for compaction management.

Behaviour (F1)

The sub-functions include:

- *F1a*: non-dimensional value (for example 0 to 1000). The scale is chosen to widen the variations obtained from accelerometers. This type of indication is very sensitive to small heterogeneities on sites. In the case of significant change (from a reference test), it may indicate a problem, the cause of which still needs to be analysed with other indicators like moisture conditions or support, to obtain suitable corrections.
- *F1b*: dimensional value: modulus, stiffness. A preliminary correlation between a standardized test and the compaction value for numerous different cases had to be established and introduced in the device. The modulus remains dependent on the depth action of the compactor on which the device is mounted. So, the correlation is easy to use on successive layers of the same characteristics, but sometimes, when large differences exist from one layer to the other, dimensional values on compactors may vary from those with specified tests.

  The depth action in the dynamic response is two or three times the operational depth action (a vibratory roller capable of compacting a 0.80 m layer provides a dynamic influence to a depth of about 2 m).

  Dimensional values concern the top of embankments. Modulus specification must be considered as complementary to compaction or densification achieved. Some non-qualities in compaction may occur when too great reliance is made to modulus: it is easy to understand when a soil dries, that more energy (number of passes or size of compactor) has to be used for the same result in density. But on the contrary, the required modulus is obtained sooner. The adjustment of the compaction process using only modulus is then subject to under-compaction, which will cause further settlement after moisture equilibrium during the service life of the resulting structure.
- *F1c*: qualitative observations: that is double jump, anarchic (chaotic) behaviour, etc. The double jump itself is not incorrect behaviour. It commonly occurs after a certain number of passes on difficult cases of compaction, but if either the amplitude or frequency is reduced to eliminate it, the efficiency decreases. An alert for anarchic (chaotic) behaviour, which is an extreme behaviour, is useful. The roller may be non-adapted.

End of compaction (F2)

'Compactomers' automatically compare the behaviour (F1) between two successive passes, generally in the same sense forward or backward, since the dynamic responses differ slightly dependent upon the direction of travel. Complete description will mention the method of comparison:

- *F2a, b*: comparison with an absolute target value. This target is non-dimensional or dimensional, depending on F1. It has to be fixed on a preliminary trial area. The condition of conservation of material characteristics is necessary to use the sub-functions properly, and the average value must be calculated on the same surface.
- *F2c*: comparison with relative evolution: for instance evolution $< 5-10\%$ of the dynamic parameter, non-dimensional or dimensional. This means that the dynamic response reaches an asymptote, and that further efficiency would be low. It is useful, but the most important aspect is to be sure that the compactor is of an appropriate size for the jobsite. In terms of sustainable development, a large number of passes with too small a roller uses a lot of energy and gives a poor result. It is

Natalie Dudley-Burton 3427

PSD-laser

Mark Davies

07468 742644

Paul Isaac (Pontarddulais)

07466 742583

50-75. BS1: 40.

grading 25kg. Roger.matthew@[illegible].co.uk.

Cyfoeth Naturiol Cymru
Natural Resources Wales

Mines on Afon Maethlon

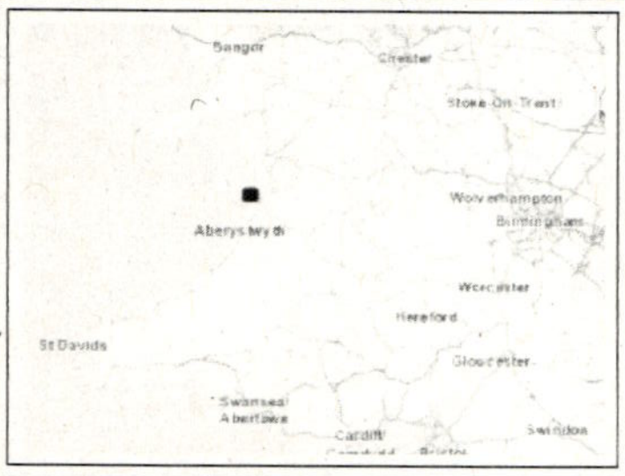

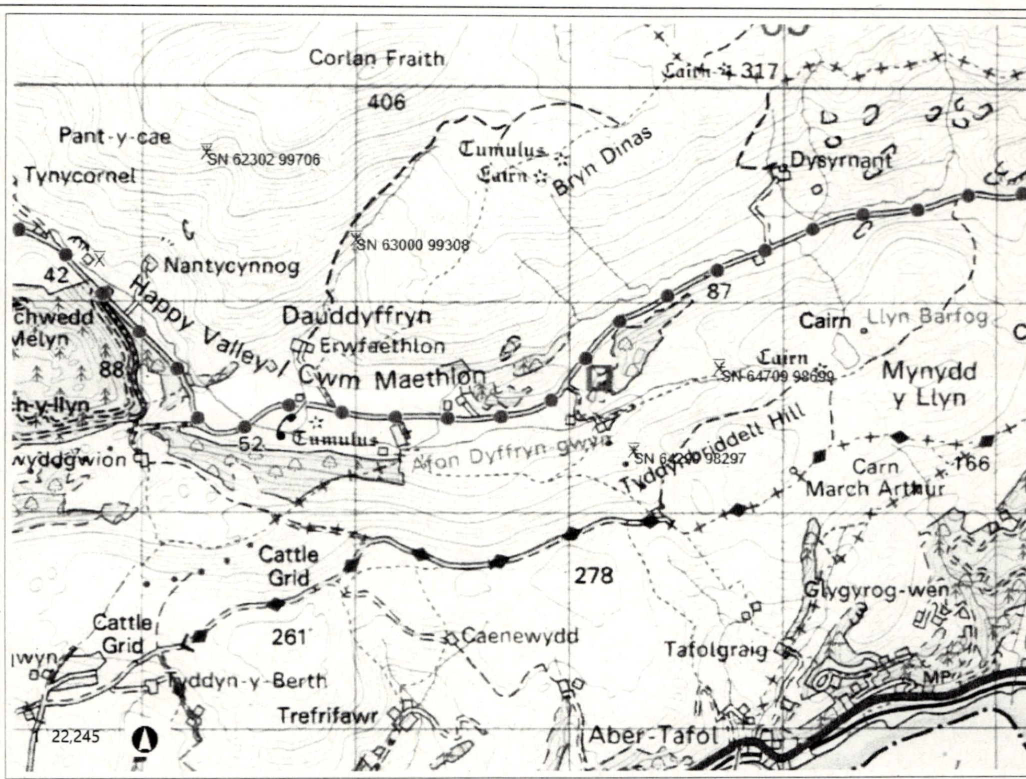

1.1 | 0 | 0.57 | 1.1 Kilometers

British_National_Grid

Legend

X Metal Mines

Notes

Nantycynnog SN62302 99706
Ffridd Hafodwen SN63000 99308
Llyn Barfog (E) SN640709 98699
Tyddyn Briddell SN 64299 98297

therefore a poor solution compared to a heavy roller making a few passes.

- *F2d*: comparison between the actual and a targeted number of passes. This sub-function is in accordance with positioning function F3. When operational tables of compaction pre-exist for the different cases of soils and equipment, like for the technical guide for earthworks in France (GTR 1992), the F2d method is the best to avoid under-compaction or over-compaction that can result in F2a,b,c when minor changes occur.

Positioning and traceability during compaction process (F3)
This sub-function describes the ability of the machine to monitor the roller position continuously:

- *F3a*: distance 1D: no satellite positioning exists, but a single presentation of result (dynamic response) *v.* length. To be sure of a good representation of the compacted area, the scheme must be strictly applied;
- *F3b*: 2D localization on board: the driver is helped by seeing the position of the roller during the process. This sub-function allows mapping of characteristics (see F3f) and progress of the percentage of the number of passes compared to the target one;
- *F3c*: 3D localization on board: the axle depth is available in addition to the sub-function F3b. Under-layers may be seen if necessary during the compaction process (weak points);
- *F3d, e*: 2D and 3D without GPS: other methods to obtain relative or absolute co-ordinates;
- *F3f*: mapping: presentation of results (dynamic response, number of passes) in selected areas.

On board operational information (F4)
The operator must have a certain number of indicators to manage the compaction or to identify that all the parameters are in accordance with the decisions taken for compaction and have the following details:

- *F4a*: actual speed V: seeing the actual travelling speed is important because the gradient of density increases and the evolution of density decreases when V is beyond a certain value, depending on soils and size of equipment. The operational tables of compaction give these optimum conditions;
- *F4b*: actual amplitude $A0$: several values may be available, or continuous variations can exist. The efficiency is related to $A0$. Correct choice or survey is therefore important;
- *F4c*: actual frequency: generally, compaction is made at maximum frequency. A lower frequency may indicate fatigue of hydraulic components. It can induce a decrease of efficiency. If adjustments are made because of anarchic (chaotic) vibratory behaviour at maximal frequency, the operator knows the corrective effect;
- *F4d*: actual number of passes: can clearly show the progress at the current position, in addition to mapping of the area;
- *F4e*: closed-loop mode: when the device can automatically adjust the parameters of the roller (amplitude and others), the operator can see which mode was used;
- *F4f*: asphalt temperature: the operator can adjust the position of the roller in priority on the lower warm areas, or closer to the finished layer, when this indication exists.

Control report, documentation (F5)
After compaction, various presentations of results may be included with the device, such as:

- *F5a*: identification of machine and continuous compaction control (CCC) device: to recall which equipment produced the results (traceability);
- *F5b*: results *v.* distance: related to F3a with 1D positioning;
- *F5c*: results by surface plot: related to F3b to F3e sub-functions;
- *F5d*: results by sequence of work (day, half day): according to global compaction coverage S when it is compared to volume of soil Q supplied during the same period. This method (Q/S) has been applied for many years with 'controlographs' that also show speed and vibratory conditions. New electronic versions of these devices may be integrated into intelligent compaction equipment;
- *F5e*: results as histograms and statistics: the software shows the population of recorded results on given areas, homogeneity, average, max, min, percentage of under-compacted or over-compacted zones, among others;
- *F5f*: data post treatment facilities, and additional information: adding separate spot-measurements, positioning landmarks, climatic conditions, among others.

Communication, others (F6)
The sub-functions deal with the transfer of information to other machines or to the supervisor's office, to have more complete management of the job carried out by several compactors or all using the same equipment. As these sub-functions are new, their description is not yet fixed and they may be developed further in the future, but they may include:

- *F6a*: data exchange from office: exchange 'on the fly'; wireless data to office software;
- *F6b*: data exchange between machines: for instance actual temperature from the finisher to compactors;
- *F6c*: set of parameters introduced in the machine (electronic card): the operator starts the job with suitable parameters controlled from the card;
- *F6d*: external control of parameters: the office can control 'on the fly' the set of parameters at every moment.

Overall view of functions and sub-functions of IC rollers

The six main functions (F1–F6) lead to 29 possible sub-functions. Each of which is related to one or several goals for compaction management (this term seems more appropriate than intelligent compaction). The general goals are:

- to help the operator during the compaction process (75% of sub-functions are concerned);

		Mapping on board F3f	Asphalt temp. F4f	Data post-treatment facilities, additional information F5f	
		3D location No GPS F3e	Closed-loop mode F4e	Result by histogram and statistics F5e	
	Number of passes (actual / absolute target) F2d	2d location No GPS F3d	Actual number of passes F4d	Results by work sequence (day, half-day) F5d	External control of parameters F6d
Qualitative observation (double-jump) F1c	Relative evolut. (% target with Dim. or Ndim.) F2c	GPS 3D on board F3c	Actual frequency F4c	Result buy surface plot F5c	Parameters introduced by card F6c
Dimensional (modulus, stiffness) F1b	Dimensional (actual / absolute target) F2b	GPS 2D on board F3b	Actual amplitude A0 F4b	Result versus distance F5b	Data exchange between machines F6b
Non-dimensional value F1a	Non-dimensional (actual/absolute target) F2a	Distance 1D F3a	Actual speed V F4a	Identification of machine and CCC device F5a	Data exchange from office F6a
Behaviour Dynamic response	End of compaction (Comparison)	Positioning, Traceability	Operational information	Control report, Documentation	Communication, Others

Fig. 1. Example of the overall matrix of IC rollers: embarked devices allowing 9 of the 29 sub-functions.

- to check the application of the compaction requirements (50% of sub-functions are concerned);
- to control the compaction automatically (10% of sub-functions are concerned). Some authors consider that the terminology 'intelligent compaction' is convenient only for this target, but intelligence of the workforce is still of great interest on sites, to avoid counter-performance in compaction even with on-board devices.

Each device can be described according to the synthetic matrix shown in Figure 1 that is unique for all domains of compaction, both earthworks and road-bases. It shows the types of functionality allowed. As a general agreement to use such a matrix already exists, it is then easier for contractors to compare different equipment and to select those that correspond to their needs.

Assessment of on-board devices

The assessment of on-board devices is not contained in the description, which is only the basis of terminology. Assessment is more difficult. It requires the answers to two questions:

- what is the value (interest) of each sub-function? For instance it is sure that F1b: dynamic response by a modulus is of a greater interest than F1a: dynamic response by non-dimensional value,
- what is the real value of the answer given by the device for this sub-function? For instance, when F1b is proposed, what kind of calibration between standardized modulus and 'compactor modulus' has been made and on what scale?

The first question is dependent on the requirements for use in each country. Harmonization of practices across Europe would permit a similar assessment method of equipment. The second question requires assessment by an independent committee, which would benefit from the manufacturers research efforts and improve the pertinence of solutions and accuracy of software in the equipment.

Impact of the new equipment for IC

We are in a period of transition. Without on-board devices, contractors have to apply constant conditions of compaction for a given job: amplitude, frequency, number of passes. In some countries, for example, France, a very large number of compaction tests that have been carried out over many years at the 'Centre d'Expérimentation Routière' (CER) in

Rouen, from which compaction tables for every case (nature of material, moisture conditions, roller parameters) have been produced. This knowledge base has been set out in technical guides like GTR (GTR 1992), for operational applications in earthworks, and it remains an important source. The latest generation of vibratory rollers allows multiple combinations of efficiency, manually or automatically controlled, and we also must take these innovations into account.

Experiments carried out at CER have shown that the best compaction effect is obtained when maximal energy is exerted on the loosened soil, and after the first passes when the vertical applied force is maximum. A forced amplitude at the first pass reduces the number of passes, and this offers an energy saving (per cubic metre). Furthermore the feasibility of single pass compaction (to current embankment compaction specifications, with moderately thick layers and high roller parameters: e.g. amplitude, frequency) has been established.

Contractors may legitimately expect to decrease the number of passes with new rollers (and then the number of rollers on big sites). However, trial test areas with density measurements are difficult to manage, and measurements are commonly too scattered to make correct interpretations. In a first approach with these new machines, it is recommended to avoid numerous and costly trial areas in earthworks (SETRA 2005). The best way is to refer to the manual and constant (highest) parameters on the machine, and to extract conditions from the tables of compaction presented in the GTR guide. To complete the process and allow extrapolations for machines with very high parameters, the compaction software simulation used to establish the tables would be of interest, and it is expected that this will be distributed when documentation is complete (Quibel 1994).

In addition to these recommendations, the modulation of parameters during the process is rarely needed in earthworks, except if anarchic (chaotic) behaviour occurs on difficult cases of soils (dry, rocky). But the adjustment of direction of applied force where possible may be useful especially on treated capping layers, when laminated surface phenomena are present, due to shear stresses induced by the drum.

The impact of intelligent compaction can be analysed from the point of view of the main causes of under-compaction on earthwork sites. Over-thickness is one cause: a thickness equal to twice the recommended thickness from the compaction tables leads on average to 10% reduction in density in the bottom of layer, and sometimes more. The number of passes cannot in anyway compensate for this, since density evolution is only about 1% when the number of passes is multiplied by 1.5. In addition to the comments about the energy of the first pass and heavy equipment able to reduce the number of passes, specific attention should be given to 3D positioning in devices, to ensure both correct thickness management, and homogeneity of compaction on every layer.

The use of devices that deliver modulus information, in addition to density, is anticipated. Weak points in the lower layers may then be detected and improved prior to the placement of subsequent layers.

IC pre-standardization work

The standardization work on intelligent compaction rollers at European national level started in the early 1990s (ZTVE StB/TP BF-StB 1994; RVS 8S.02.6, 1999; ATB Väg 2004). The ISSMGE TC3 activities during the 2001–2005 period culminated with a unified document (pre-standard) establishing technical contractual provisions and recommendations in order to prove continuously the compaction in different layers and planes in earthworks, roads, railways and airfield structures, by means of roller-integrated (RI) measuring systems (Adam 2007). This allows the operator to control and check the quality of the volume of compacted layers during the compaction process performed by vibratory rollers, saving time and reducing costs. Adam (2007) gives an introduction to intelligent compaction and the continuous compaction control principles, components and values. It includes the following sections:

- application
- definitions
- principles of RI-CCC with vibratory rollers
- application of RI-CCC
- prerequisites and requirements
- influences on the measured values
- calibration by compaction on a test field
- conduction of RI-CCC
- tests
- costs.

There are also appendices with important notes concerning measurements, operating conditions of the drum, linear regression rules for correlations and contractual terms and recommendations for tender texts for CCC application. Figure 2 shows the evolution of CCC values and defined limits as control criteria for standardization application (Adam 2007).

CCC can also be carried out using a test device, not a compactor, which is rolled on the ground while vibrating like a roller, so-called 'Portancemeter' (Fig. 3). This test device also allows continuous measurements of the load-bearing capacity of a platform, identifying weak points and their precise locations become straightforward (Quibel 1999). It can also give the variation of modulus continuously during rolling on the ground and defined limits as control criteria for standardization application. Refined between 1998 and 2001, this device has been standardized and resulted in a patent; moreover, it was awarded the innovation prize at the INTERMAT 2000 international trade show, where ISSMGE TC3 also held a workshop devoted to compaction (Gomes Correia & Quibel 2000).

Both these CCC methods, roller integrated CCC (RI-CCC) and 'Portancemeter', aim to provide a higher quality embankment or road layer by implementing a reliable quality and interactive assurance system, which in the case of RI-CCC starts from the very beginning of construction.

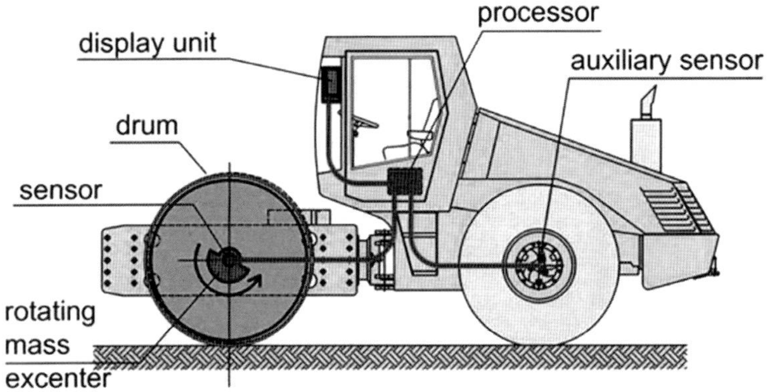

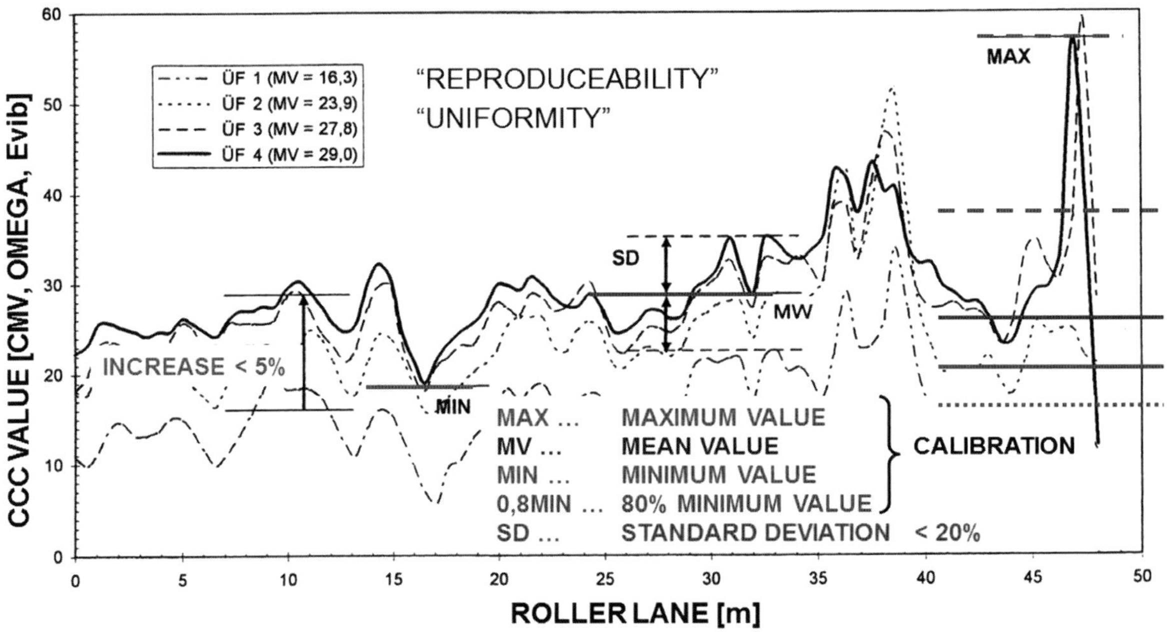

Fig. 2. Evolution of CCC values and control criteria for standardized applications (Adam 2007).

Materials and techniques in earthworks

There are a number of innovative materials and techniques that are currently available and/or emerging for use in earthworks.

The materials generally fall into four categories:

- reused and recycled geomaterials,
- non-traditional geomaterials such as marble,
- non-geomaterials such as tyre-derived products (shred, chip, crumb), and
- hydraulically-bound materials.

The techniques may be classified as follows:

- *in-situ* tests used for compaction control and performance based specification (see section on Intelligent compaction);
- continuous compaction control (see section on Impact of new equipment for IC); and
- compactor technologies such as impact rollers.

Materials

The growing demand for more sustainable solutions and the better use of natural materials, many of which are irreplaceable, creates a strong demand for the use of recycled

Fig. 3. 'Portancemeter'.

geomaterials, non-traditional geomaterials, non-geomaterials, and hydraulically bound materials (Winter & Henderson 2003).

Typical amongst the recycled category are the recycled asphalt planings (RAP) that were once viewed as a waste but now occupy a strong position in most national specifications (Winter & Henderson 2001). Indeed, in the UK a Quality Protocol (QP) allows such materials that have been processed in accordance with its requirements to be treated as primary materials once processing is complete. This is in contrast with those materials that have not been processed in accordance with the requirements of a QP and which still fall within the legal definition of waste, with the associated administrative burden that this status implies, until they are incorporated into permanently constructed works. In parallel the more efficient use of excavated earthworks materials is a priority (Winter 2004).

Less-traditional geomaterials include burnt colliery spoil (Winter 1998), spent oil shale (Winter 2001) (Fig. 4), and pulverized fuel-ash (PFA), or fly ash (Winter & Clarke 2002) (Fig. 5).

Non-geomaterials are typified by tyre-derived products and tyre bales in particular have seen widespread use in the UK and USA (Prikryl *et al.* 2005; Winter *et al.* 2005, 2006, 2009), not least as a foundation for roads over soft ground (Winter 2008). In the UK, a British Standard has been issued for the production of tyre bales for use in construction (Anon 2007; Simm *et al.* 2008).

Hydraulically-bound materials have come to the fore in recent years and now take a prominent place in the UK Specification for Highway Works for both road foundations and structural layers. In the late-1990s an experimental recycled road (Robinson *et al.* (2001) made substantial use of recycled materials (Fig. 6). Winter & Henderson (2001) estimated the proportion of recycled materials to be around 75% and Ghazireh & Robinson (2001) estimated substantial energy savings.

Tyre bales

Tyre bales comprise between 100 and 115 car/light goods vehicle tyres compressed into a lightweight block of mass around 800 kg and density *c.* 0.5 Mg m^{-3}, indicating a

Fig. 4. Placement of spent oil shale sub-base on top of a geosynthetic separator.

Fig. 6. Mixing lime into a clay subgrade material.

truly lightweight material. The bales measure approximately 1.3 × 1.55 × 0.8 m and are secured by five galvanized steel tie-wires running around the length and depth of the bale (Fig. 7). They have considerable potential for use in construction particularly where their low density and ease of handling places them at a premium. A porosity of around 62% and permeability of approximately 0.02 m s^{-1} through the length and 0.2 m s^{-1} through the depth (Simm *et al.* 2005) makes them ideal for drainage applications. The bale-to-bale friction angle is around 35° in dry conditions and stiffness in the vertical direction of Figure 4 is up to around 1 GPa (Winter *et al.* 2006; Freilich & Zornberg 2009). Furthermore, the process of tyre bale manufacture consumes around 1/16 of the energy required to shred a similar mass of tyres (Winter *et al.* 2006).

Fig. 5. PFA used to construct storage lagoon dams. Colliery spoil is used to dress the side slopes and minimize the potential for erosion due to runoff.

Substances which could potentially leach from tyres are already present in groundwater in developed areas. Studies suggest that leachate levels generally fall well below allowable regulatory limits and will have negligible impact on the water quality in close proximity to tyres (Hylands & Shulman 2003) and that rates of release decrease with time (Collins *et al.* 2002).

Spontaneous fires in whole tyre dumps are not known to the authors. In the USA, while combustion due to sparks from

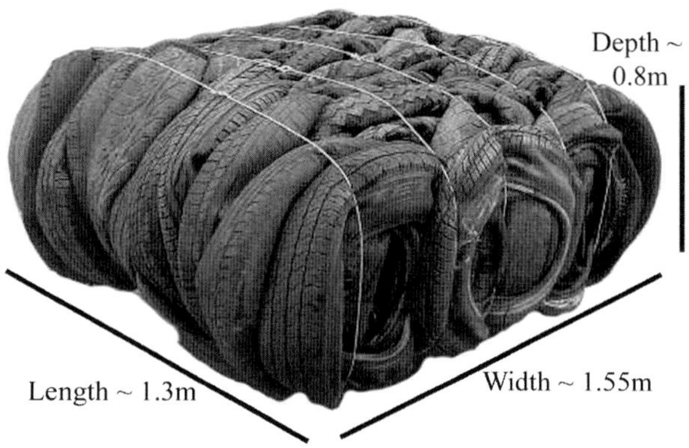

Fig. 7. A typical tyre bale with dimensions.

agricultural machinery and lightning have been reported, most observers suspect some form of arson in almost all cases. Baling piles of whole uncompressed tyres reduces the volume by a factor of four to five greatly reducing the available oxygen within the tyre mass as well as the exposed rubber surface area as tyre-to-tyre contacts are formed, without exposing any steel reinforcing in the tyres. The exothermic oxidation reaction potential is thus significantly reduced compared to whole tyres and the risk of spontaneous combustion from tyre bales is thus generally viewed as extremely low. A modelled storage condition for a 17.5 × 6.0 × 3.0 m volume of bales needed to reach and maintain a temperature of 188 °C for 39 days before spontaneous combustion became possible. In contrast reports have been made of internal heating of tyre shred and of apparently spontaneously combusted fires in large volumes in the USA (Sonti *et al.* 2000). Further details of tyre bale properties and behaviours are available (Simm *et al.* 2005; Winter *et al.* 2006; Anon. 2007; Freilich & Zornberg 2009).

Successful applications involving the construction of tyre bale road foundations have been achieved in both the USA (New York State) and the UK (Winter *et al.* 2005). Chautauqua County Department of Public Facilities completed five projects using tyre bales as a lightweight subgrade replacement for roads over soft ground (Fig. 8). The tyres result from the clean-up of a tyre dump and from a tyre amnesty programme. Future projects of this nature will depend upon the now limited availability of tyres.

The geology of Chautauqua County is characterized by sands and gravels in the river valleys with glacially deposited fine silty clays elsewhere, primarily on the hilltops which are often depressed forming high level swamps. These materials are stable if dry but are sensitive to moisture and to the freeze–thaw cycle which can turn them into a material like 'pottery slip' (used to manufacture clay utensils). Conventional roads constructed on them can turn into impassable quagmires. Tyre bale road construction was targeted on these roads.

To date, with the roads having been in service for up to nine years, no major signs of distress have been observed that could be attributed to the presence of tyre bales. In the case of County Road 342 (CR342) the traffic levels have been greatly increased (up to around 1600 vehicles per day Annual Average Daily Traffic (AADT) since construction due to local development.

A public road has been constructed by Highland Council in the far north of the UK (Winter *et al.* 2005). It was completed in late-2002 and performance to date has been broadly satisfactory despite extreme loadings imposed by a very high proportion of heavy logging trucks using the route.

One of the largest recent applications involved the construction of a lightweight embankment over an in-filled lake (Fig. 9). The construction involved the use of over 6000 bales (0.6–0.7 M tyres) and forms part of the A421 trunk road scheme promoted by the Highways Agency in southern England. Monitoring of the embankment indicates that it is performing entirely as expected approximately 18 months after construction. This project was the recipient of the British Geotechnical Association's prestigious Fleming Award for 2010.

Techniques

Techniques related to *in-situ* tests for compaction control, performance-based specification and continuous compaction control are discussed in earlier sections.

Compactor technologies related to the input of energy to the compacted material are relatively static in terms of their development. However, impact rollers, which have been developed in nominally triangular, square and pentagonal shapes allow high levels of energy to be applied to the

Fig. 8. Construction: bulldozing sand to fill voids, County Road 342 (CR342) in 2000 (left); vibrating sand into inter-bale voids, CR647 in 1999. (Photographs courtesy of Ken Smith, Chautauqua County Department of Public Facilities, New York State.)

Fig. 9. Construction of the A421 embankment.

material under compaction in a very rapid fashion; these have been developed in both South Africa and Australia.

The total compactive effort applied to the ground by the impact roller comprises three components:

- gravitational kinetic energy imparted to the ground by the fall of the impact mass;
- energy stored within torque arm springs during the lifting phase which is discharged during the falling phase; and
- rotational kinetic energy derived from the forward motion of the impact roller. The maintenance of optimum speed (between 9 and 11 km h^{-1} is generally quoted) is essential if the full effectiveness of the device is to be maintained.

Unpublished work by M. Winter suggests that typically a square-wheeled roller will achieve a higher dry density than an equivalent vibrating roller. With outputs of up to 1000 m^3 h^{-1}, this may be achieved rapidly. Layers of around 500 mm tend to be most suitable for compaction with an impact roller, generally thicker than those addressed by a vibratory machine, and that the compaction of subsequent layers can have a substantial densifying effect on lower layers. Allied to this is the tendency for the surface of the compacted layer to be corrugated and broken up (poorly compacted) and finishing of the final layer with a vibratory roller is usually recommended.

The high output of impact rollers is a great benefit but may limit their applicability. It can be difficult to supply material sufficiently rapidly to keep such devices gainfully employed in, for example, a road construction context. Where impact rollers seem to be able to make a very positive contribution to construction is in situations where very large amounts of fill need to be placed rapidly including, for example, filling disused quarries, and coal and discard stockpiling. Impact rollers may also be effective for proof rolling to identify residual weak spots after the compaction of large areas. The South African version of the square impact roller was originally developed to compact the dry, cemented aeolian (collapsing) sands of the Kalahari. A blanket of well-graded, partly cohesive material was normally placed on top of the sands as they would not generally support the weight of the impact roller. These devices were effective in ensuring that such sands collapsed prior to construction activities being commenced. A more generalized form of this application is the *in-situ* subgrade compaction prior to construction, and again impact rollers have been found to be effective in this arena.

Future developments

The agreement between the main manufacturers concerned with intelligent compaction devices for a harmonized and consistent approach to the description of the main functions of such devices is an important step and should be continued. This will, in the near future, provide the main framework for the worldwide implementation of a harmonized approach to compaction management (which is a broader and more appropriate term than is intelligent compaction).

ISSMGE TC3 activities will continue the work of promoting liaison at CEN (European Committee for Standardization) level to standardize continuous compaction control and also measurements carried out with other performance based devices. Its work includes the promotion of alternative materials. In this paper the use of non-traditional materials such as recycled asphalt planings (RAP), PFA (fly ash) and spent oil shale has been highlighted. RAP in particular was only recently viewed as a waste but now occupies a clear position in most national specifications. Similarly a formal specification for the use of PFA in earthworks has allowed its use in such applications with greater confidence.

Although environmental regulators and related bodies often speak of promoting the use of waste in construction, the industry itself often articulates a somewhat different view. As a highly competitive sector, in order for waste materials to be adopted for use on a regular basis a number of criteria must be met. These include *inter alia* the existence of a suitable specification, a favourable regulatory position, cost competitiveness and that the properties of the material are suitable for the application at hand. Tyre bales have been highlighted as a good example of a non-geomaterial that is subject to a robust specification, has lightweight and free-draining properties that benefit many applications and which are cost comparable or less costly than the more traditionally used materials. In addition, the ease of placement and construction often mean that construction with tyre bales confers significant time and plant savings.

The construction industry has an obligation to ensure that its activities are sustainable. However, it is important that sustainability is addressed in its broadest sense and that inappropriate and costly materials are not employed simply for the sake of increasing the use of recycled materials. There are many opportunities to take such appropriate action.

References

Adam, D. 2007. Roller-Integrated continuous compaction control (CCC): technical contractual provisions, recommendations. *In*: Gomes Correia, A., Momoya, Y. & Tatsuoka, F. (eds) *Design and Construction of Pavements and Rail Tracks – Geotechnical Aspects and Processed Materials*. Swets & Zeitlinger, Rotterdam, 111–138.

Anon. 2007. *Specification for the Production of Tyre Bales for use in Construction, PAS 108*. British Standards Institution, London.

ATB VÄG. 2004. Kapitel E – Obundna Material VV Publikation 2004:111. *General technical construction specification for roads*. Road and Traffic Division, Sweden.

Collins, K. C., Jensen, A. C., Mallinson, J. J., Roenelle, V. & Smith, I. P. 2002. Environmental impact assessment of a scrap tyre artificial reef. *ICES Journal of Marine Science*, **59**, 243–249.

Dussaugey, C. 2009. Les constructeurs européens de compactage et le compactage intelligent. *Revue Générale des Routes et des Aérodromes*, **875**, avril 2009.

Freilich, B. & Zornberg, J. 2009. *Mechanical Properties of Tire Bales for Highway Applications*. Report No. FHWA/TX-10/0-5517-1. The University of Texas at Austin.

Ghazireh, N. & Robinson, H. 2001. The Echline road construction project: environmental assessment. *In*: Yong, R. N. & Thomas, H. R. (eds) *Proceedings, Geoenvironmental Engineering: Geoenvironmental Impact Management*. Thomas Telford, 519–524.

Gomes Correia, A. 2007. ISSMGE TC3 Geotechnics of pavements, workshop on management of materials for infrastructures in urban environments. Geotechnical Engineering in Urban Environments. *In*: Cuéllar, V., Dapena, E. *et al.* (eds) *Proceedings of the 14th European Conference on Soil Mechanics and Geotechnical Engineering*, 24–27 September 2007, Madrid. Millpress, Rotterdam, **5**, 619–620.

Gomes Correia, A. & Brandl, H. (eds) 2001. *Geotechnics for Roads, Rail Tracks and Earth Structures*. Swets & Zeitlinger, Rotterdam.

Gomes Correia, A. & Loizos, A. (eds) 2004. *Geotechnics in Pavement and Railway Design and Construction*. MillPress, Rotterdam.

Gomes Correia, A. & Quibel, A. (eds) 2000. *Compaction of Soils and Granular Materials*. Presses de l'Ecole Nationale des Ponts et Chaussées, Paris.

Gomes Correia, A., Momoya, Y. & Tatsuoka, F. (eds) 2007. *Design and Construction of Pavements and Rail Tracks - Geotechnical Aspects and Processed Materials*. Swets & Zeitlinger, Rotterdam.

GTR. 1992. *Guide technique: Réalisation des remblais et des couches de forme*. Fascicules 1 et 2, Ref 9233 – LCPC/SETRA, Septembre 1992.

Hylands, K. N. & Shulman, V. 2003. *Civil Engineering Applications of Tyres*. Viridis Report VR5. TRL Limited, Crowthorne.

Prikryl, W., Williammee, R. & Winter, M. G. 2005. Slope failure repair using tyre bales at interstate highway 30, Tarrant County, Texas, USA. *Quarterly Journal of Engineering Geology and Hydrogeology*, **38**, 377–386.

Quibel, A. 1994. Modélisation pratique du compactage par vibration. *Bulletin de liaison des Ponts et Chaussées*, **192**, réf 3835.

Quibel, A. 1999. New in situ devices to evaluate bearing capacity and compaction of unbound granular materials. *In*: Gomes Correia, A. (ed.) *Unbound Granular Materials. Laboratory Testing, In-situ Testing and Modelling*. A. A. Balkema, Rotterdam, 141–151.

Robinson, H., Ghazireh, N. & Jeffrey-Wright, J. 2001. The Echline road construction project using suitable design and construction techniques. *In*: Yong, R. N. & Thomas, H. R. (eds) *Proceedings, Geoenvironmental Engineering: Geoenvironmental Impact Management*. Thomas Telford, 543–549.

RVS 8S.02.6. 1999. *Continuous compactor integrated compaction – Proof (proof of compaction)*. Technical Contract Stipulations RVS 8S.02.6 – Earthworks, Federal Ministry for Economic Affairs, Vienna.

SETRA. 2005. *Note d'information, Eléments techniques pour la conception et la réalisation de planches d'essais de compactage dans les chantiers de terrassements*. SETRA, août 2005.

Simm, J. D., Wallis, M. J. & Collins, K. (eds) 2005. *Sustainable re-use of tyres in port, coastal and river engineering: guidance for planning, implementation and maintenance*. SR 669. HR Wallingford, Wallingford.

Simm, J. D., Winter, M. G. & Waite, S. 2008. Design and specification of tyre bales in construction. *Proceedings, Institution of Civil Engineers (Waste and Resource Management)*, **161**, 67–76.

Sonti, K., Senadheera, S., Jayawickrama, P. W., Nash, P. T. & Gransberg, D. D. 2000. *Evaluate the Uses for Scrap Tires in Transportation Facilities. Research Study 0-1808, Centre for Multidisciplinary Research in Transportation*. Texas Tech University, Lubbock, TX.

Winter, M. G. 1998. *The Determination of the Acceptability of Selected Fragmenting Materials for Earthworks Compaction*. TRL Report 308. Transport Research Laboratory, Crowthorne.

Winter, M. G. 2001. Spent oil shale use in earthwork construction. *Engineering Geology*, **60**, 285–294.

Winter, M. G. 2004. Determination of the acceptability of glacial tills for earthworks. *Quarterly Journal of Engineering Geology and Hydrogeology*, **37**, 187–204.

Winter, M. G. 2008. Tyre bale foundations for roads on soft ground. *In*: Brown, M. J., Bransby, F. M., Brennan, A. J. & Knappett, J. A. (eds) *Proceedings, Second BGA International Conference on Foundations, ICOF 2008*. IHS BRE Press, Bracknell, **2**, 1437–1448.

Winter, M. G. & Clarke, B. G. 2002. Improved use of pulverised fuel ash as general fill. *Proceedings, Institution of Civil Engineers (Geotechnical Engineering)*, **155**, 133–141.

Winter, M. G. & Henderson, C. 2001. *Recycled Aggregates in Scotland*. Central Research Unit, Published Research Report Series. The Scottish Executive, Edinburgh.

Winter, M. G. & Henderson, C. 2003. Estimates of the quantities of recycled aggregates in Scotland. *Engineering Geology*, **70**, 205–215.

Winter, M. G., Reid, J. M. & Griffiths, P. I. J. 2005. *Tyre Bales in Construction: Case Studies*. TRL Published Project Report PPR045. Transport Research Laboratory, Crowthorne.

Winter, M. G., Watts, G. R. A. & Johnson, P. E. 2006. *Tyre bales in Construction*. TRL Published Project Report PPR080. Transport Research Laboratory, Crowthorne.

Winter, M. G., Williammee, R. & Prikryl, W. 2009. The application of tyre bales to the repair of slope failures. *Proceedings, Institution of Civil Engineers (Engineering Sustainability)*, **162**, 145–153.

ZTVE STB/TP BF-STB. 1994. Surface Covering Dynamic Compaction Control Methods. German Specifications and Regulations, Additional Technical Contractual Conditions and Guidelines for Earthwork in Road Construction and Technical Testing Instructions for Soil and Rock in Road Construction, Research Society of Road and Traffic, Germany.

Optimum use of material: selection of limits for suitable earthworks fill: Irish experience

P. C. Rutty[1,2]* & T. P. Johnston[3]

[1] Mott Macdonald Group Ltd, 8–10 Sydenham Road, Croydon CR0 2EE, UK
[2] Previous address: AGEC Ltd, The Grainstore, Singletons Lane, Bagenalstown, Co. Carlow, Ireland
[3] AGEC Ltd, The Grainstore, Singletons Lane, Bagenalstown, Co. Carlow, Ireland

Corresponding author (e-mail: peter.rutty@mottmac.com)

Abstract: Most Irish earthworks fills consist of glacially derived soils with some excavated rock. Excavated rock and granular glacial soils will nearly always provide acceptable fill. Cohesive glacial soils are generally well-graded tills, characterized by low plasticity and relatively high undrained strength. These cohesive tills provide acceptable fills providing they can be placed and compacted to provide sufficient strength and stiffness, and achieve low air voids.

In standard earthworks specifications, acceptability limits for cohesive fills may be specified by moisture content, moisture condition value or undrained shear strength. The lower limit of strength or moisture content adopted ensures acceptable trafficability, compressibility and stability. Laboratory relationship testing is carried out to establish the relationship between the various test properties.

Experience has shown that simple measurements and descriptions of strength derived from data collected routinely during ground investigations are adequate for estimation at tender. For construction control, moisture condition value testing has been widely adopted, is suitable for the range of soils encountered and provides quick results allowing timely decision-making. Irish tills are moisture susceptible and on-site results allow marginal materials to be processed on site for re-use, providing care is taken to work only in dry weather and protect placed fill.

Over the past ten years or so there has been substantial road building activity in Ireland to improve highway infrastructure with motorways or high-quality dual carriageways connecting the major commercial centres of the Republic of Ireland to Dublin, Dublin to Northern Ireland, and also Belfast to other parts of the north. Generally these projects have been procured using design and build or design, build, finance and operate (Public–Private Partnership or PPP) contracts. The contracts have placed most of the ground risk with the contractors but also encouraged value engineering by way of using excavated marginal material for earthworks construction where appropriate.

Most Irish earthworks fills consist of glacially derived soils with some excavated rock. Excavated rock and granular glacial soils will nearly always provide acceptable fill. The cohesive glacial soils generally are over-consolidated well-graded tills, characterized by low or medium plasticity, and with relatively high strengths (e.g. Clarke *et al.* 2001). These cohesive tills normally meet the classification of general cohesive fill under the National Roads Authority (2000) of Ireland Specification for Road Works (NRA SRW) and UK Highways Agency Specification for Highway Works (HA SHW). Acceptability of cohesive till for use as general earthworks fill depends on its strength. However, due to their low plasticity, Irish cohesive tills are extremely sensitive to minor changes in moisture content, with a significant strength reduction resulting from a small increase in moisture content. This sensitivity often makes the use of criteria based on moisture content limits difficult in the assessment and classification of earthworks materials for general fill. Over time the material properties used to assess and specify earthworks acceptability have changed and developed allowing better use to be made of site arisings.

Within the published specifications, grading limits are predetermined, and for general cohesive fills the designer has the option of specifying acceptability limits by moisture content or moisture condition value (MCV) or undrained shear strength. These limits are decided based on factors such as earthworks trafficability, fill compressibility and slope stability. Originally moisture content was used, later in combination with plastic limit (PL), to set the acceptability limits of fill. For some cohesive materials, including cohesive glacial soils, moisture content alone or with PL has been found to provide poor guidance to acceptability and other tests have been employed.

For monitoring construction acceptability of cohesive tills, the MCV test (Matheson & Winter 1997) has gained

From: Radford T. A. (ed.) 2012. *Earthworks in Europe*. Geological Society, London, Engineering Geology Special Publications, **26**, 79–92, http://dx.doi.org/10.1144/EGSP26.10

Fig. 1. Ireland showing locations of case history projects.

wide acceptance; however, MCV testing is not always carried out as part of ground investigation work, and when testing has been undertaken, results have commonly correlated poorly with other material acceptability parameters. The paper describes other means of assessing the strength of cohesive tills, particularly for design, and presents summary case studies from three road construction contracts in Ireland (Fig. 1).

Specifications and acceptability limits

Specifications

The earthworks specification used for road construction in the Republic of Ireland is the National Roads Authority (2000) (NRA) Specification for Road Works (SRW). The

SRW has been adapted for use in Ireland from the UK Highways Agency Specification for Highway Works (SHW). The SHW is employed for projects in Northern Ireland with amendments incorporated by the Roads Service of Northern Ireland. Throughout Ireland, available materials for earthworks typically comprise soils that have resulted from glacial activity, together with excavated rock. Series 600 is the section within both the SRW and the SHW that specifically deals with earthworks and appendix 1/5 is used to specify the frequency of testing during construction for classification purposes.

Earthworks materials are classified in table 6/1 of Series 600, with grading requirements set in table 6/2. Class 1 general granular materials are defined as those with less than 15% fine material (i.e. clay and silt size particles, $<63\ \mu m$). The specifications further sub-divide Class 1 materials into well-graded and uniformly-graded materials on the basis of uniformity coefficient, and to coarse granular material which may include particle sizes up to 500 mm and for which a minimum 10% fines value is required. Unlike UK SHW, there is no requirement in SRW to specify other acceptability limits, such as moisture content, for Class 1 materials.

Class 2 general cohesive materials as those with more than 15% fine material. The specification sub-divides Class 2 into four sub-classes: Class 2A and 2B, wet and dry cohesive materials respectively, comprise predominantly fine material with no more than 20% >2 mm, and are differentiated on the basis of moisture content relative to plastic limit; Class 2C, stony cohesive material, has a wider grading allowing up to 80% >2 mm which generally includes most excavated cohesive tills; and Class 2D, silty cohesive material occurs rarely in Ireland. Using the Irish SRW (and similar to UK SHW) it is the responsibility of the earthworks designer to decide which acceptability limits to use and to set the limits of acceptability.

Granular fills

Most Irish glacial sands and gravels, are well-graded, containing a wide range of particle sizes and are classified as Class 1A. In the Republic of Ireland it has been found that these fills drain readily and there is no need to specify moisture contents to achieve adequate compaction, though the requirement is retained in UK SHW. Generally as-dug rock fill material also will be well-graded and is classified as either Class 1A or Class 1C and is rarely processed unless required to be used as a 'selected fill'.

Uniformly graded silty fine sands, that could be classified as Class 1B material, occur more rarely and these can be more difficult to work as earthworks materials as they require careful management of water content, placement, compaction and subsequent trafficking. Selection of acceptability limits for granular materials for earthworks is covered in the literature and reference might be made to for example Trenter (2001) or UK HA44/91. Use of granular fills in earthworks is not considered further in this paper.

Cohesive fills

Typically Class 2A or 2B materials are rarely found in excavations on Irish roads projects and are not considered further.

Generally most cohesive tills meet the grading requirements for Class 2C 'Stony cohesive material' and provide the majority of earthworks fill material in Ireland. To employ such materials, a design should ensure that they are capable of being excavated, placed, compacted by appropriate roller, and subsequently trafficked (but only by earthmoving plant not as a general haul route) to allow succeeding layers to be placed and compacted. Earthworks management should avoid over-trafficking of placed material as this will cause rutting and material deterioration, perhaps leaving shear surfaces within the fill. To allow these materials to provide an adequate road pavement foundation, it will usually be necessary to employ a capping layer of good quality granular material. In the long term, design should ensure no undue settlement of the fill and stable slopes.

To achieve these requirements the fill requires appropriate strength and stiffness, both during construction and long term. The long term requirement for stability depends on the effective friction strength and requires the designer to select a safe slope angle using conventional design methods (see e.g. Trenter 2001 or UK HA44/91). Adequate construction and long term stiffness are achieved by proper earthworks management including compaction to achieve low air voids. Note that over-compaction should be avoided as this can lead to 'mattressing' when high pore pressures are generated in lower layers of wet clay fill, preventing proper compaction of subsequent layers. Short term strength is required to enable earthmoving and compaction plant to traffic the newly placed layers and construct further layers of fill. This strength might be ensured by different means; traditionally in the Ireland moisture content was used; however more recently, more direct measurements of short term strength have been used. The strength requirement for earthworks fills is no more than this and for most Class 2C fill it will be necessary to form bespoke haul roads to allow more intensive trafficking.

Class 2D material is not commonly encountered in Ireland and is not considered further.

Moisture content: historical use, limitations, suitability and relationship testing

Historical use

Early methods of assessing soil acceptability were based on either visual recognition of unsuitable soil types or by the establishment of an upper limit of moisture content beyond which the soil was deemed to be unworkable (see e.g. Winter 2001). Over the last fifty years, the earthworks specification (in both the UK and Ireland) has been altered several times, partly reflecting the difficulties in adequately specifying cohesive fills.

In the 3rd edition of UK *Specification for Road and Bridge Works* (HMSO 1963) all excavated material was classified as either suitable or unsuitable, with suitable material defined as that which was acceptable for use in the works. For the purpose of use in earthworks all excavated material was classified as either plastic or non-plastic. Non-plastic materials were defined as those on which it was impossible to carry out a plasticity index test. Plastic materials were defined as all other acceptable materials described in CP 2001 as 'fine grained cohesive materials'. The amount of compaction used was decided by the contractor and confirmed by site trials, to achieve an end product dry density corresponding to an air voids content not greater than 10% at the measured moisture content (5% in topmost 2 feet below formation). For 'plastic materials', compaction was only permitted when the moisture content was not greater than 2% above the PL, though the accompanying notes for guidance indicated that higher values could be used for some clays. These specification requirements would appear to exclude most Irish tills as, although they are typically cohesive (i.e. have a plasticity index) they could not be described as fine grained cohesive material as defined in CP 2001 because of their gravel content.

The 4th edition of *UK Specification for Road and Bridge Works* (HMSO 1969) implemented method compaction. Suitable excavated soil materials acceptable for use in earthworks, capable of compaction and of forming stable specified slopes were classified as either:

- 'Cohesive soil', included clays and marls with up to 20% gravel and moisture content not less than PL–4;
- 'Well-graded granular and dry cohesive soils', includes clays and marls with more than 20% gravel and/or having a moisture content less than PL–4; this would include typical Irish cohesive tills together with most sands and gravels;
- 'Uniformly-graded material', includes sands and gravels with uniformity coefficient less than 10, and silts and pulverised fuel ash (PFA).

In this 4th edition, the upper moisture content limit was to be decided by the designer, and the notes for guidance suggested a value of $1.2 \times$ PL.

The 5th edition of UK *Specification for Road and Bridge Works* (HMSO 1976, 1978), generally referred to as the 'Blue Book', used similar classifications for material suitability and compaction methods. The accompanying notes for guidance stated that to achieve maximum economy, the specification had been drawn up to encourage best use of all materials on the site and that the design principle should be that if materials can be excavated, transported, and compacted they were suitable for most earthworks. The notes reported the TRRL Report LR 406 conclusion that in certain instances cohesive soils can be used with moisture content up to $1.3 \times$ PL, although this may depend on the final design requirements for the earthworks, and also may restrict the plant deployed by the contractor. In practice, moisture content limits between 0.9 and $1.3 \times$ PL were selected by designers based on trials and experience. The Irish Department of the Environment adapted this specification for Irish conditions and published the *Specification for Road Works* in 1978 (Davitt 2001). It was commonly referred to as the 'Green Book'.

The 6th edition of the *UK Specification for Highway Works* was published in 1986 and was referred to as the 'Brown Book'. This edition introduced a change to the system for classifying excavated materials, introduced a wider range of definitions of suitable materials that would be acceptable for use in earthworks and moved away from reliance on moisture content to specify acceptability by allowing the designer to adopt other measures, such as strength. Typical Irish tills will fall into Class 2C stony cohesive material.

The 7th edition of the *UK Specification for Highway Works* (SHW, Highways Agency) was published in 1998 and continued the classification system adopted in the 'Brown Book'. To advise designers on the design and preparation of contract documents, the UK Highways Agency produced advice note HA44/91 (1991). This provided advice on selecting appropriate strength limits to define acceptability of the various earthworks classes and for Class 2 general cohesive fill it suggested that a lower limit could be 35–50 kPa and an upper limit 150–200 kPa. The option of establishing equivalent values of moisture content or MCV to these strength limits was suggested.

In 2000, the NRA published a new *Specification for Road Works* (NRA SRW), based on UK HA SHW 7th edition (1991), and this has been used for all trunk road and motorway projects constructed more recently in Ireland. The UK specifications are used in Northern Ireland.

Limitations

Traditionally, moisture content has been employed as a measure of acceptability for cohesive soils, as it is simple and relatively quick to measure; however it cannot easily be measured in the field, and usually requires laboratory testing. The use of a moisture content relationship with PL to determine earthworks suitability is based on the change in consistency of a cohesive material as its moisture content increases. The PL is the point at which the soil changes from a 'semi-solid' state to a 'plastic' state. At PL undrained shear strength has been assessed variously as 110 kPa to 170 kPa (Wroth & Wood 1978; Whyte 1982). The liquid limit (LL) is the point at which the material becomes liquid. As the moisture content of a cohesive material approaches its LL, the undrained shear strength of the soil reduces until the strength is approximately 1.5 kPa. Typically, moisture content of soils encountered in earthworks ranges from below PL to less than LL, generally in the 'plastic' range.

Relationships with PL require PL to be determined for each sample, increasing the number of tests for each sample and the time taken. PL is measured on the fraction of soil less than 425 μm and includes some sand in the sample, while moisture content is measured on that part of the sample with sizes less than 20 mm (BS1377 1990). Hence there is a discrepancy between the material used to measure PL (particle sizes less than 425 μm) and the fraction used for moisture content determination (less than 20 mm). If a moisture content relationship with PL (or liquid limit) is used then the moisture content should be adjusted to give a 'matrix moisture content' that represents the moisture content of the fraction less than 425 μm. Dumbleton & West (1966) showed PL varies with the granular content of the sample and the nature of the sand (shape, grading, etc). A number of workers have shown that determination of PL is subject to operator skill and may not be repeatable (Sherwood 1970; Arrowsmith 1979; Whyte 1982), perhaps partly because there is a subjective element in rolling the clay sample. Barnes & Staples (1988) showed that as the stone content of a till increases then the multiplicand applied to PL should decrease to maintain constant minimum shear strength. For these reasons moisture content, and its relationship to PL, are no longer widely used to assess acceptability of cohesive soils.

Acceptability

Materials that are acceptable for use in earthworks must be capable of being placed, compacted, and subsequently trafficked by earth-moving plant to allow succeeding layers to be placed and compacted. Overtrafficking must be avoided by earthworks/site management and any short-term slopes in the fill should be stable. Providing an adequate road pavement foundation is not a necessary requirement, a capping layer of good quality granular material can be provided. In the long term the fill should not settle unduly and its slopes should be stable. These criteria determine the lower strength limits for acceptable fill material.

Experiencc in Ireland has shown that the critical requirements are that compacted fill material will be stable, not settle and be trafficable. During construction these requirements are largely determined by the short term or undrained strength of the compacted fill. Slope stability can be assessed using the method described by Taylor (1948). Acceptable settlement of the fill is produced by proper compaction, which can only be achieved if the material is trafficable.

Over the years a number of attempts have been made to assess the trafficability of compacted cohesive fills and relate trafficability by different plant to a measure of soil strength. Arrowsmith (1979) reported on experience with boulder clay/glacial till in NW England and observed that caterpillar tractors and scrapers required a minimum shear strength of 35 kPa to operate but for large rubber-tyred scrapers the minimum strength was 50 kPa. Dennehy (1979) described observations and *in-situ* strength testing carried out at a number of sites where 'wet cohesive fill' was being placed. He concluded that for light scrapers with low tyre pressures the lower limit of soil strength was between 40 kPa and 60 kPa, while for medium to heavy scrapers at high tyre pressures the limiting strength was between 60 kPa and 80 kPa. Kennard *et al.* (1979) reviewing construction of a series of embankment dams with clay cores, observed that using a specification based on shear strength had advantages compared to one based on moisture content and that the lower strength limit is usually governed by plant trafficability rather than fill stability. They found that the lower strength limit varied from 42 kPa to 70 kPa.

Parsons & Darley (1982) carried out an extensive study including observations at 25 motorway and all-purpose road construction sites in the UK. They recorded the performance of different types of earthmoving plant in terms of various factors affecting productivity compared to soil conditions measured using MCV. The work showed that heavier plant required soil with a higher MCV to operate and that even the most lightly loaded plant would not operate with an MCV lower than 6, while most plant would operate with MCV of 10 or more. Their results showed that a high proportion of twin-engined scrapers could operate on soils at MCVs less than 8, equivalent to the usual criterion of suitability then adopted.

Advice note HA44/91 recommends relationship testing. It indicates that a minimum strength limit of 30–50 kPa would be typical and that the lower limit would depend on trafficability and stability. It is suggested that lower MCVs of between 7 and 9 could be used as a basis for deriving alternative lower limits of acceptability in terms of 'an equivalent value' of moisture content, California Bearing Ratio test (CBR) or undrained shear strength. To apply an equivalent value, calibration testing is critical as both the lower and upper bound limits vary for differing 'fines' contents and differing plasticity index (PI) values. This applies particularly to the more moisture sensitive materials such as Class 2C. HA44/91 references work carried out the Transport Research Laboratory, Arrowsmith (1979), Al Shaikh-Ali (1979) and Snedker (1973). HA44/91 also provides useful information on working with glacial till and notes that behaviour depends on the amount of fines and that high silt content makes till very weather susceptible.

The lower strength limit for acceptable cohesive fill controls trafficability and short-term stability of fills. A low strength fill with undrained strength of 30 kPa and a slope of 1:2 will be acceptably stable up to more than 8 m height. Trafficability will therefore govern highway embankment construction, and this will depend on the type of plant deployed for the bulk earthworks operations. Although this might be difficult to predict for a traditional contract, for the 'design and construct' contract the designer should be able to select an appropriate value based on discussion with the constructor subsequent to site specific calibration tests and optional site trials. In Ireland, material is typically hauled using large pneumatic-tyred dumpers, placed by tracked excavators and dozers, and compacted using heavy smooth-wheeled, often vibratory, rollers.

Relationship testing

Parsons & Boden (1979) showed that MCV was related to undrained strength of a soil, and they presented a series of correlations for different types of clays and silts (Fig. 2). They published relationships for low, intermediate and high plasticity soils – for low plasticity clays, Log $S_u = 0.91 + 0.112$ MCV and for low plasticity silts, Log $S_u = 0.91 + 0.120$ MCV. The relationships show that for any given soil strength the MCV is higher for high and intermediate plasticity clays and silts, when compared to low

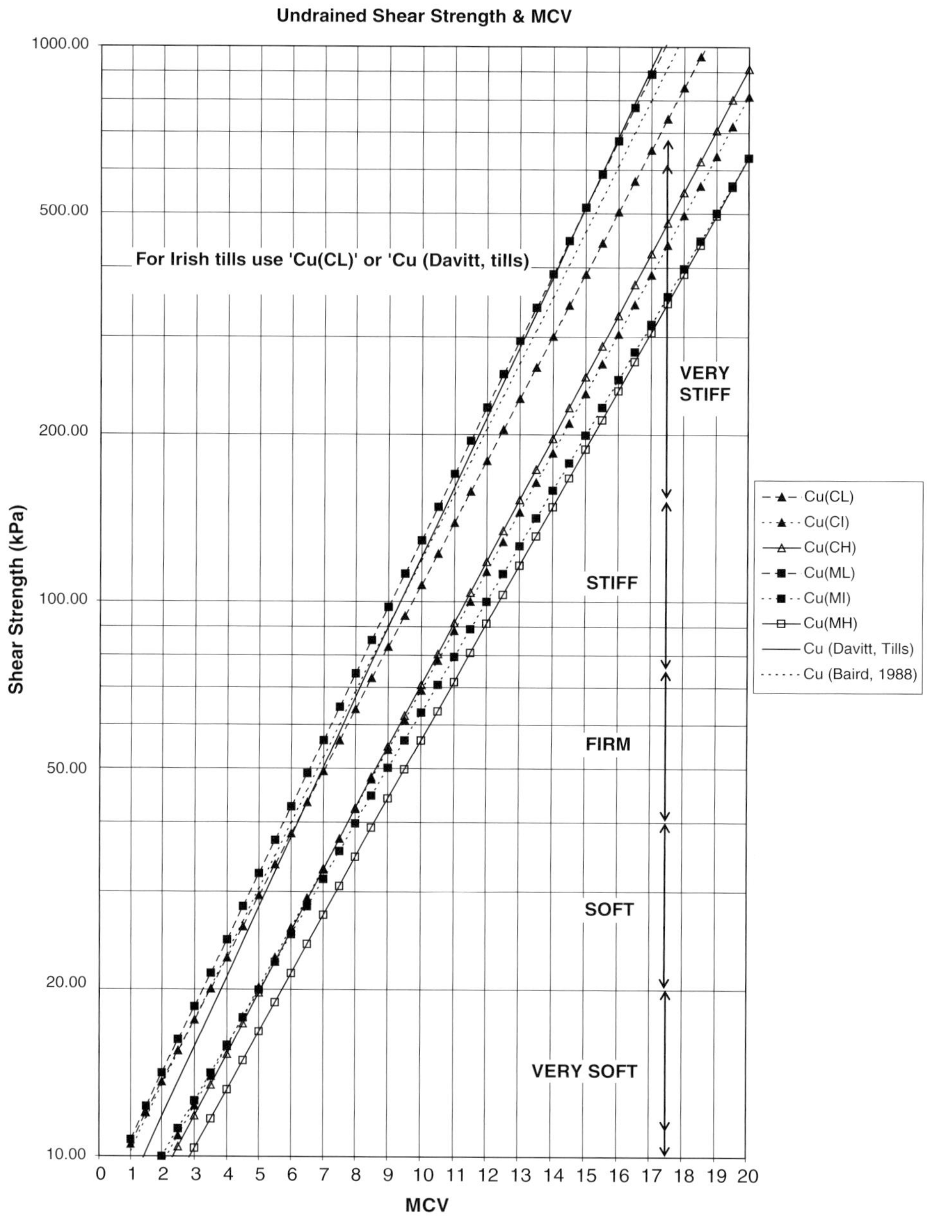

Fig. 2. Correlation between undrained strength and MCV for different soil types.

plasticity clays and silts. The correlations are straight line relationships when undrained strength is plotted on a logarithmic scale.

Baird (1988) contended that suitable fill will have stable side slopes, support construction plant and allow construction of the pavement layers, and that for cohesive soils this depended only on the compacted fill's undrained shear strength. The experience was based on testing ten samples of glacial clay from Northumberland (with PI values typically between 11 and 15) and a relationship between undrained shear strength and MCV was derived (Log $S_u = 0.891 + 0.1148$ MCV). This correlation is similar to that produced by Parsons & Boden (1979) for low plasticity cohesive clays and silts.

Davitt (1984) reviewed published work from the UK, and undertook studies of the relationship between MCV and undrained strength for Irish glacial tills (with PI values of 11–14) from three roadwork sites. Soil strength was measured using quick undrained triaxial tests on remoulded samples of 100 mm diameter by 200 mm high and by laboratory CBR tests, and these results were compared to MCVs. The correlation between MCV and shear strength for Irish tills (Log $S_u = 0.82 + 0.126$ MCV) is similar to that produced by Baird for Northumberland tills, and to Parsons & Boden (1979) for low plasticity cohesive materials (Fig. 2).

For many years the relationship between the moisture content of a cohesive material and its shear strength was considered to be one of the most important for earthworks. However, the shear strength of low plasticity clay soils is very sensitive to change in moisture content, whereas the moisture content for a given strength is dependent on the clay mineral range and can therefore cover a wide range of values (Baird 1988). For tills, the situation is further complicated by the common intermixing of cohesive tills with layers or lenses of water bearing granular material, which has the twin adverse effects of reducing PL (by the addition of sand to the matrix) and typically releasing pore water to soften the fines matrix. Relationship testing has demonstrated that moisture content typically provides a poor indicator of strength.

Unambiguous relationships between undrained strength, CBR and MCV have been established by the referenced workers for cohesive tills, and together with the earthworks experience with these soils it is possible to:

- Use a range of data from ground investigation including laboratory undrained strength, undrained strength derived from Standard Penetration Test (SPT) N-values, site and laboratory MCV testing, site and laboratory CBR testing, and soil strength descriptions following BS 5930 1999 (and now BS EN ISO 14688-2:2004), to assess the proportion of excavated soils likely to be acceptable for use in earthworks; and
- Reliably select an acceptable lower strength criterion, as undrained strength or MCV, as required by SRW/SHW, for Class 2C material based on the proposed earthworks plant and continued monitoring of the trafficability of the engineered fill during construction.

Irish experience

Assessment of acceptability

There are two basic activities which require assessment of acceptability for design of earthworks:

(1) Assessment of available volume of acceptable material from proposed excavations (cuts) that will be available for use as fill in embankments. This is essential to plan the works, to design a vertical alignment that balances cut and fill and consider the logistics of the earthworks operations. This activity is carried out relatively early in the design process – indeed a preliminary assessment would usually be required as part of the design used as the basis for pricing and tendering the works. The assessment will be based on the ground investigation and other geotechnical information available at the time together with any additional local knowledge. The process should consider all of the data available, identify anomalous results and warn the constructor of any geotechnical risks associated with the assessment. To undertake this assessment the earthworks designer and the constructor can agree acceptability limits based on the proposed earthmoving plant. For cohesive materials these limits might be any, or all, of the various properties: moisture content, shear strength, CBR, MCV, as mentioned above.

(2) Compilation of a specification of acceptability criteria for the works in line with the appropriate specification (SRW or SHW). Both specifications require acceptability limits for cohesive fills and these should be one of moisture content, undrained shear strength or MCV. For design and construct contracts, the specified limits should suit the constructor's proposed earthworks plant but this can only be established properly with calibration tests and site trials, or from experience with similar materials.

Typically Irish cohesive tills have a vertical profile comprising a near surface layer of 1–3 m thickness with variable moisture content controlled by rainfall. This zone has low confining stresses and in wet conditions without soil suction, can sustain only low strengths although in dry conditions soils may become stiff or very stiff. As depth and confining stresses increase, soil strength typically is greater and exceeds 'stiff' consistency.

Data readily available from most ground investigations includes strength descriptions to BS5930 and SPT N-values. These can both be readily correlated with *in situ* undrained shear strength and have been found to provide good guidance to acceptability for use as fill. From BS5930, 1999, a description of soft or very soft indicates a strength of less than 40 kPa and will always represent unacceptably weak clay soil, while stiff or very stiff indicates a strength of 75 kPa or more, which will be acceptable. Cohesive soils described as 'firm' (BS EN ISO 14688-2:2004 'medium') should have strength in the range 40–75 kPa and can be considered marginal, dependent on the plant and experience of the constructor; the descriptor firm to stiff can be particularly useful as an indicator that strength is towards the upper end of the range and more likely to be acceptable. (N.B. BS5930 1999 has been amended subsequently to its use for these projects and Amendment No. 1 (2007) incorporates changes from BS N ISO 14688-2:2004. Values of undrained strength are no longer associated with the consistency descriptions very soft through to very stiff. Where undrained strength has been measured in the field then descriptors extremely low, through to extremely high can be used. 'Low' represents strength of 20–40 kPa, 'medium' 40–75 kPa and 'high' 75–150 kPa.) In Ireland, to assess undrained strength, it has been found that using Stroud's correlation a value of $f_1 = 6.0$ is generally applicable to low plasticity tills, hence $c_u = 6 \times N$ (see e.g. Clayton 1995). Hence an SPT N value of 10, indicating strength of 60 kPa, can often provide a useful indicator of the lower acceptable shear strength.

The remainder of the paper describes earthworks design and construction using cohesive till soils on three road projects in Ireland. In each case the materials were low plasticity cohesive tills and ground investigation used traditional methods of sampling and testing. The case histories illustrate how simple assessments of strength from tactile tests and SPTs can be used as a basis for quantifying available volumes of acceptable materials. Anomalous results from some other tests are described. Low plasticity cohesive tills are very moisture sensitive and Ireland can be a wet place to carry out earthworks construction. The importance of following specified construction procedures to maintain material properties during the works is noted, and strength measurement during construction by MCV is shown to provide a quick, reliable basis for assessing acceptability.

Case studies

N8 Cashel to Mitchelstown

The N8 Cashel to Mitchelstown road improvement was the first early contractor involvement (ECI) road scheme in Ireland. The route runs for some 37 km SW of Cashel to west of Mitchelstown. A series of ground investigations were carried out over a period of three years prior to construction. These investigations were designed and supervised by the NRA's consultant. The ECI contract was awarded to Roadbridge-Sisk JV, and design was carried out by their consultants McCarthy-Hyder-CarlBro with AGEC Ltd as sub-consultant responsible for geotechnical design. Construction started in March 2006 and was completed in July 2008.

The ground investigations showed that most of the cohesive till tested was well-graded with a range of particle sizes from clay to cobbles. Generally more than 15% of the material was fines (clay and silt sized) and hence meeting the SRW requirements for Class 2C material (Fig. 3a). Descriptions from trial pits, and reports of refusal in cable tool boreholes indicated the presence of boulders. Plasticity testing shows that the fines have low plasticity (mostly low plasticity clay soils, CL) with liquid limit (LL) = 20 to 35 and PL = 12–20 (Fig. 3b). Test data shows little variation of plasticity with depth and that natural moisture content was higher in the upper 1–2 m (Fig. 3c).

The ground investigations data revealed poor correlation between laboratory MCVs and natural moisture content (Fig. 4), and poor correlation between laboratory MCVs and likely *in-situ* strengths as described in borehole and trial pit records and revealed by SPTs. The correlation between laboratory CBR testing and *in-situ* strength was also poor. More than 50% of MCV results were less than 8, indicating more than half of the excavated material would likely be unacceptable for re-use as fill; there were few values of MCV ≥ 9 (Figs 4 & 5). Review of the SPT values and shear strength descriptions indicated that most of the material (around 75%) would be suitable for re-use (Fig. 6).

As a check, the relationship between plasticity and natural moisture content was considered for individual soil samples of cohesive till. Generally the results indicated that natural moisture content was less than or close to the PL of soil samples, which would suggest that generally samples of cohesive till would have a strength descriptor of 'stiff' or stronger (Fig. 3c). These results correlated reasonably well with strength descriptions and strength derived from *in-situ* SPT N-values.

It was agreed with the constructor that a minimum MCV of 7, equivalent to strength of *c*. 50 kPa, would be specified for recompacted cohesive till (SRW Class 2C material) bearing in mind the moisture sensitive nature of the material and this value was used to assess quantities for earthworks planning. Approximately 90% of material was consequently determined as acceptable. MCV was selected as the on-site control, with a lower acceptability limit of 7 noted above. This is at the lower end of the range suggested in HA44, by correlation with undrained strength short term embankment stability could be demonstrated, and the contractor was happy that the value would suit his plant and methods. In particular, the contractor did not intend to use compacted fills for construction traffic. Higher strength fill with MCV = 9 was specified immediately beneath capping to limit capping depth.

Recognizing the moisture sensitivity of the cohesive till material, the particular specification was varied to allow acceptability testing at the point of deposition (as compared to point of excavation generally in a traditional contract). This enabled any near surface wetter, weaker soils to be processed, for example by air drying, to render them acceptable for re-use as general fill. Trials had indicated air drying

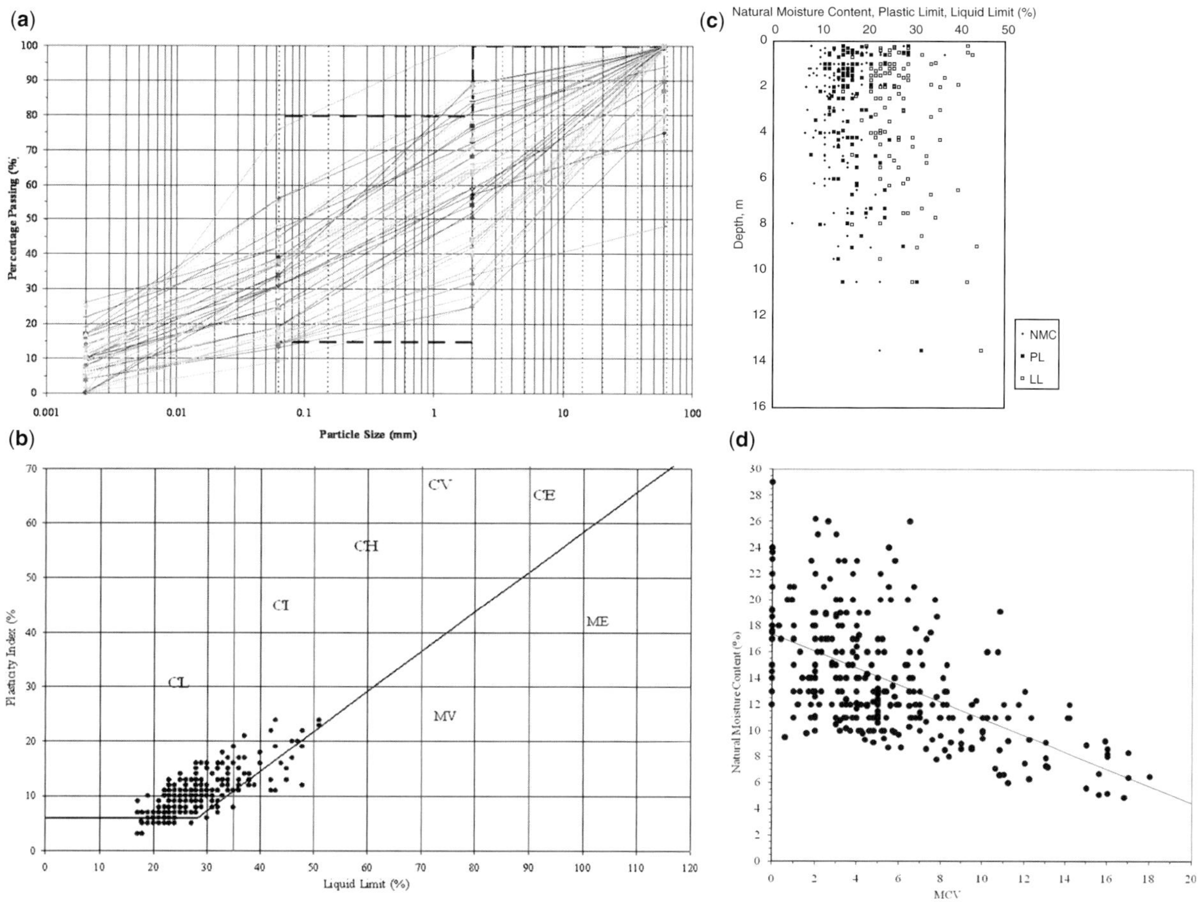

Fig. 3. N8 Cashel to Mitchelstown. (**a**) Typical cohesive till gradings; (**b**) plasticity data for cohesive till; (**c**) natural moisture content, plastic limit and liquid limit for cohesive till; and (**d**) natural moisture content and MCV for cohesive till.

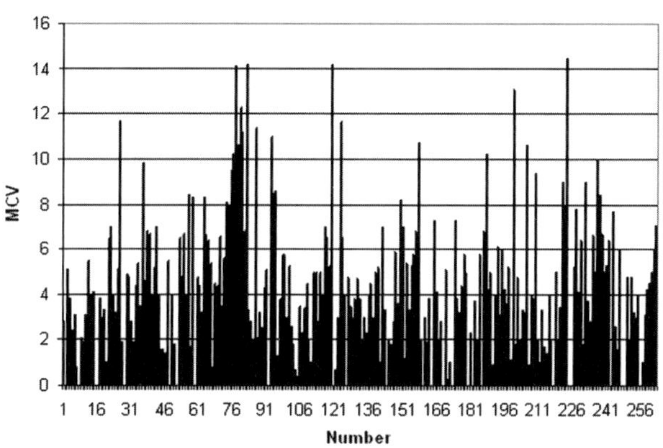

Fig. 4. Number of MCV results from ground investigations: N8 Cashel to Mitchelstown.

could allow low plasticity till soils to gain strength from MCV of 5 or 6 to more than 9 in around 6 hours.

During construction more than 95% of excavated cohesive till soils were used for earthworks construction confirming the assessment based on strength description and SPT was cautious. Earthworks required excavation and placing of around four million cubic metres of fill and this was continued throughout the year. It is considered that this was possible because the contractor paid attention to the details specified in SHW (clause 602) requiring him to keep earthworks free from water and protect constructed fills by sealing and shaping to falls. He also had extensive Irish experience. Typically, he refrained from bulk earthworks operations during periods of wet weather when it would be difficult to maintain the condition of moisture-sensitive materials and did not traffic constructed fills but made use of well-constructed haul roads to maintain productivity; he

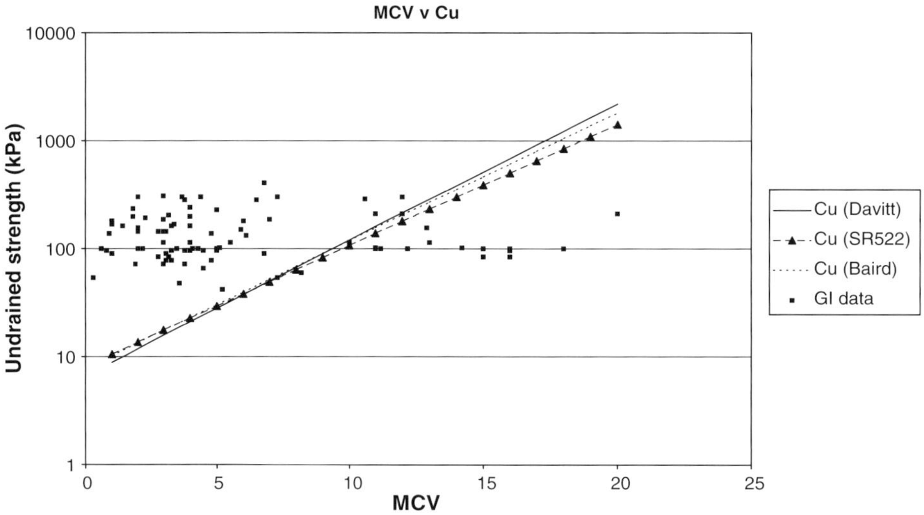

Fig. 5. Undrained strength and MCV from ground investigations: N8 Cashel to Mitchelstown.

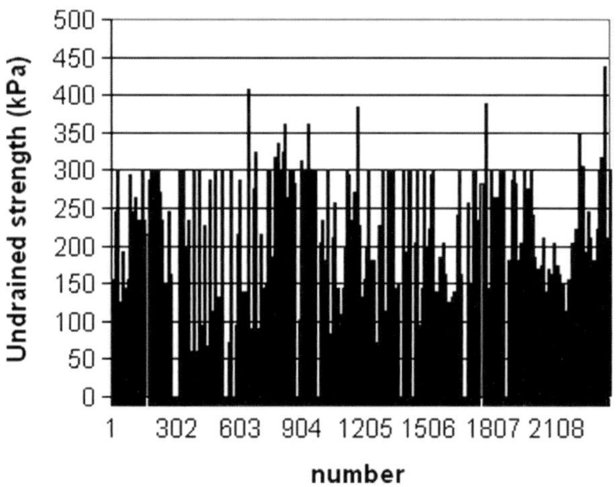

Fig. 6. Number of undrained strength results from ground investigations: N8 Cashel to Mitchelstown.

exploited periods of dry weather to progress the works. The lower MCV limit of 7 allowed economical use of site-won fill and was successful.

M3 Clonee to north of Kells

In March 2007, Eurolink Motorway Operations (M3) Ltd, a consortium formed by Cintra Concesiones de Infraestructuras de Transporte, S.A. ('Cintra') and SIAC Construction Ltd. ('SIAC'), was awarded the contract to develop the M3 Clonee to North of Kells Scheme in Ireland. This project in County Meath is part of the NRA's plans to improve road links to Dublin from the regions and the consortium entered into a PPP (Public Private Partnership) contract with M3 Motorway JV (M3 JV), a Joint Venture contracting entity formed by Ferrovial-Agroman, S.A. (through Ferrovial-Agroman Ireland Ltd) and SIAC Construction Ltd for the design and construction of the new motorway.

Prior to the award of the construction contract a series of ground investigations were carried out for the NRA, which were designed and supervised by their consultants. Detailed design was by McCarthy-Hyder-CarlBro, with geotechnical design by AGEC Ltd, Eptisa (Spain) and Webber Associates (UK) all acting as sub-consultants to McCarthy-Hyder-CarlBro. AGEC Ltd was engaged as geotechnical project design manager for the whole route. Detailed design was carried out between November 2006 and December 2007, allowing construction to start in April 2007.

Ground investigation data showed a typical Irish cohesive till grading, with a substantial proportion of coarse material, but in most cases more than 20% fines. The material generally meets the grading requirements for Class 2C stony cohesive material (Fig. 7a). Plasticity of the matrix material is low (CL) with typically LL = 20–35 and PL = 12–25 (Fig. 7b). Test data showed little variation of plasticity with depth, and again revealed that natural moisture content was higher in the upper 1–2 m (Fig. 7c).

Data from the ground investigations showed a general correlation between moisture content and MCV (Fig. 7d), but the laboratory results indicated that, based on MCV testing, up to 40% of samples tested would be unsuitable for re-use. By comparison, examination of SPT results and soil

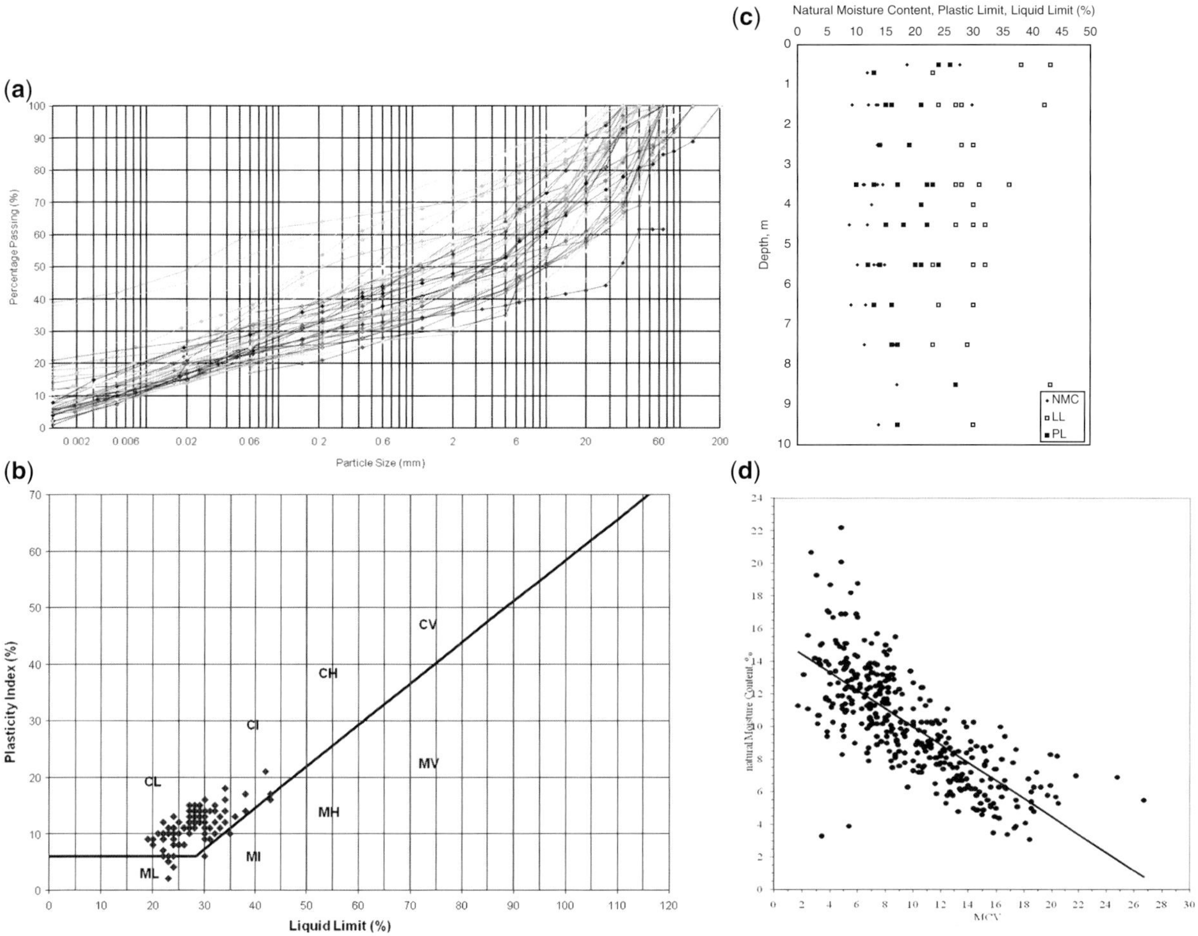

Fig. 7. M3 Clonee to Kells. (**a**) Typical cohesive till gradings; (**b**) plasticity data for cohesive till; (**c**) natural moisture content, plastic limit and liquid limit for cohesive till; and (**d**) natural moisture content and MCV for cohesive till.

strength descriptions indicated that most of the material would be acceptable. Natural moisture content data was generally less than or equal to the PL, which confirmed the *in situ* SPT and strength descriptions were likely to be correct. Similarly to the N8 project, the volume of acceptable site-won fill was estimated based largely on SPT and strength descriptions. MCV results from the ground investigation were not consistent with this data or other data, and would have severely underestimated the acceptable material quantity.

MCV limits of 7 (lower) and 15 (upper) were specified for construction. The Irish partner in the consortium was experienced at working with moisture sensitive cohesive tills in Irish weather conditions and this enabled earthworks to progress without a seasonal break. Ireland might be considered relatively rainy with from 750–1500 mm rainfall and 150–225 rain days per year. Rainfall is spread throughout the year with no obvious wet or dry seasons, though spring and summer are commonly drier than autumn and winter. There can be substantial variations from one year to another. However it does not rain all the time and good drying conditions are frequent as there is often a breeze or stronger wind.

A4/A5 Dungannon to Ballygawley

The 27 km long A4/A5 Dungannon to Ballygawley DBFO (Design, Build, Finance and Operate) Scheme 3, in County Tyrone, Northern Ireland, was constructed by joint venture contractor, Lagan Ferrovial-Agroman JV, for Amey Lagan Roads Limited on behalf of the Roads Service of Northern Ireland (RSNI). Ground investigation was procured for RSNI and designed by its consultants. Detailed design for the project was carried out by Arup between January 2008 and June 2009, with AGEC Ltd as geotechnical sub-consultant, responsible for earthworks and other geotechnical design. Construction started in February 2008 and the road was opened in December 2010.

Cohesive till material available from proposed excavations along the route typically was well-graded and includes more than 20% fines (Fig. 8a). The matrix has low plasticity, LL = 25– 40; PL = 12–20 (Fig. 8b), with some variation with depth, and again moisture content is more variable in the upper 1–2 m of the ground profile (Fig. 8c). The material available for fill was generally described as stiff gravelly clay, with a poor correlation between moisture content and MCV from laboratory testing (Fig. 8d). MCV results indicated that a substantial percentage (around 40%) of material would not be acceptable, despite descriptions of material as stiff and very stiff.

Strength derived from SPT results along with strength descriptions were examined and indicated that around 80% of material in cuttings would be acceptable for re-use, and again this is significantly different from an assessment based on MCV results. From Figure 8c it can be seen that the natural moisture content results are typically less than or equal to the PL, and are consistent with the SPT results and descriptions indicating stiff and better strength. Again, and reflecting the relative consistency of the low plasticity tills, MCV limits of 7 and 15 were specified for construction.

The joint venture included an Irish partner, and Ferrovial had Irish experience from a series of road construction projects. Through the implementation of the specified

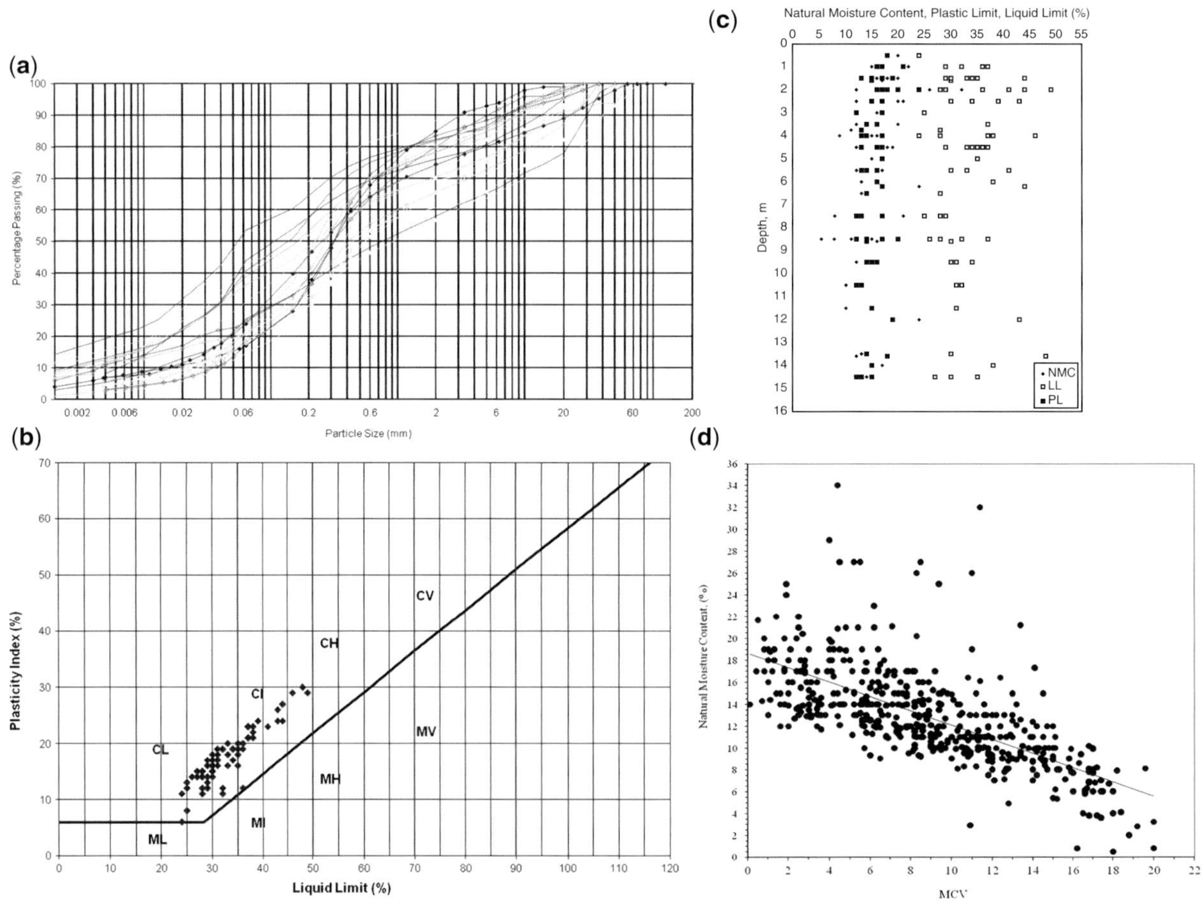

Fig. 8. **A4/A5** Dungannon to Ballygawley. (**a**) Typical cohesive till gradings; (**b**) plasticity data for cohesive till; (**c**)natural moisture content, plastic limit and liquid limit for cohesive till; and (**d**) natural moisture content and MCV for cohesive till.

requirements and adoption of good construction practice, the contractor was able to work throughout the year.

Summary

In Ireland, the NRA SRW allows all Class 1A well-graded granular material to be used regardless of the moisture content as experience has shown these soils generally are free-draining. This practice contrasts with that adopted in Northern Ireland, where moisture content limits are generally specified. The majority of earthworks fill in Ireland is Class 2C stony cohesive material, derived from glacial cohesive tills, and both NRA and HA earthworks specifications require the designer to choose lower limits of acceptability from one of moisture content, undrained strength or MCV. Work in the UK has established that a lower limit based on trafficability of earthworks plant generally governs this selection, and correlations between MCV and strength have been established, both in general terms (with variations for different plasticity soils) and specifically for glacial clay soils. Work by Davitt has confirmed that a similar relationship to that for UK glacial clay soils and those for low plasticity silts and clays is valid for low plasticity cohesive tills in Ireland.

Irish experience with cohesive tills soils has shown that strength is a good indicator of acceptability for use in earthworks construction. In design a range of measures of strength are available from ground investigations and include field material description, moisture content, SPT N-value, CBR value, MCV value and laboratory undrained strength triaxial testing. Of these, moisture content (and matrix moisture content) did not provide good indicators of likely material strength. Results from laboratory CBR testing and MCV testing did not correlate well with other tests and appeared to underestimate the *in-situ* strength of soils. It is unclear why laboratory MCV and CBR testing provide unrepresentative indicators of field strength of cohesive tills. Field descriptions and strength derived from SPT, assuming $c_u = 6 \times N$, provided the most reliable and consistent indicators of strength, and these have been the most useful guide to assessing the likely volumes of acceptable fill material from sections of highway cut. For most contractors and generally for short term stability an acceptable lower limit of strength is 50 kPa. This strength value can be compared with MCV using published work on similar low plasticity tills and will usually correlate with MCV = 7. For construction acceptability limits the MCV test is an appropriate test and provides results quickly and straightforwardly.

Critically the susceptibility of earthworks materials to moisture must be recognized and managed in order to ensure optimum use of site arisings for bulk earthworks operations. In this regard implementation of the requirement to keep earthworks free of water described in SRW 602.15 is critical, and measures such as proper drainage, shaping placed fills to provide falls and ensuring fills are sealed against water ingress make a significant difference. Cohesive tills of only medium (firm) strength are generally unsuited to general construction traffic and proper haul routes should be provided.

This work is based on experience of design and construction of roads in Ireland and the authors would like to thank their colleagues at Applied Ground Engineering Consultants Limited for their help over the years when this work was underway. They would also like to extend their thanks to the contractors who constructed the three schemes described in the case studies: Roadbridge-Sisk JV, M3 Motorway JV and Lagan Ferrovial-Agroman JV.

References

AL SHAIKH-ALI, M. M. H. 1979. The behaviour of Cheshire Basin lodgement till in motorway construction. *Proceedings of the Conference on Clay Fills*, 14–15 November 1978, Institution of Civil Engineers, London. Institution of Civil Engineers, London, 15–24.

ARROWSMITH, E. J. 1979. Roadwork fills – a materials engineer's viewpoint. *Proceedings of the Conference on Clay Fills*. Institution of Civil Engineers, London, 25–36.

BAIRD, H. G. 1998. Earthworks control – assessment of 'suitability'. *Ground Engineering*, May, 21–28.

BARNES, G. E. & STAPLES, S. G. 1988. The acceptability of clay fill as affected by stone content. *Ground Engineering*, January, 22–30.

BS1377. 1990. *British Standard Methods of Test for Soils for Civil Engineering Purposes, Parts 1 to 9*. British Standards Institution, 1990.

BS5930. 1999. *Code of Practice for Site Investigations*. British Standards Institution, 1999.

CLARKE, B. G., HUGHES, D. B. & HASHEMI, S. 2001. Characteristic parameters of tills in relation to earthworks, *Proceedings on the Seminar on Earthworks*, 12 December 2001, Institution of Engineers of Ireland/Geotechnical Society of Ireland, Paper 6.

CLAYTON, C. R. I. 1995 The Standard Penetration Test (SPT): Methods and Use. CIRIA Report 143. CIRIA, London.

CP. 2001 *Code of practice on site investigations*. British Standards Institution, 1957.

DAVITT, S. 1984. The use of the moisture condition test in assessing soils for roadworks, Report RC.278. Environmental Research Unit, November.

DAVITT, S. 2001. Comparison of Specification of Earthworks – NRA Specification for Road Works (2000) and DOE Specification for Road Works (1978). *Proceedings of the Seminar on Earthworks*, 12 December 2001, Institution of Engineers of Ireland/Geotechnical Society of Ireland, Paper 6.

DENNEHY, J. P. 1979. The remoulded undrained shear strength of cohesive soils and its influence on the suitability of embankment fill. *Proceedings of the Conference on Clay Fills*, 14–15 November 1978. Institution of Civil Engineers, London, 87–94.

DUMBLETON, M. J. & WEST, G. 1966. Some factors affecting the relation between the clay minerals in soils and their plasticity. *Clay Minerals*, **6**, 179–193.

HMSO. 1963. *Specification for Road and Bridge Works*. 3rd edn.

HMSO. 1969. *Specification for Road and Bridge Works*. 4th edn. Ministry of Transport, Scottish Development Department & Welsh Office. HMSO, London.

HMSO. 1976. *Specification for Road and Bridge Works*. 5th edn. Department of Transport, Scottish Development Department & Welsh Office. HMSO, London.

HMSO. 1978. *Specification for Road and Bridge Works, Supplement no 1*. Department of Transport, Scottish Development Department & Welsh Office. HMSO, London.

HIGHWAYS AGENCY. 1986. *Specification for Highway Works*. 6th edn.

HIGHWAYS AGENCY. 1991. *Design Manual for Roads and Bridges*, Volume 4, Section 1, HA44/91 Design and Preparation of Contract Documents [Incorporating Amendment No. 1 dated April 1995] (http://www.standardsforhighways.co.uk/dmrb/vol4/section1.htm)

HIGHWAYS AGENCY. 1998. *Manual of Contract Documents for Highway Works*, Volume 1, Specification for Highway Works. 7th edn (http://www.standardsforhighways.co.uk/mchw/vol1/index.htm)

KENNARD, M. F., LOVENBURY, H. T., CHARTRES, F. R. D. & HOSKINS, C. G. 1979. Shear strength specification for clay fills. *Proceedings of the Conference on Clay Fills*, 14–15 November 1978. Institution of Civil Engineers, London, 143–147.

MATHESON, G. D. & WINTER, M. G. 1997. *Use and Application of the MCA with particular reference to Glacial Tills*. TRL Report 273, Transport Research Laboratory.

NATIONAL ROADS AUTHORITY. 2001/2002*. Manual of Contract Documents for Road Works, Volume 1 Specification for Road Works.

PARSONS, A. W. & BODEN, J. B. 1979. *The Moisture Condition Test and its Potential Applications in Earthworks*, Report SR522, Transport and Road Research Laboratory.

PARSONS, A. W. & DARLEY, P. 1982. *The Effect of Soil Conditions on the Operation of Earthmoving Plant. Report* LR1034, Transport and Road Research Laboratory.

SHERWOOD, P. T. 1970. *The Reproducibility of the Results of Soil Classification and Compaction Tests*. Report LR 339. Road Research Laboratory.

SNEDKER, E. A. 1973. Choice of an upper limit of moisture content for highway earthworks. *Highways Design and Construction*, January 1973, 2–5.

TAYLOR, D. W. 1948. *Fundamentals of Soil Mechanics*. John Wiley & Sons, New York.

TRENTER, N. A. 2001. *Earthworks: A Guide*. Thomas Telford Limited, London.

WINTER, M. G. 2001. Earthworks acceptability for glacial tills. *Proceedings of the Seminar on Earthworks*, 12 December 2001. Institution of Engineers of Ireland/Geotechnical Society of Ireland, Paper 5.

WHYTE, I. L. 1982. Soil plasticity and strength – a new approach using extrusion. *Ground Engineering*, **15**, 16–24.

WROTH, C. P. & WOOD, D. M. 1978. The correlation of index properties with some basic engineering properties of soils. *Canadian Geotechnical Journal*, **15**, 137–145.

Overconsolidated clays and flysch for embankments in Slovenia

Ana Petkovšek* & Bojan Majes

Faculty of Civil and Geodetic Engineering, University of Ljubljana, Jamova 2, Ljubljana, Slovenia

**Corresponding author (e-mail: ana.petkovsek@fgg.uni-lj.si)*

Abstract: In Slovenia 512 km of motorways have been constructed since 1994. Most of them run over hilly terrain, where the bedrock is stiff clay and soft argillaceous sedimentary rock. In order to improve the understanding of stiff clays/soft rock behaviour, series of conventional and advanced tests were performed. The inherent suction of freshly compacted material was up to 2000 kPa whereas after construction, it equilibrated at between 100 and 500 kPa. The decrease in suction resulted in material wetting and softening. Therefore, the Proctor compaction diagrams were complemented with the corresponding Soil Water Characteristic Curve (SWCC). The fundamental issue was that during compaction, inherent suction had to be maintained below a certain level and the fills had to be constructed as soil fills. For quality control, the control of soil suction was found to be important as the relative density control. The conventional monitoring with settlement plates, inclinometers and piezometers was complemented with suction and TDR probes and the seasonal water content changes were measured continuously both during and after construction. Several millions of tons of excavated flysch, shales and stiff overconsolidated clays were used for high embankments using this newly developed approach.

Highly overconsolidated stiff clays, marls, shales and other types of argillaceous soft sedimentary rocks have represented the most challenging geomaterial for geotechnical engineers for decades. They are between rocks and soils in their behaviour and tend to change from rock-like into a soil-like material within a relatively short time period on a construction site. There is no laboratory test method available to describe this softening process, which is why it is difficult to predict the magnitude and the time frame of softening. Many authors have reported premature failures of cut slopes and embankments, constructed in stiff clays and shales (McGuffey 1982; Oakland & Lovell 1982; Šuklje 1980, among others). There have been a number of studies on the evolution of the mechanical behaviour of materials such as shales and stiff clays (Leroucil & Vaughan 1990; Cafaro & Cotecchia 2001; Sorensen *et al.* 2007). However, very few studies deal with the behaviour of the material in actual embankments. There is also a lack of professional guidelines and technical specification on the use of material for embankments. Consequently, engineers are faced with a lot of questions when deciding about how to treat such materials for high embankments.

In 1994, Slovenia started with its national motorway construction programme, which involved the construction of 500 km of new motorways. The programme was later adjusted to the new political situation in the Balkans and extended to the total of 713 km of motorways with the estimated costs of the 5425.89 million EUR. Since 1994, 512 km of motorways and access roads to the motorways have been constructed.

Slovenia's surface area is fairly small, 20 273 km^2, but it is characterized by its morphological diversity, controlled by the geology, tectonics and high amount of rainfall. As a consequence, the construction of motorways requires many deep cuts, high embankments and numerous structures such as bridges, viaducts, tunnels and retaining walls (Fig. 1).

The lack of good quality fill materials for high embankments was recognized as one of the main challenges for geotechnical engineers. On a number of motorway sections the only available material was from cuts in highly overconsolidated clays, marls, shales and flysch, mostly of Tertiary age (Fig. 2). Past experience of constructing older structures in comparable environments was not good. The lack of good quality material and increasingly strict regulations on the protection of natural resources led to a decision that even in the cases of high embankments locally excavated material should be used as a priority.

Experiences in Slovenia before 1994

Before 1994, only a limited number of large earthwork projects (deep excavation, high embankments) were undertaken

From: Radford T. A. (ed.) 2012. *Earthworks in Europe*. Geological Society, London, Engineering Geology Special Publications, **26**, 93–100, http://dx.doi.org/10.1144/EGSP26.11

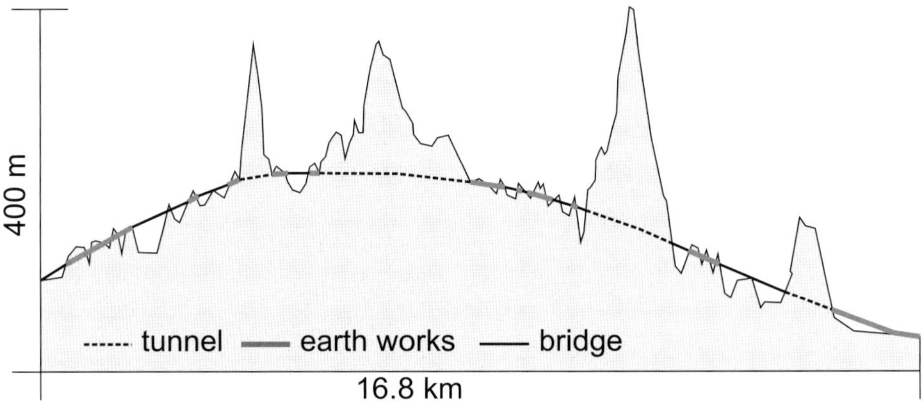

Fig. 1. Typical longitudinal cross section of the 17 km long motorway Vransko–Blagovica (courtesy of DDC).

in the highly overconsolidated clays and flysch in Slovenia. The unpredictable behaviour of the materials encountered during the construction of these existing structures sounded a note of caution for the future. As part of the construction of the Moste power plant, a 20 m deep excavation for the underground power station was made in 1949 in the stiff fissured Oligocene clay. During the excavation, the material behaved as a rock. Soon after the construction, fissures appeared in the concrete walls and since 1950 the concrete walls have had to be reinforced several times by additional anchoring to help maintain the stability of the walls. The main reasons for the damage were defined by

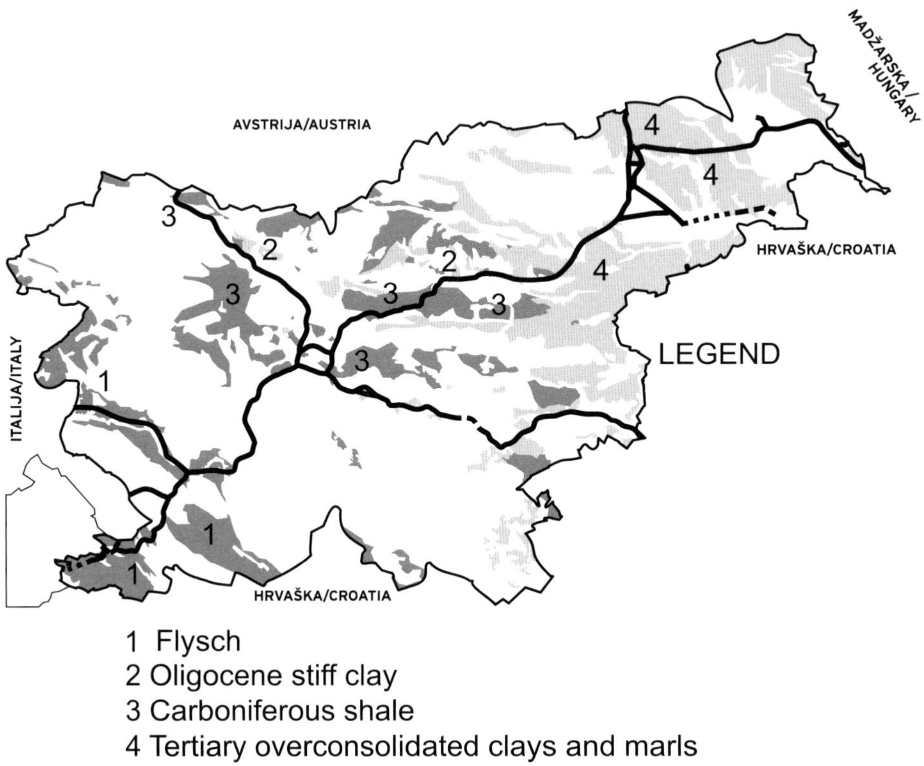

Fig. 2. Slovenian motorway network, showing the types of geological materials.

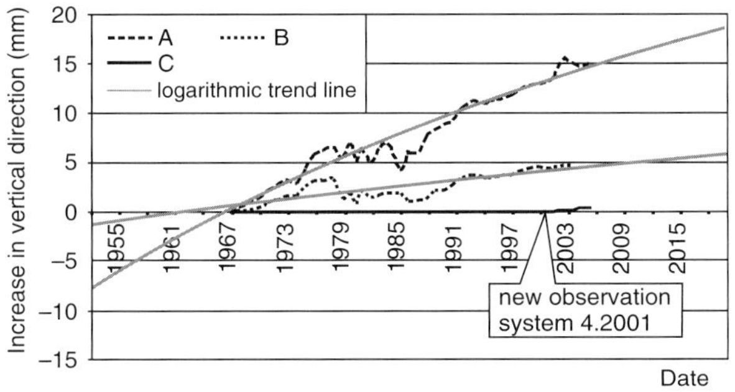

Fig. 3. Results of measurements of cracks in the Moste engine room. Deformations progress despite additional anchoring.

Šuklje (1980) as the consequence of swelling and slow creep of the Oligocene clay. Similar interdependencies between water content and swelling pressures in Oligocene clays were confirmed by other experts for different structures, such as tunnel Ljubno, where pavement heave was observed, Moste 2, etc. (Figs 3 & 4).

The impervious core of the 35 m high Vogrščec earth dam was made of weathered Eocene flysch, consisting of silty clays and some coarse-grained flysch clasts. In 1987, only one year after its completion, inspection boreholes showed that inside the dam core very soft, almost liquid lenses, a few metres thick, had developed. The measured excessive pore pressures inside the dam core due to the weight of the dam were significant, up to 130–285 kPa; however, the measured pore pressures in the adjacent piezometers may vary by a factor of up to 2 or 3. It was feared that if the lenses became continuous, there was a risk of hydraulic failure of the dam, so remedial measures were introduced. An analysis showed that the most intensive softening of compacted core developed in those layers that had become compacted during the dry summer period of August 1986, at a water content

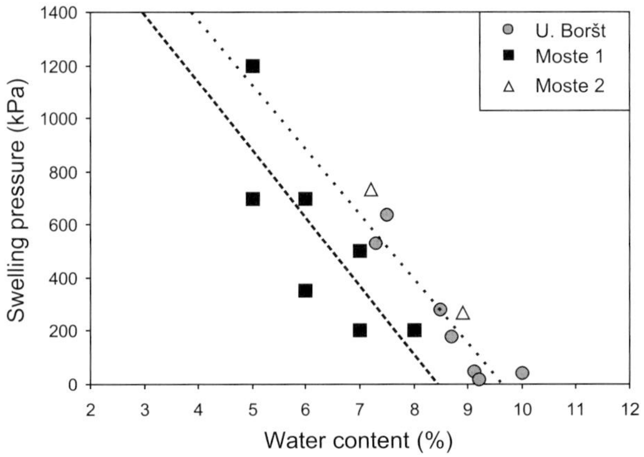

Fig. 4. Swelling pressure depending on water content.

well below the optimum water content. When comparing the data from the inspection boreholes and the construction period, it was found that the water content in the deteriorated core lenses increased from the average of 15–20% to the average of 24–30%, while the dry density decreased from the average of 1650 kg m^{-3} down to 1430 kg m^{-3}. The lenses consisted of a liquid matrix of fines with rare coarse flysch fragments.

As a consequence, the geological names of these materials are commonly supplemented by the adjective 'swelling', that is Oligocene swelling marine stiff clay, Miocene swelling marl, etc.

Marls and crushed flysch have also been used as fill material in numerous national roads and railway embankments. Although the structures were built using conventional compaction procedures, different types of failure and defect were observed soon after the construction:

(1) progressive and premature failures of asphalt pavements,
(2) heave, settlements and unevenness of pavements,
(3) reduction of subgrade stiffness and strength, and
(4) pumping of fines from the subgrade into the unbound aggregate layers which lost permeability and became more sensitive to frost action.

Slovenian technical specification and practice for earthworks on roads

Quality of fill material

Slovenian Technical Conditions for Earthworks on Roads was published in 1989 (PTP 1989). Amongst other instructions and requirements it lays down the quality of fill materials, conditions for compaction and compaction control for road embankments. Fill materials are divided in three main categories: soil fill material, rock fill material and fly ash.

For the soil fill material there are almost no limitations foreseen for a specific soil type. According to the technical conditions, all types of soil can be used as fill material, even when the liquid limit exceeds 65%. The only limitation is given for the content of organic matter, which must be checked using the colorimetric method.

The limitations for rock fill material are:

- coefficient of uniformity greater than 8
- maximum grain size below 300 mm or 1/3 of layer thickness, whichever is the smaller, and
- rocks, prone to rapid weathering (like flysch) must be protected against the influence of weather immediately after compaction.

According to the present state of knowledge, the technical conditions cannot compete with modern European technical specifications for earthworks. They also do not satisfy the needs of engineers, who have to use marginal materials and

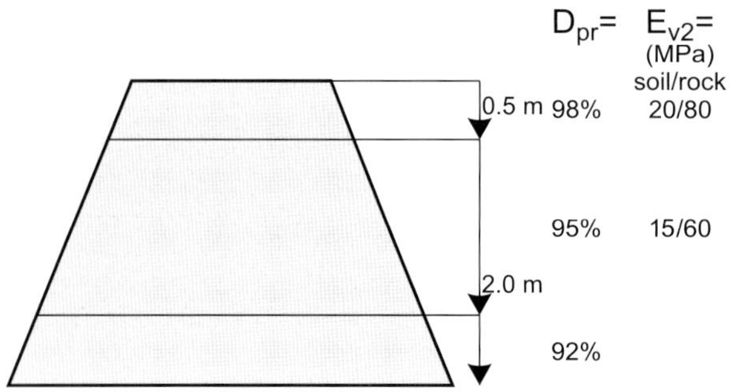

Fig. 5. Embankment sections and compaction degree.

materials with less than tolerable characteristics in earth structures. To overcome the lack of relevant national technical regulations, a special permanent 'Commission for earthworks' was established in the framework of the National Motorway programme. This Commission was authorized to observe and review the specific geomaterials and site conditions at different motorway sections and recommend specific criteria if necessary. For some motorway sections, these criteria were included in the main design and became a part of the tender applications. Some of these typical criteria were:

(1) clays with liquid limit greater than 85% are not allowed to be used as fill material;
(2) clays with liquid limit between 65 and 85% must be modified (improved) using lime or fly ash;
(3) uniform silts and sands with the coefficient of uniformity less than 8 are not allowed for structural embankments without improvement by stabilizing agents etc.

Compaction degree and layer stiffness

Technical conditions classify road embankments in three different zones (Fig. 5). For soil fills, the reference density is given by the standard Proctor test and for rock fills by the modified Proctor test. The water content, at which the fill material is allowed to be compacted is defined by the limits of optimum water content $\pm 2\%$. The stiffness of compacted layers has to be tested using the static circular plate. Deformation modulus, E_{v2}, defined after conducting the second load cycle is given for the purpose of approval. After 1994, the time consuming E_{v2} tests were partly replaced by plate tests using the light falling weight.

Compaction control

The density, water content and stiffness need to be determined for compaction control. For the national motorway project, a programme of quality control was prepared for each motorway section, for both compaction control and all the other types of works. In 2005 the National Technical specification (TSC 06) for continuous compaction control (CCC) was published. In 2007 the first embankment dams using the CCC as the main tool for the compaction control were constructed.

Overconsolidated clays as fill material and the compaction approach in the past

Before 1994 the general recommendation was to use stiff, overconsolidated clays and flysch as fill material immediately after excavation. Temporary deposits of the excavated materials were not allowed by the engineers on the job sites.

The reason for such an approach was that freshly excavated stiff clays, marls and flysch, compacted at natural moisture content, behave like a rock fill and the achieved layer stiffness may exceed the values of $E_{v2} > 60$ MPa and the dry density of 1900 kg m^{-3}. However, when the natural moisture content increases or exceeds the optimum moisture content, the stiffness of a layer progressively decreases and the effect of 'rubber pillow' can be observed during compaction. When the already compacted 'dry' layer is exposed to rain or frost during winter, it deteriorates completely down to 0.7 m depth and must be removed and replaced with new material.

To protect the embankments from the effect of delayed deterioration the technical specifications require that the layer must be protected immediately after compaction. According to some experiences, stiff clays, marls, shales and flysch were compacted only on the dry side of the Proctor curve, using the modified Proctor test as a reference density.

Table 1. *Physical properties of some typical stiff soils/soft rocks*

	w (%)	w_L (%)	w_P (%)	ρ_d (t/m^3)	q_u (kPa)	w_A (%)	MBf (g/kg)
Flysch	4–10	42–77	11–20	2.0–2.1	100–40000	55–105	32–60
Oligocene clay	6–8	35–50	13–15	2.2–2.3	1000–20000	50–70	32–60
Pliocene clay	20–35	50–90	19–26	1.5–1.7	30–1000	50–110	61
Miocene marl	8–20	39–53	14–26	1.9–2.1	600–15000	50–90	28–30

w, natural water content; w_L, liquid limit; w_P, plastic limit; ρ_d, dry density; q_u, unconfined compression strength; w_A, Enslin Neff water adsorption; MBf, methlene blue value on fines.

Experiences from different job sites indicated that the old principles were not helpful and that new test methods and new compaction conditions had to be defined.

Overconsolidated clays, shales and flysch as fill material for the motorway programme

Laboratory tests

During the motorway programme different types of stiff overconsolidated clays and soft, argillaceous sedimentary rocks were encountered in quantities of several millions tons. They differed in their geological age and origin, type of structure, etc. The properties of some typical deposits are given in Table 1. Common to all types of materials are their plastic nature, low natural water content and relatively high unconfined compression strength, which indicates, that materials in their natural state could be classified as rock.

As an example, a comparison between the virgin compression line and the compression lines of natural Oligocene clay is given in Figure 6. Figure 7(a, b) shows the relationships between the soil suction and the unconfined compression strength together with a virgin soil water characteristic curve (SWCC). More than 130 core samples were tested. The unconfined compression strength–water content relationship is described using the same fitted equation as was used for the SWCC. In natural state, for the water content of 5–8%, the clay/claystone possesses high suction up to 20 MPa.

On sites, soon after excavation, fissures appear and then widen until single blocks fall down from the slope (Fig. 8). The fragmentation and degradation of blocks then continues. Rain and cycles of wetting and drying accelerate the fragmentation (Fig. 9).

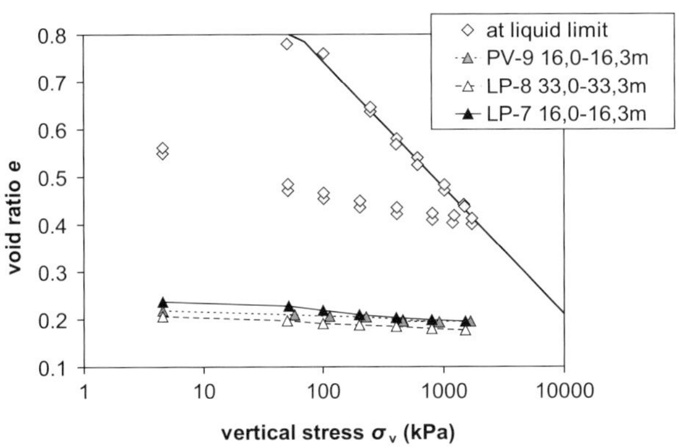

Fig. 6. Virgin compression line and the compression lines of stiff Oligocene clay.

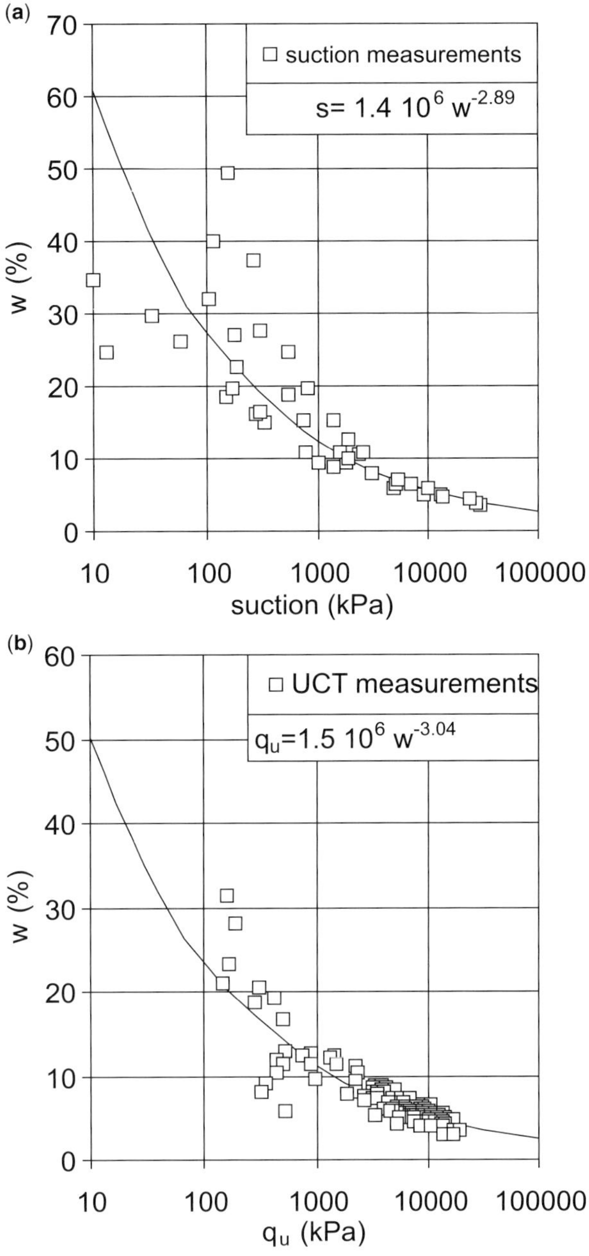

Fig. 7. Relationship between unconfined compression strength, water content and suction for Oligocene stiff clay.

Swelling tests on Oligocene stiff clay soft rock samples in oedometric conditions show that the swelling stops at water contents around 12%. At the same water content compacted samples still draw in water and equilibrate at the water content of around 18–20%. The behaviour of samples

Fig. 8. Fissures on cut slopes appear and widen soon after excavation in Oligocene clay.

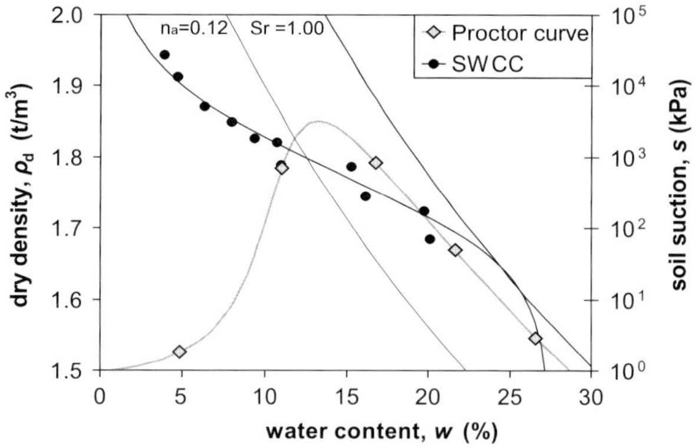

Fig. 10. Proctor curve for stiff Oligocene clay.

could be divided into two groups: poorly compacted samples collapse, whereas well compacted samples swell. The most stable volumetric behaviour is shown by those samples that are well compacted at the water content close to the plastic limit or at the saturation degree of 85–90% and at the impressed suction lower than 800 kPa (Fig. 10). It was also found that the breakage of the coarse grains was higher for drier fragments, which agrees well with the results from the old sites, where failures were observed and studied. Similar mechanisms of rock degradation and softening of the compacted material were found for all types of tested materials; only the magnitude was different. The main concern for the investigation was how to protect the compacted layers with coarse grains against the degradation of fragments and softening due to the loss of suction. Therefore, to protect the embankments against uncontrolled deformations that originate from deterioration of the coarse grained fraction, the fill material must be treated as soil fill material and not as rock fill material. After excavation, material must additionally be broken down into smaller pieces, wetted and placed in a temporary location, where the water content and the suction can equilibrate. After the resting period the material must be compacted as a soil fill at a degree of saturation of around 85% and a compaction degree higher than 95% of Proctor density.

In the Pliocene formation there were stiff, plastic clays which possessed high suction even at a very high degree of saturation. For these materials, lime improvement was proposed and used on job sites. To improve the control over

Fig. 9. Breakage of flysch block.

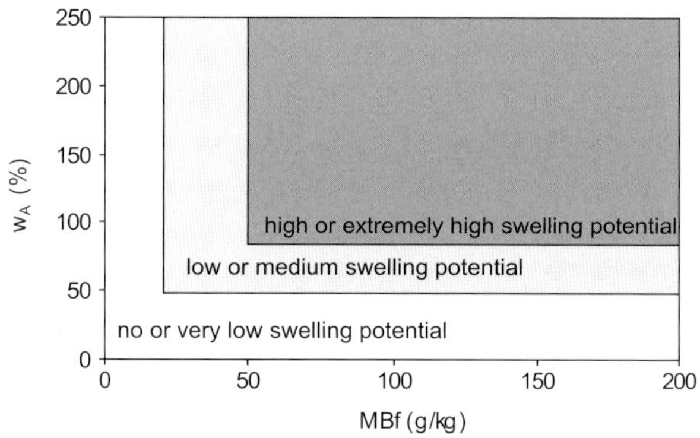

Fig. 11. Enslin-Neff and methylene blue test relationship as an indicator of swelling (Petkovšek 2002).

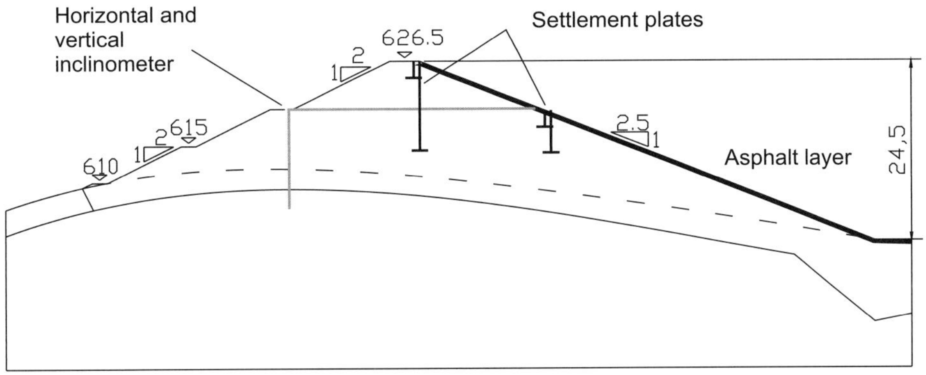

Fig. 12. Typical monitoring profile on embankment dam, constructed from flysch.

these highly plastic and swelling materials, two new index methods were introduced: the Enslin–Neff water adsorption test and the methylene blue test. Both methods are simple and fast, and they offer a good insight into the activities of the soil. An empirical diagram was proposed by Petkovšek (2002) for fast identification of swelling potential of all type of clay soils (Fig. 11).

Construction and monitoring during construction

For all fills, constructed from stiff clays and soft argillaceous rocks, the compaction procedure is defined using an experimental embankment at the start of the works. Settlement plates, vertical and horizontal inclinometers are installed at different levels of the embankment to control the deformation. In the case of indications of swelling/shrinkage, suction probes are installed within the upper 3–4 m. Figure 12 shows a typical monitoring profile on an earth embankment dam, constructed in 2007, using flysch, and Figure 13 presents the results of monitoring of suction inside the finished embankment. The construction of pavement layers starts soon after the monitoring results have confirmed that the deformation of the embankments has been completed. The resting time depends on the type of fill material and the height of the embankment. Untreated stiff clays and flysch must not be used for embankments which are lower than 2 m.

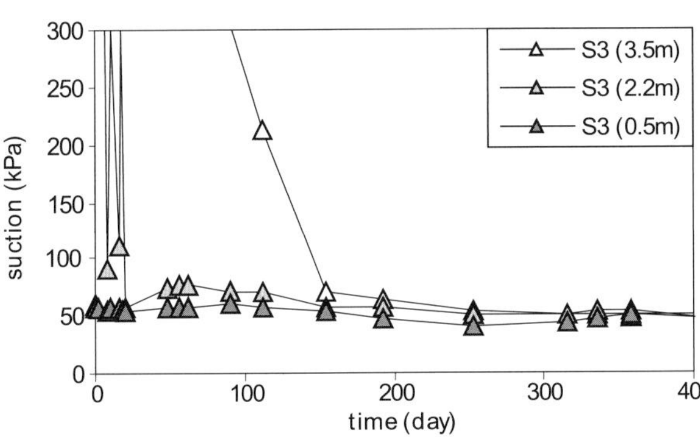

Fig. 13. Soil suction monitoring results on embankment, constructed using Pliocene clay.

Conclusions

The national programme for construction of the Slovenian motorway network was a tremendous challenge for geologists and civil engineers. In the field of earthworks and construction of embankments, the understanding of those processes that control the behaviour of stiff clays and shales in deep cuttings and embankments has been significantly improved and adopted for operational use.

The material, excavated as rock, must be treated as a soil fill material. Soil suction was found to play a key role in the behaviour of coarse fragments that remain inside the clayey matrix, as well as in the compacted material as a whole. Before the material is allowed to be compacted in the embankment, it must be placed in a temporary location while the water content equilibrates with the new surrounding and the coarse fragments deteriorate. For compaction control, the degree of saturation is very important and must not be below 85%.

Monitoring has shown that inside embankments, the suction equilibrates at 80–300 kPa, rarely up to 500 kPa. When layers are compacted at suctions higher than 1500 kPa, the heterogeneous deformation due to softening may occur.

Post-construction behaviour of compacted stiff clays and soft clay-bearing rocks strongly depends on the impressed matrix suction during compaction. Therefore the results of Proctor tests must be complemented with the SWCC. In cases when at optimum water content the sample still has high suction the material must be modified by chemical agents such as lime. Two fast and simple index tests have proved to be very useful, the Enslin–Neff water adsorption test and the methylene blue test. Both tests indicate the activity of the clay fraction and excellently complement the SWCC.

For the retaining walls in deep cuttings, the design soil parameters must be derived from the tests, performed under the conditions of controlled suction. When the results of *in-situ* tests, or the parameters defined in the laboratory on the intact samples were used in the design, the strength and deformation parameters were found to be too optimistic and during or after construction the retaining structures exhibited higher deformation than predicted.

Several million tonnes of stiff clays, shales and flysch have been used as fill material for motorway embankments within the last 15 years in Slovenia. So far, no unpredicted damage or failure has been observed where these procedures have been used. There are some short sections where the importance of suction and water content control was underestimated and the dry density and stiffness have been used as acceptance criteria according to the old principles.

New Slovenian technical specifications for earthworks are now in preparation and the knowledge gathered within the last 15 years on the motorway programme will be included in the new edition.

References

CAFARO, F. & COTECCHIA, F. 2001. Structural degradation and changes in the mechanical behaviour of a stiff clay due to weathering. *Géotechnique*, **51/5**, 441–453.

LEROUEIL, S. & VAUGHAN, P. 1990. The general and congruent effects of the structure in natural soils and weak rocks. *Géotechnique*, **40/3**, 467–488.

MC GUFFEY, V. C. 1982. Design of cut slopes in overconsolidated clays. *Transportation Research Record*, **873**, 8–11.

OAKLAND, M. W. & LOVELL, C. W. 1982. Standarized tests for compacted shale highway embankments. *Transportation Research Record*, **873**, 15–22.

PETKOVŠEK, A. 2002. Methylene blue test – a simple and fast method for the recognition of swelling soils. *In*: VILHAR, M. (ed.) 6. slovenski kongres o cestah in prometu, Portorož, 23.–25.oktobra 2002. *Zbornik referatov*. Družba za raziskave v cestni in prometni stroki Slovenije, Ljubljana, 197–205. [In Slovene]

PTP. 1989. Posebni tehnični pogoji za zemeljska dela na cestah (Special Technical Conditions for Earth Works on Roads). SCS Ljubljana, Slovenia.

SORENSEN, K. K., BAUDET, B. A. & SIMPSON, B. 2007. Inflence of structure on the time- dependent behaviour of a stiff sedimentary clay. *Géotechnique*, **57/1**, 113–124.

ŠUKLJE, L. 1980. Earth pressures on the underground power station moste. *Gradbeni vestnik*, **29**, 201–212.

TSC 06.713. 2005. Continuous compaction control (in Slovene). Technical specification. Ministry of Transportation.

Use of tyre bales as lightweight fill for the A421 improvements scheme near Bedford, UK

Ivan F. Hodgson[1]*, Stephen P. Beales[2] & Martin J. Curd[3]

[1] URS (formerly Scott Wilson), Scott House, Alençon Link, Basingstoke RG21 7PP, UK
[2] URS (formerly Scott Wilson), 15th Floor, McLaren Building, 35 Dale End, Birmingham B4 7LN, UK
[3] Le Quereau, 86400 St Macoux, France

*Corresponding author (e-mail: ivan.hodgson@urs.com)

Abstract: The A421 improvements scheme near Bedford in England requires construction of the new dual carriageway and a realigned side road over two former borrow pits that have been partially infilled with up to 20 m of soft clay. The side road was constructed using lightweight fill comprising 'Maxit' expanded clay aggregate and tyre bales manufactured from old vehicle tyres under a Standard that produces an engineering material from a former waste product.

A deskstop study and a series of ground investigations were carried out to establish the history of the site and the current condition of the infill and to provide parameters for geotechnical design. The solution was chosen after consideration of various options against a number of constraints. The bulk of the embankment was formed from layers of tyre bales with alternate thin levelling layers of 'Maxit'. A period of surcharge with wick drains was incorporated to reduce settlements in the long term and a temporary toe berm provided for stability during construction. The use of tyre bales on this project provided a number of environmental benefits as well as practical benefits including cost and ease of storage, handling and placing.

The A421 Improvements (M1 Junction 13 to Bedford) is being carried out under an Early Contractor Involvement (ECI) agreement by Balfour Beatty/Scott Wilson (BBSW) for the Highways Agency. The scheme comprises 14 km of new dual carriageway in the English East Midlands. The new road generally follows the alignment of the existing A421 from the M1 through the Marston Vale to join the Bedford Southern Bypass. Oxford Clay dominates the geology, locally with glacial clay cover. The terrain is flat with the exception of a prominent hill close to the M1.

At Brogborough the route follows the alignment of the existing A421 along a narrow strip of *in-situ* ground between two former clay borrow pits. The clay pit to the south of the route is partially backfilled and flooded to form Brogborough Lake, the pit to the north is an active landfill. In this area, in addition to the new dual carriageway, the improvements require side roads on each side of the new road: an access road to the landfill on the north side and a realigned section of the existing A421 to the south to serve local access. The width of the works means that the new construction would extend beyond the *in-situ* Oxford Clay between the pits and be founded partly on the infilled former pits. The alignment of the new road requires the road to be supported up to 7 m above the existing ground level in this area.

At the time of writing, the lightweight embankment has been constructed and is being monitored under surcharge loading. It is intended that further publications will be produced in due course to describe the monitoring results and longer term performance of the lightweight embankment.

Desk study and investigation

The steep sides of the former pit on the lake side were identified in the desk studies and walk over surveys. Figure 1 shows the tell-tale hummocky ground that indicates the potential for soft infill to the pit.

The desk study, which included examination of old maps and aerial photographs, revealed the extent of the former borrow pits, provided an understanding of the method by which they were worked and indicated the way in which the lakeside pit was partly infilled. Figure 2 is an aerial photograph from 1962 in which it can be seen that exploitation of the lakeside pit is complete with prominent dragline traces in the base. The pit to the north is being worked and waste is being conveyed across the A421 road and discharged into the lakeside pit. The material appears to be a soft slurry which is able to flow and form alluvial-style fans.

Following the desk study a ground investigation was carried out to provide data for feasibility assessments and optioneering during the preliminary design phase. Ground investigations carried out comprised cable tool boreholes,

From: Radford T. A. (ed.) 2012. *Earthworks in Europe*. Geological Society, London, Engineering Geology Special Publications, **26**, 101–108, http://dx.doi.org/10.1144/EGSP26.12

Fig. 1. Site of the new alignment viewed from the high ground outside the former pit.

windowless sampling, cone penetration testing using a piezocone (CPTu) and trial pits. The results indicated that the fill has an undrained shear strength (Cu) of between 25 kPa and 35 kPa and is, in places, up to 21.5 m thick.

Design options and solution

Optioneering

An extensive optioneering exercise reached the following conclusions.

Conventional earthfill embankments were not feasible due to:

- The requirement for staged construction to ensure short term stability. This raised problems of programming, especially as the timings of the stages are difficult to predict.
- Settlement of up to 1 m was predicted for a 5 m high embankment over only 12 m of made ground. This leads to maintenance risks.
- Risk of large differential settlement at the edges of the borrow pits, also leading to risk of increased maintenance.

Replacement methods were not feasible owing to the excessive depth of soft fill and the proximity of the lake which would require very large coffer dams.

Ground treatment was ruled out because mix volumes for lime or cement treatment would be large and the low strength and significant depth of soft soil was marginal for the use of vibro replacement or lime/cement columns. Settlements after treatment would be difficult to predict.

A piled concrete raft could be used but stricter settlement limits would have been required than for an embankment to avoid damage to the raft itself. This would lead to a risk of differential settlements at the contacts with *in-situ* Oxford Clay.

The solution

A pile-supported geotextile load transfer platform was considered the most suitable solution for the new dual carriageway sections where it passes over the corners of the two backfilled pits.

The new landfill access road is on a very low embankment which can be built of earthfill as the infill to the landfill side borrow pit is slightly stiffer and of more varied

Fig. 2. Aerial photograph taken in 1962 showing the former A421 between the two borrow pits with slurry flowing into the lakeside pit from three discharge points forming fans that obscure the dragline traces.

composition than the slurry in the lakeside pit. The domestic waste landfill itself is not affected as the works are restricted to the margin of the pit which was backfilled with mixed inert materials prior to the start of commercial landfill operations.

The solution adopted for the realigned A421 across the former pit on the lakeside was a lightweight embankment. Over most of this area the design included wick drains and embankment surcharge, with 3 m of earthfill for up to 6 months, in order to reduce the risk of long term settlements that would require additional maintenance. The juxtaposition of the pile-supported load transfer platform and lightweight embankment is shown in Figure 3. In the central part of the embankment, where the embankment is low, excavation and replacement with lightweight material can be used to achieve negligible additional load and hence reduce the predicted settlement to acceptable levels.

The various areas of construction in the lake area described above are shown in plan in Figure 4.

Lightweight fill embankment

Materials

Pulverised fuel ash (PFA) is one of the most commonly used lightweight fill materials, but for this application its bulk density of between 1500 and 1700 kg m^{-3} meant that staged construction would still be required for short term stability. The preliminary design of the side road embankment was therefore carried out using a proprietary expanded clay aggregate called 'Maxit' (now known as Leca) which has a density of between 300 kg m^{-3} (dry), 600 kg m^{-3} (wet) and kg m^{-3} (saturated).

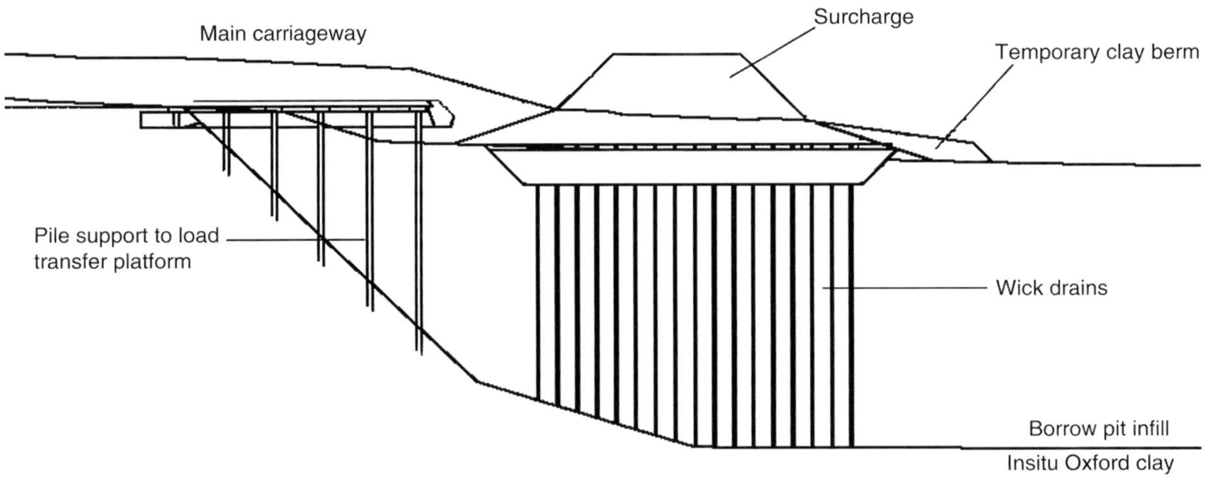

Fig. 3. Cross section through the new main carriageway and realigned side road with surcharge and in place.

The concept of tyre bales was originally introduced to the design team for consideration for the construction of a drainage chamber in another area of the scheme. This use was not adopted but it was realized that tyre bales might be used to reduce the volumes of Maxit needed for the lightweight embankment at Brogborough Lake.

Baling used tyres was originally employed to enable more efficient transport and storage of used tyres by the waste disposal industry. Each bale is formed by lacing together between 100 and 120 ordinary car or van tyres and compressing them into a rectangular bale with nominal dimensions of 1.33 m × 1.55 m × 0.83 m. The bale is secured with galvanized wire straps whose primary function is to retain the shape of the bale during storage and handling as in the long term the tyres in the bale adopt a permanent compressed shape. Bales ready for placing are shown in Figure 5. The

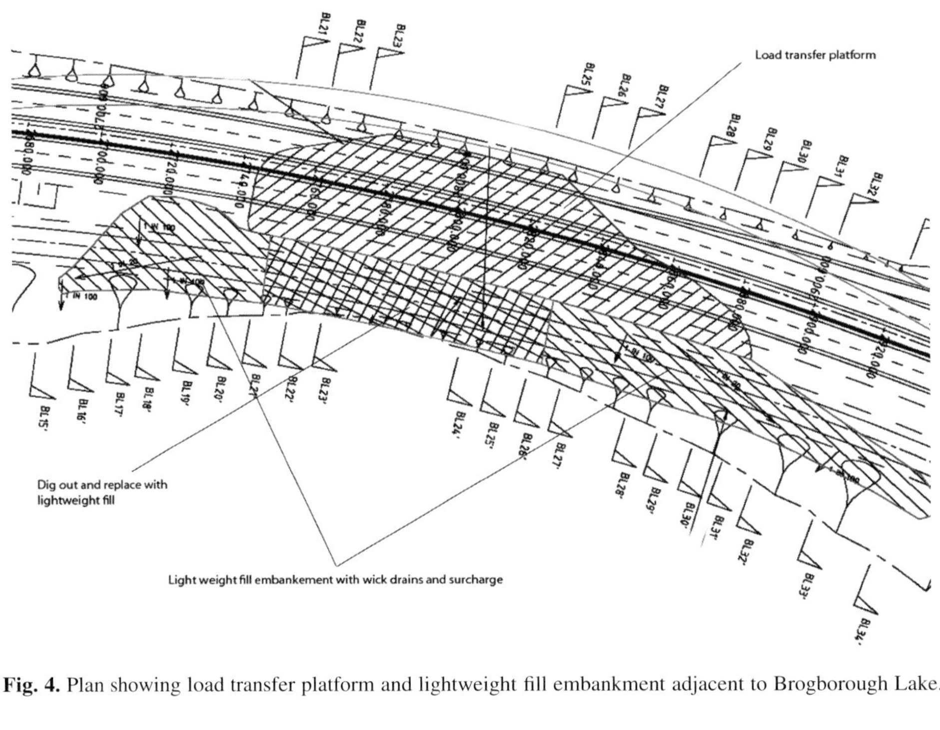

Fig. 4. Plan showing load transfer platform and lightweight fill embankment adjacent to Brogborough Lake.

Fig. 5. Tyre bales being stacked ready for placement.

transformation of tyre bales from a waste product to a construction material is largely due to the production of Publicly Available Standard (PAS) No.108 (2007) by the British Standards Institution.

Bales constructed according to this standard are not permitted to contain damaged or dirty tyres that would constitute a risk of contamination. According to HR Wallingford (2005), research carried out by the Building Research Establishment into the use of tyre bales for construction included tests that showed that not only were the bales not susceptible to spontaneous combustion in typical construction works, but they were also much more difficult to ignite than other recycled tyre products and propagation of fire within a mass of tyre bales is virtually impossible.

Tyre bales have been used in road construction in the USA and in Scotland, but only for low embankments comprising a single layer of bales. Monitoring from these sources and observation of a tyre bale structure at Pevensey on the south coast of England provided useful data on their stiffness and long term creep performance. This data is included in PAS 108 as shown in Figure 6.

Embankment design

The initial design using Maxit was relatively easily modified to accommodate tyre bales because the inter-bale friction is similar to the internal friction angle of Maxit, and the mass density of the bales is similar to Maxit in the long term. These are some of the most critical properties for design of a lightweight embankment. Regarding the long term performance of the road, it was possible to demonstrate that the use of tyre bales resulted in predicted consolidation settlement and secondary compression of the embankment, that were similar to Maxit-only construction.

The lightweight embankment comprised the following elements, as shown in Figures 7 and 8:

- Granular fill working platform built of Maxit with 300 mm of conventional granular fill as an operating surface for the wick drain installation.
- Layers of tyre bales placed in staggered 'bricks and mortar' style with reducing layer width to produce a stepped embankment with an equivalent overall side slope of about 1v:3h.
- Levelling layers of Maxit between the tyre bale layers, vibrated into the gaps between adjacent bales using a vibratory roller.
- Maxit is also used where the embankment is too low for a single layer of tyre bales to be used.
- The top layer of tyre bales is covered with a woven geotextile separator layer to prevent the upper layers of fill migrating down if voids open between the tyre bales.
- The upper 1 m of the embankment and the outer minimum 1 m is composed of conventional cohesive earthfill. This is to accommodate buried services in the verges, for aesthetic reasons and protect against UV light and vandalism. The finished side slopes are formed at 1v:3h and vegetated.
- During the application of the surcharge a temporary toe berm of earthfill is required to ensure stability of the

Property	Value(s)	Comments
Nominal mass density	470 kg/m^3 (± 50 kg/m^3)	Mass density of the cuboid enclosing the tyre bale (see Annex B).
True mass density	500 kg/m^3 (± 70 kg/m^3)	Mass density of bale only (see Clause C.2.3)
Porosity	62 % (± 5 %)	
Shear strength: angle of inter-bale friction, ϕ'	35° to 36°	Tests performed on dry bales. Cohesion is small and can be ignored in design.
Stiffness (expressed as Young's Modulus), M	800 MPa to 1,000 MPa	Values increase within range with increasing degree of confinement. Values based on a preliminary interpretation of USA data for arrangements of two and three bales with no joint filling.
Total Creep (35 months)	Up to 1.1 %	Based on measurements of a bale stack in a beach at Pevensey. Measurements indicate creep process now substantially slower and long term creep not expected to exceed 1.5 %.
Permeability through depth (see Figure 1)	0.1 m/s to 0.2 m/s	
Permeability through length (see Figure 1)	0.02 m/s to 0.04 m/s	

Fig. 6. Physical properties of tyre bales from PAS108 (2007).

embankment against deep rotational failures of the soft clay foundation.

Embankment monitoring

Monitoring of the lightweight fill embankment is required primarily to confirm stability during construction (piezometers and inclinometers) and to determine that the required settlement of the underlying soil has taken place (equivalent to the predicted settlement under the load of the lightweight fill alone) to allow the surcharge to be removed. To this end BBSW has installed vibrating wire settlement gauges/cells (4no.), magnetic extensometers (4no.), vibrating wire piezometers (18no. at 6 locations), inclinometers (7no.), surface settlement markers (45no.) and rod and plate settlement gauges (8no.).

In addition to the above, the Transport Research Laboratory has installed instrumentation to monitor the performance of the tyre bales within the embankment. The results will serve to confirm the anticipated long term performance of tyre bales in the embankment and to enlarge the database that will be available for design of similar structures in the future. This monitoring includes recording the compression of the tyre bales themselves and the temperatures within the tyre mass.

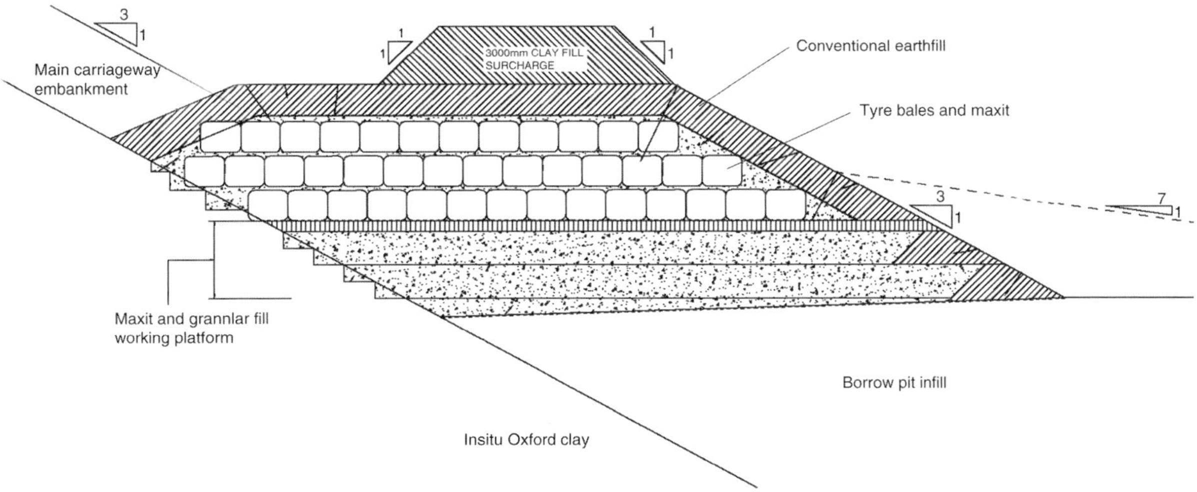

Fig. 7. Cross section of the lightweight embankment during construction with surcharge in place.

Fig. 8. Tyre bale embankment under construction at Brogborough.

Advantages of the use of tyre bales

A wide range of benefits arises from the use of tyre bales for the lightweight embankment on the A421 scheme at Brogborough:

- the delivered cost of the tyre bales is significantly less than that of Maxit;
- tyre bales can be easily stockpiled on site (subject to Environment Agency restrictions) and can be placed with standard construction machinery;
- tyre bales are sourced within the UK whereas Maxit is imported from mainland Europe;
- the rate of embankment construction compares favourably with other earthworks and will probably improve with experience;
- construction with tyre bales is not significantly affected by wet weather.

There are also a number of environmental advantages in the use of tyre bales over the use of conventional fill materials and manufactured lightweight materials:

- transport distances are reduced compared with Maxit which is shipped from mainland Europe;
- used vehicle tyres are no longer permitted in landfill in the UK and must be incinerated. The use of tyres in construction represents an environmental advantage over disposal; and
- tyre bales represent a recycled resource compared with use of primary fill materials.

The use of lightweight fill for the side road embankment at Brogborough provided a cost effective solution to the problems of construction and long term performance of embankments up to 7 m high over soft borrow pit infill. The inclusion of tyre bales as a lightweight fill added value in terms of cost savings and environmental benefits increasing the attractiveness of this solution.

The authors are indebted to the Highways Agency and Balfour Beatty/Scott Wilson for permission to publish this paper. David Baker of Balfour Beatty Civil Engineering Ltd introduced the concept of tyre bales to the design team. Mike Winter of TRL and Steve Mendes of Anglo Environmental provided advice and information during the feasibility studies.

References

PAS 108. 2007. *Specification for the Production of Tyre Bales for use in Construction*. British Standards Institution.

HR WALLINGFORD. 2005. *Report SR669 Sustainable Re-use of Tyres in Port, Coastal and River Engineering*. Guidance for planning, implementation and maintenance.

Use of industrial by-products in earthworks in the Czech Republic

František Krest

ARCADIS Geotechnika a.s., branch Ostrava, 28.října 150, 702 00 Ostrava – Moravská Ostrava, Czech Republic (e-mail: kresta@arcadisgt.cz)

Abstract: The use of industrial by-products as a construction fill is set to increase in order to minimize the environmental impact of large infrastructure construction schemes. The Czech Republic is one of a number of countries where these products are now used. This paper gives a brief background to the use of colliery spoil, fly ash, blast furnace slag and PRESTAB, with particular reference to the construction of the D47 motorway and R48 expressway near Ostrava.

Industrial by-products represent alternative sources of material which can be used as fill. The Czech Republic is one of a number of countries where these products are now widely used (Fig. 1), mostly in traffic engineering, for example in more than 90% of new embankments. Small quantities of these products are used in earthworks connected with site preparation for construction of new logistic centres or production facilities in industrial zones.

In the Czech Republic, the industrial by-products are primarily from colliery spoil (mine stone), fly ash and blast furnace slag. However, other by-products from industry (like industrial gypsum PRESTAB) have been tested to be used as fill or a part of a binder.

The use of industrial by-products is concentrated in the industrial parts of Czech Republic with coal deposits and steelworks: the Northern Moravia (colliery spoil and blast furnace

Fig. 1. Use of industrial by-products in Czech Republic.

From: Radford T. A. (ed.) 2012. *Earthworks in Europe*. Geological Society, London, Engineering Geology Special Publications, **26**, 109–113, http://dx.doi.org/10.1144/EGSP26.13

Fig. 2. The Paskov 'D' tip.

Fig. 4. Sample sizes for tests of susceptibility to spontaneous combustion using the isotherm method. (cubes with edge measurements: 50, 100, 300, 450 and 600 mm).

slag) and NW Bohemia (fly ash). Industrial by-products are very cheap; however, each of them has their own specific properties which must be considered prior to their use.

Colliery spoil

Colliery spoil (mine stone) is one of the most important sources of alternative materials that are available for possible use as fill in embankments. In the construction of the D47 motorway and the R48 expressway near Ostrava over 3 million m^3 of the colliery spoil was used as a bulk fill, instead of natural materials. The best material is located on the Paskov 'D' tip (Fig. 2). It is represented by colliery spoil of a 0–180 mm grading after processing. It is homogenous and contains low amounts of organic matter. Reserves of this tip are over than 14 million m^3.

Fig. 3. The Heřmanice tip, burning colliery spoil. Steam rises from the material burning underground.

It is better to use burnt colliery spoil than unburnt (as has previously been used); there is a lack of burnt colliery spoil. Burning has occurred in several tips in the Ostrava region and the susceptibility to spontaneous combustion of colliery spoil has already been analysed (Fig. 3).

Spontaneous combustion tests carried out by the isotherm method showed that large size samples of 600 mm are not susceptible to spontaneous combustion (see Fig. 4, Table 1) and that relationships between sample size, time of combustion and temperature could be calculated. These results have been used to calculate the safe heights of embankments. For a 30 m high embankment, with ambient temperature of 85 °C, a 10 °C rise in the internal temperature of the embankment is possible (see Fig. 5) without risk of combustion.

Colliery spoil in the Ostrava region is not susceptible to spontaneous combustion in the naturally occurring conditions and embankment heights (Fig. 6).

Table 1. *Results of susceptibility to spontaneous combustion tests using the isotherm method of large size samples*

Cube	*600 (mm)*	*450 (mm)*	*300 (mm)*
Max. temperature of surroundings (°C)	87	87	87
Max. temperature of sample (°C)	87	87	86
Time necessary to achieve maximal sample temperature (hours)	270	130	95
Loss of oxygen	Not measured		

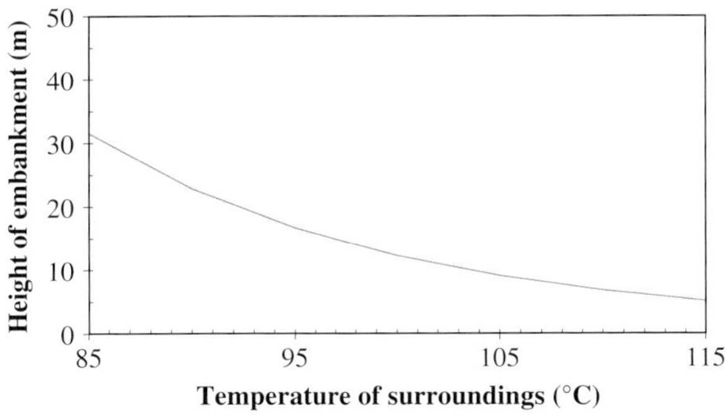

Fig. 5. Relationship between the embankment height and surrounding ground temperature during tests in a closed adiabatic chamber.

Fig. 6. The D47 motorway, section 4708, the Ostrava– Rudná overpass, aerial view of the embankment construction. The embankments were built from colliery spoil.

Fig. 8. Mixture of fly ash from the Třinec power station using lime used as a binder in the soil treatment trial section.

Fly ash

Fly ash is used as fill in the Czech Republic in areas with brown coal power stations (NW Bohemia; Fig. 7). The main advantages of fly ash are its high shear strength and constant quality. It is available in large quantities in tips near the power stations. It is a light material and therefore used especially in embankments located on soft subgrade soil. Fly ash can be used in the mixture, with lime as a binder or as a component in new mixtures used as a binder in soil treatment (Figs 8 & 9). However, it contains sulphur and has unpleasant environmental properties as a result of ettringite formation, which is susceptible to swelling.

Fig. 7. The embankment of D11 motorway section Osíčky–Hradec Králové where more than 2 000 000 m^3 of fly ash was used as fill.

Fig. 9. Lime-rich fluidized ash bed product beyond the bridge abutment (the R6 expressway).

Fig. 10. Blast furnace slag from the Třinec steelwork being incorporated into the I/11 Jablunkov by-pass.

Fig. 12. Mixing of PRESTAB, fly ash and lime (2004).

Blast furnace slag

Blast furnace slag is used mostly in earthworks in Northern Moravia where iron and steel is produced (Accelor Mittal, Tž Třinec). Potentially, slag from the (now closed) steelwork tips located near Prague could also be used. Blast furnace slag is used as bulk fill, in active zones (0–90 mm) or as the construction layer (0–32 mm) (Fig. 10).

Blast furnace slag is a very good fill material; however, it is brittle and is crushed during compaction . It can be substituted by steel slag or steelwork waste (dinase bricks) (Fig. 11), but in general, the heavy metal content of steel slag is too high and steelwork waste has the potential to

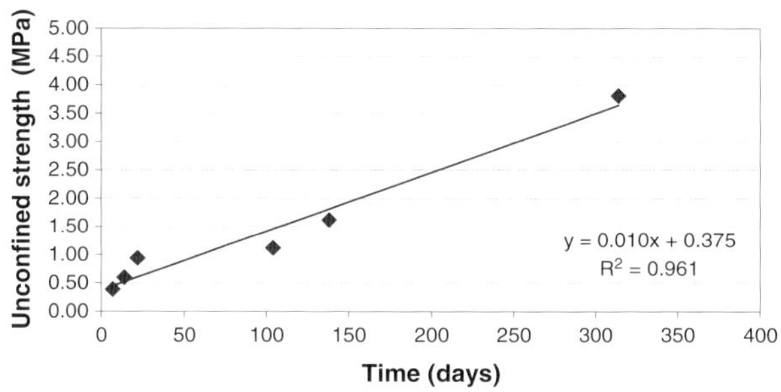

Fig. 13. Relationship of the unconfined strength of the mixture (PRESTAB, fly ash, lime) prepared *in situ* with time.

Fig. 11. Decomposition, and accompanying expansion, of the dinase bricks to secondary minerals (white) .

Fig. 14. Testing area 3 years after its construction in 2007.

expand causing the deformation of the construction work; subsequent remedial works are typically very expensive.

Prestab

In the D1 motorway near Přerov there is a lack of suitable material for embankment fill and a mixture of PRESTAB (industrial gypsum from titanium white (TiO_2) production), fly ash and quick lime has been used (Fig. 12). Large trials with accompanying laboratory and field tests and monitoring have approved its use as a fill (Figs 13 & 14). Since it can be difficult to manipulate PRESTAB, it is recommended that PRESTAB, fly ash and lime are combined in mixing centres or used in layers of 15 cm thickness.

Conclusions

In spite of the various difficulties, the use of industrial by-products as fills will increase in the near future. Construction of new embankments of the D47 motorway near Ostrava (a section with length of 15 km) used more than 3 million m^3 of colliery spoil and more than 0.5 million m^3 of blast furnace slag, with all fills being sourced from the built environment.

The use of deep counterfort drains as an effective method of stabilizing cuttings constructed in overconsolidated clays

G. J. Macdonald, A. R. Vooght* & S. Parkin

Mouchel Ground Engineering, Export House, Cawsey Way, Woking, Surrey, GU21 6QX, UK
*Corresponding author (e-mail: alan.vooght@mouchel.com)

Abstract: Slope failures in cuttings constructed in heavily overconsolidated clays often occur after a period of 20 to 30 years due to the gradual dissipation of negative pore pressures and the effects of water ingress into the slopes. These slope failures are often stabilized with the installation of deep counterfort drains, which reduce the pore water pressures on the potential critical slip planes and act as a reinforcing element in the slope. Counterfort drains are normally designed using empirical methods. However there is little published data on drain performance. This paper summarizes the findings of a research project undertaken by Mouchel on behalf of the Highways Agency to assess the performance of counterfort drains installed in two highway cutting slopes that have been monitored over a number of years. The study has shown that measured pore water pressures can exceed the theoretical pressures, due mainly to what is believed to be the poor performance of the counterfort drains and that the apparent improved slope stability may be primarily due to the reinforcing effect of the drains within the slope. Further studies are required to determine the effects of construction techniques and materials used on the performance of counterfort drains before the efficiency of the existing empirical design methods can be fully assessed.

The stability of highway cuttings, over a particular height and slope angle, in overconsolidated clays is usually dependent on the magnitude and distribution of pore water pressure within the clay. The stress relief on excavation of the cutting, results in the development of negative pore water pressures which will normally ensure short to medium-term stability. Over a period of time, these negative pore water pressures dissipate as the clay becomes wetter. The rate of dissipation depends on the permeability of the clay, potential recharge along the cutting boundaries, flow conditions within the cutting and the effects of vegetation which will abstract moisture from the clay. Seasonal variations in pore water pressure may also occur in the slope possibly leading to cyclic swelling and shrinkage of the clay. All of these and many other factors, such as soil homogeneity, evapotranspiration and surface suction, affect the surface hydraulic boundary conditions which influence the stability of the cutting, making it extremely difficult to predict if and when slopes will fail.

Counterfort drains have been employed for many years as a cost-effective remedial solution to remediate and/or stabilize cuttings with high groundwater. The effectiveness of the counterfort drains is dependent on the amount by which the groundwater table is to be reduced and maintained. Counterfort drains usually run down the slope, perpendicular to the crest, and provide drainage to a depth of 2–7 m, although the deeper drains are very often difficult to construct. When installed at between 5–10 m centres along the line of a cutting the counterfort drains not only reduce the pore water pressures but also provide a buttressing effect and enhance the overall stability of the slope.

Counterfort drains are normally between 0.3–0.6 m in width and are filled with stone or granular backfill, or sometimes no-fines concrete (Fig. 1). Drains constructed with coarse fill material are normally lined with a geotextile to prevent the migration of fines which may affect the drain functionality. The surface of the counterfort drains are generally sealed to prevent surface water runoff and fines from entering the top of the drains.

Due to the complexity of accurately modelling the boundary conditions and soil properties, counterfort drain design has generally been based on semi-empirical formulae. Analytical, and in the last two decades, numerical solutions have also been used to estimate flow rates and pressure distributions associated with subsurface drainage systems. The empirical solutions proposed by Bromhead (1984) and Hutchinson (1977) take the form of charts in which the piezometric head reduction varies with counterfort drain spacing and drain efficiency. However, the published data on counterfort drains suggests that there are often significant discrepancies between observed field data and predicted curves (Hutchinson 1977), which may be due to the fact that these solutions inherently adopt assumptions of flow through saturated soil. When the soil de-saturates (as may happen in trench drainage cases, involving lowering of the water table) the transient response of the system will alter.

From: Radford T. A. (ed.) 2012. *Earthworks in Europe*. Geological Society, London, Engineering Geology Special Publications, **26**, 115–124, http://dx.doi.org/10.1144/EGSP26.14

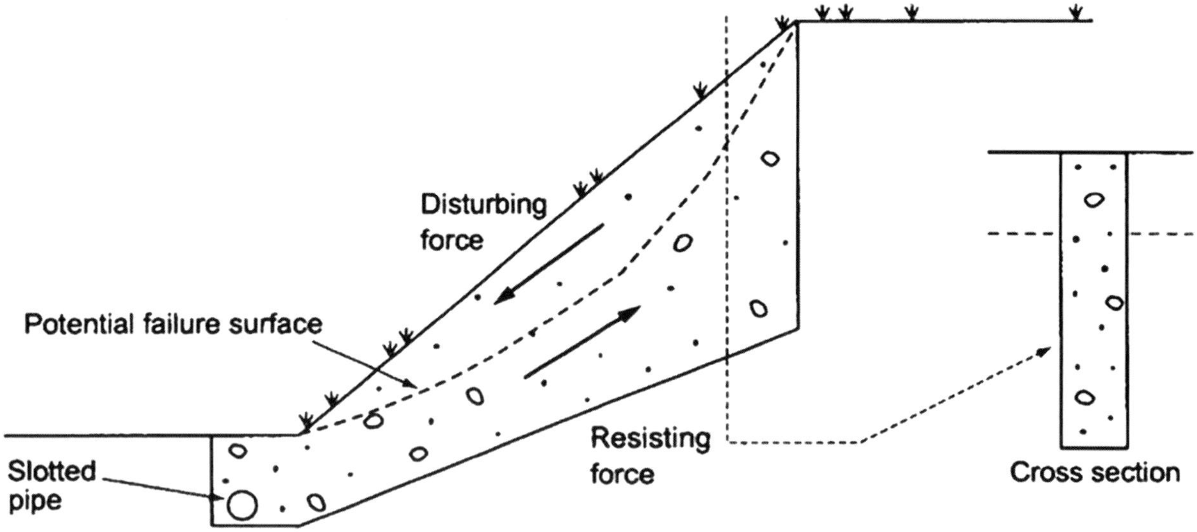

Fig. 1. Typical counterfort drain.

This paper summarizes the relationship between the theoretical drawdown of groundwater predicted between counterfort drains based on the method proposed by Hutchinson (1977) and pore water pressure measurements recorded at two remediated highway cutting failures that occurred within overconsolidated clay. The paper is based on the findings of a research study undertaken by Mouchel for the Highways Agency (Mouchel 2010) into the use of deep counterfort drains as an effective method of stabilizing cuttings in overconsolidated clays.

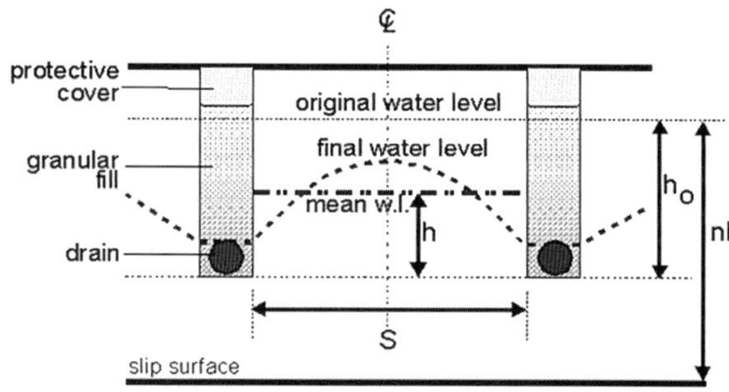

Fig. 2. Cross-section of typical counterfort drain (after Hutchinson 1977).

Design

An approximate method for designing trench or counterfort drains has been developed by Hutchinson (1977) using finite element analyses and assuming two-dimensional steady-state flow in saturated soil. Hutchinson (1977) used the results of the finite element analysis to define the efficiency ($\eta = (ho-h)/ho$) of the counterfort drains and to relate the efficiency to the ratio S/ho, where S is the spacing of the drains and ho is the depth of the drains beneath the groundwater level (Figs 2 & 3). Hutchinson (1977) showed that for maximum effectiveness the drains should penetrate to the base of the predicted critical slip surface and suggested that lines G and H in Figure 3 can be taken as reasonable lower and upper bounds for the design of counterfort drains. The required spacing and the depth of counterfort drains required to achieve the pore water pressure reduction determined from the stability analyses can therefore be determined from Figures 2 and 3.

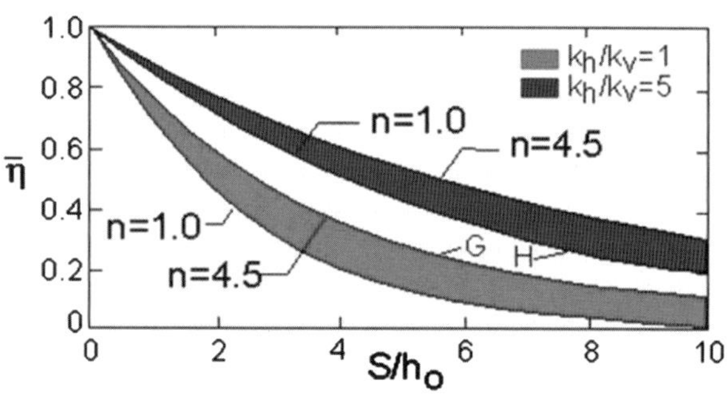

Fig. 3. Efficiency chart for counterfort drain (after Hutchinson 1977).

Filter materials for the construction of counterfort drains for highways are specified in Clause 505 of the *Manual*

Fig. 4. M11 Chigwell cutting finished drains prior to placing topsoil.

of Contract Documents for Highway Works Volume 1 (MCHW; Highways Agency 2009). Type A is fine grained general purpose filter material suitable for some, but not all, soils. Type B is coarse free draining material which is not suitable as a filter material but which could be used in counterfort drains together with geosynthetic filters in accordance with Clause 514 of the MCHW. Type C is a filter material and requires designer specification in accordance with Clause 505 of the MCHW and BS EN 13285 (BSI 2010). Typical counterfort drainage layout and construction with Type A backfill is shown on Figures 4 and 5.

Fig. 5. M26 Sevenoaks cutting placing type A drain material in counterfort drain.

Case studies

M11 Chigwell cutting

The M11 motorway cutting at Chigwell is located SW of Junction 5 on the M11 southbound carriageway with the motorway at the toe and Roding Lane at the crest of the cutting (Fig. 6).

The 12 m deep cutting was constructed through a thin layer of Terrace Gravel into the underlying London Clay with an average slope angle of 1 vertical to 3.7 horizontal (15°). The slope was mainly grassed, with trees near the crest (Fig. 6).

Slope failures occurred in the cutting in 1998 and were remediated in 2001 with the installation of 1.0–1.5 m deep counterfort drains and the 'benched' removal and replacement of soil in the upper part of the failed slope. A further failure on this remediated section occurred during the winter of 2001/2002. Investigations revealed water, possibly from the Terrace Gravels, emerging at the surface just below the tree line near the crest of the cutting.

Pore water pressure data recorded in piezometers installed prior to the remedial works indicated the presence of two roughly hydrostatic profiles. The first with water table at 0.5–1.0 m below ground level at mid-slope and the second with the water table at about 2 m depth near the crest.

Fig. 6. M11 Chigwell cutting shallow translation failure (February 2002).

Remedial work and slope monitoring

Remedial works were carried out between January and June 2005, which included the installation of thirty-one 3 m deep counterfort drains at 6 m centres, filled with Type A filter material with no geotextile separator or carrier pipe (Fig. 4). A filter geotextile, Lotrak 1800 was placed on the upper side of the cut-off trench to prevent the ingress of silt into the counterfort drainage system. The site arisings were mixed with lime and a toe berm constructed over the lower 14 m of the batter to provide additional stability and to reduce the volume of material required for off-site disposal.

Two arrays of Geotechnical Observation Limited Type 1 piezometers were installed between the counterfort drains to measure both negative and positive pore water pressures at shallow depth. One array was located near the crest and the second towards the toe of the completed slope (Fig. 7) and monitored over a 3-year period to assess the effectiveness of the drains.

Immediately following the remedial works, negative pore water pressures (suctions) were measured within the cutting, these suctions reaching a maximum in the summer of 2006, as shown for the upper array of piezometers plotted on

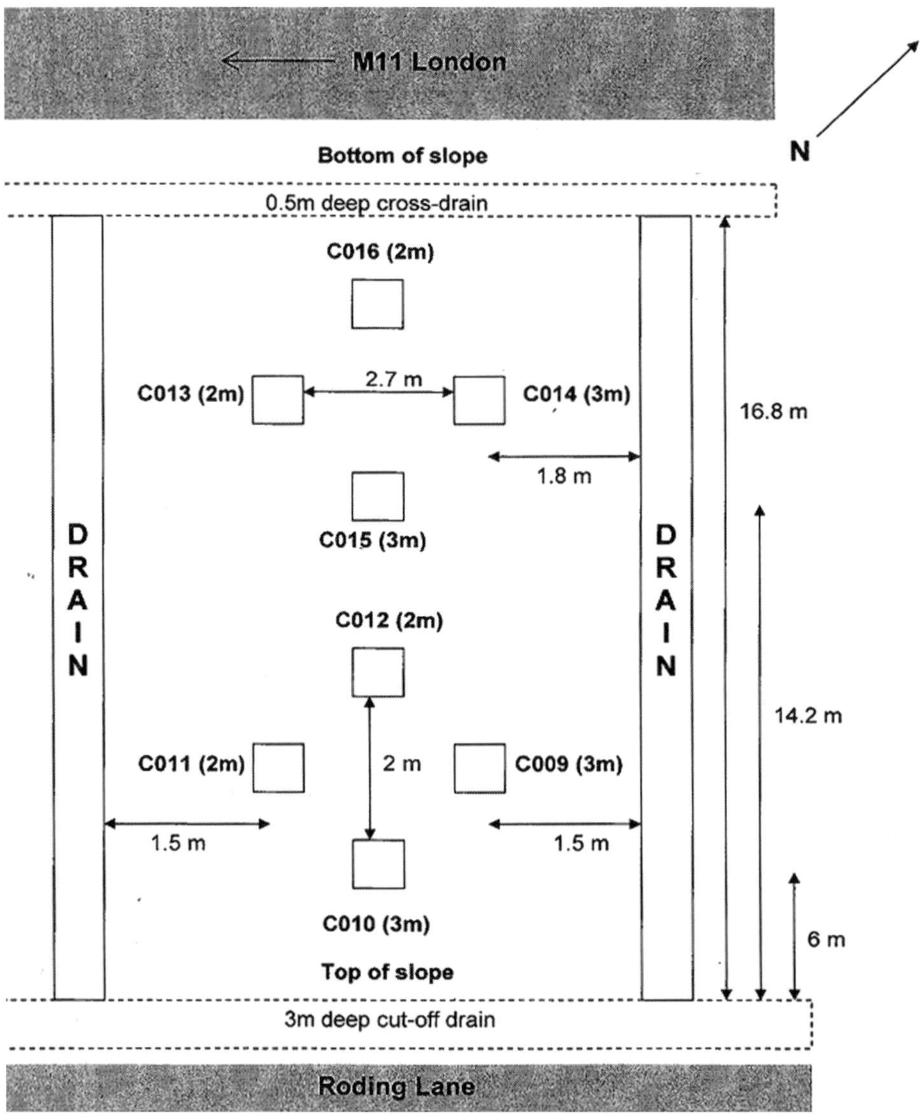

Fig. 7. M11 Chigwell cutting piezometer layout.

Figure 8. The suctions recorded in the piezometers gradually dissipated with a steady increase in pore water pressure until, at the end of 2007, positive pore water pressures were recorded in the piezometers close to the counterfort drains (C009 and C011). The positive pore water pressures in piezometers C009 and C011 continued to increase, reaching an initial peak close to the predicted average drawdown level of 25.3 m Above Ordnance Datum (AOD; that is the datum used by the Ordnance Survey in the UK to define elevation) by January 2008. Pore water pressures in the two piezometers close to the counterfort drains fluctuated seasonally above and below the predicted average drawdown level throughout 2008. By February 2009, the pore water pressure in piezometers C009 and C011 (close to the counterfort drains) increased to a maximum of 26.6 m AOD, that is 1.4 m above the predicted average drawdown level, before reducing steadily to below this level throughout the summer of 2009.

In contrast, the pore water pressures in piezometers C010 and C012 midway between the counterfort drains remained negative throughout the four-year monitoring period, although there is an apparent trend of these negative pore water pressures becoming less negative with time (Fig. 8).

Figure 9 shows the pore water pressures recorded between July 2008 and July 2009, together with the theoretical drawdown curve (based on Hutchinson 1977) required to lower the average water level (predicted average drawdown) from ground level to 2 m depth. This diagram also shows the pore water pressures recorded in July 2008 (summer level) and February 2009 (winter level). It can be seen from Figure 9 that the winter levels in piezometers C010 and C012 (midway between the counterfort drains) are similar but the summer suctions differ by about 1.3 m head of water. Figure 9 also shows that elevated pore water pressures were recorded in the piezometers closest to the counterfort drains and that suctions persist in the piezometers midway between the drains.

The results for the lower array of piezometers installed in the cutting slope (C013, C014, C015 and C016, Fig. 8), although not presented in this paper, also recorded elevated pore water pressures in the piezometers closest to the counterfort drains and in the piezometer adjacent to the mid-slope cross drain. The higher pore water pressures are equivalent to a water table close to ground level.

It is apparent from the piezometer data that transient peak pore water pressures exceed the theoretical levels close to the counterfort drains and are the equivalent in some cases to a ground water table close to ground level. The dates at which the initial suctions were lost and positive pore water pressures recorded show a progressive trend from August 2005 from the toe of the slope to January 2008 upslope, suggesting that the counterfort drains have become progressively less efficient with time. This would suggest that the

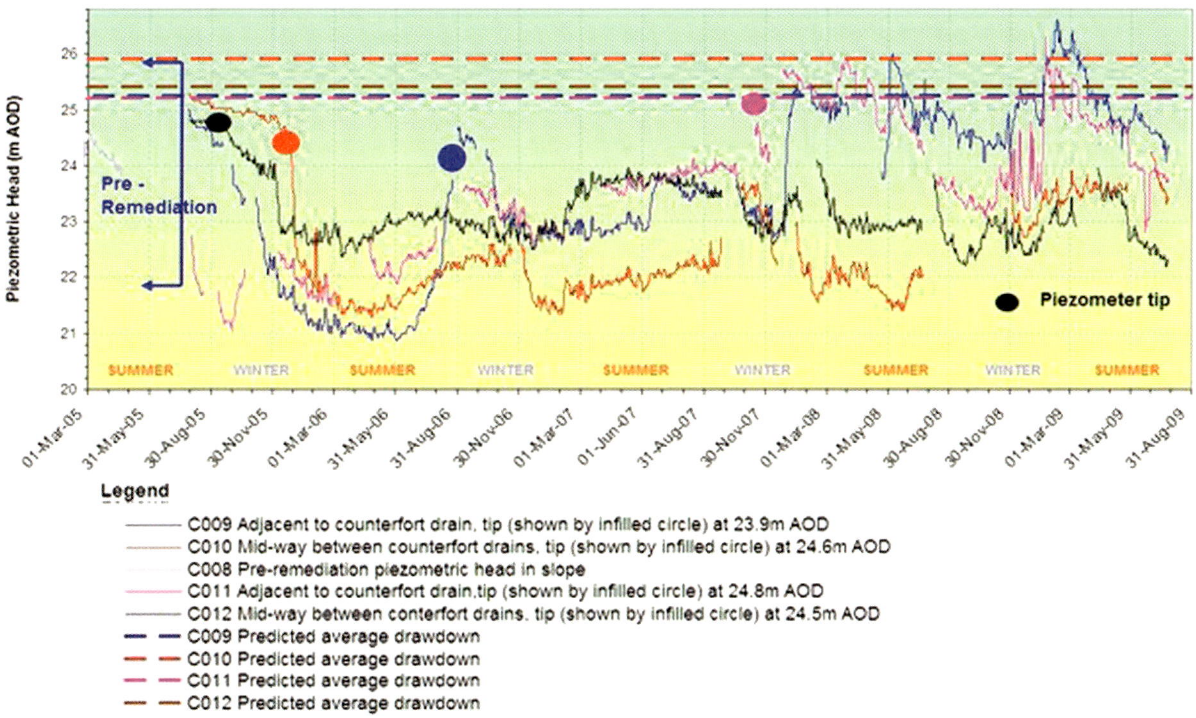

Fig. 8. M11 Chigwell cutting piezometric head *v.* time (upper piezometer array).

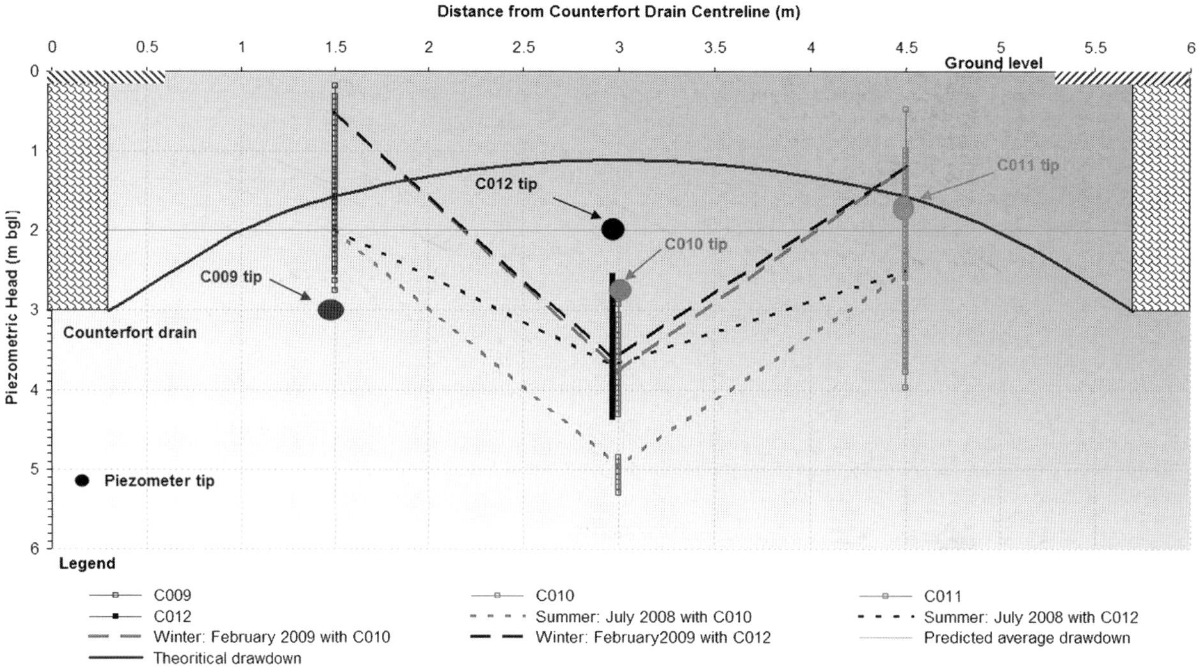

Fig. 9. M11 Chigwell cutting piezometric head variation: July 2008–February 2009 (upper piezometer array).

counterfort drains are not functioning as designed and as a consequence may be feeding water into the cutting slope. However, notwithstanding the above, the trend of increasing pore water pressure with time observed in the piezometers may reflect the response to a general trend of increasing annual rainfall, based on data provided by the Meteorological Office, which shows a significant increase in rainfall between 2006 and 2008.

M26 J2A to J1, Sevenoaks cutting, Kent

The motorway cutting is located on the westbound carriageway of the M26 motorway between Junction 2A and Junction 1 near Sevenoaks, Kent.

The earthwork comprises a 5 m deep cutting through Head Deposits (comprising clay and gravel) and the underlying Gault Clay with an average slope angle of 1 in 6 (10°). The slope had been planted with a mixture of mature trees and shrubs (Fig. 10). A slope failure was recorded in the cutting in 2003 and comprised a shallow translational movement within the batter slope with multiple backscarps, up to 0.6 m in height and a 1 m high toe bulge. Water was observed ponding within the failed mass and draining from the toe of the slope.

Remedial work and slope monitoring
Remedial works were carried out between September 2006 and December 2006 and comprised reprofiling the failed slope and the installation of fourteen 3.5 m deep counterfort filter drains 0.6 m wide at 6 m centres, including a toe drain and a cut off drain at the crest of the slope. The counterfort, cut off and toe drains were filled with Type A filter material (Fig. 5) and a filter geotextile (Lotrak 1800) was placed on top of the drainage trenches to prevent the ingress of fines into the drainage system. Carrier pipes were not included in the counterfort and cut off drains.

Piezometers and inclinometers were installed centrally between the counterfort drains (Fig. 11) and monitored between March 2007 and September 2009 to assess the effectiveness of the drains.

Piezometers P1 and P2 were installed at the base of the cutting with P1 located approximately 1 m from the counterfort drain. Piezometers P3 and P6 were positioned centrally

Fig. 10. M26 Junction 2A to J1 Sevenoaks cutting failure.

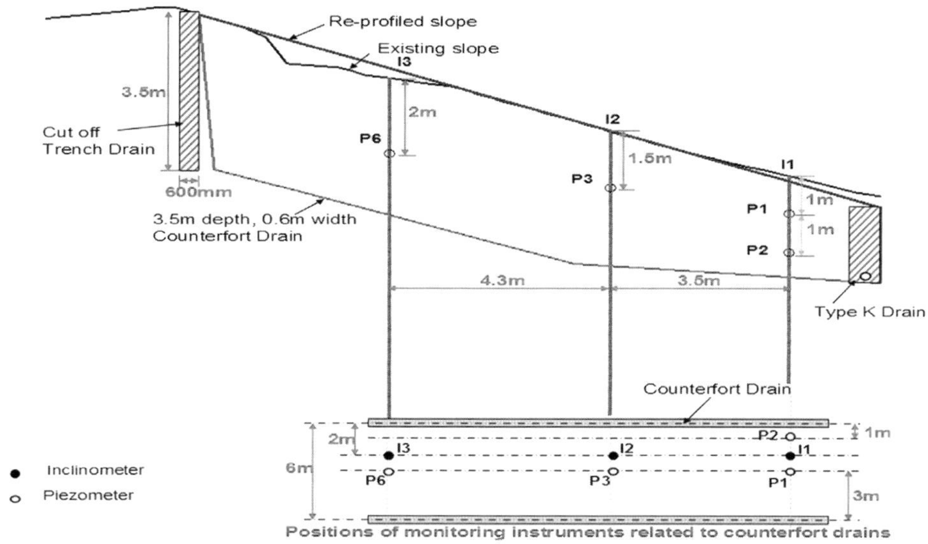

Fig. 11. M26 Sevenoaks cutting piezometer layout.

between the counterfort drains, located mid-slope and towards the crest of the cutting, respectively.

Following installation of the counterfort drains small negative pore water pressures were measured in all of the piezometers (Fig. 12). Pore water pressures began to increase at the end of March 2007 becoming positive with a small head in the piezometers installed at the base of the slope (P1 and P2) in early May 2007. Only small variations in pore water pressures were recorded in the mid to upper slope piezometers (P3 and P6) during this period.

Figure 12 shows that the pore water pressures in the slope from May 2007 varied seasonally, particularly at the toe of the cutting, with suctions in the summer and positive pore pressures during the winter. Pore water pressures become positive near the crest of the cutting (P6) and mid-slope (P3) by October 2007 and December 2007 respectively, that is approximately 10 months after the completion of the remedial works. The pore water pressure readings in piezometer P6 show only small fluctuations from March 2007 through to November 2008 following which more pronounced variations were recorded (Fig. 12).

The deep counterfort drains were designed with a predicted average drawdown of the water table to 2.2 m depth below ground level. This predicted average drawdown and the theoretical drawdown curve between the drains based on Hutchinson (1977) are shown on Figure 13. The elevation of the predicted average drawdown at the location of each piezometer has been determined for the period between August 2008 and January 2009 (Fig. 13). It can be seen from Figure 13 that in all of the piezometers except piezometer P6, the theoretical drawdown curves imply that suctions are to be maintained whereas pore water pressures near the cutting crest result in a small positive head in piezometer (P6). The pore water pressures vary seasonally by 1.6–1.7 m head above and between 1.8–3.0 m head below the theoretical drawdown levels in piezometers P1, P3 and P6 located mid-way between the counterfort drains and by 2.3 m head above and 1.0 m head below the theoretical curve for piezometer P2, located adjacent to the counterfort drain at the toe of the cutting.

It can be seen from Figure 13 that the highest measured piezometric heads significantly exceed the theoretical piezometric head between the counterfort drains and are close to being hydrostatic from near ground level, suggesting that the counterfort drains are not effective in maintaining a lower ground water level in the cutting slope.

Both the counterfort and cut off drains were backfilled with acceptable Type A drainage material laid to a fall but with no carrier pipe. The drainage material contained a large proportion of sand-sized material (Fig. 5). Whilst this material is likely to have a permeability orders of magnitude greater than the surrounding clay, it is possible that the counterfort drains have become saturated during heavy rainfall and are feeding water back into the slope leading to the temporary increase in pore water pressures recorded in the piezometers. The elevated piezometric head of water above the theoretical curve in the piezometer (P2, Fig. 13) installed adjacent to the counterfort drain is greater than those measured in the three piezometers located midway between the counterfort drains supporting the assumption that the drains are in fact feeding water into the slope during high rainfall. However, further work is required to verify this assumption.

The presence of granular layers and lenses of Head Deposits near the crest of the cutting may also have

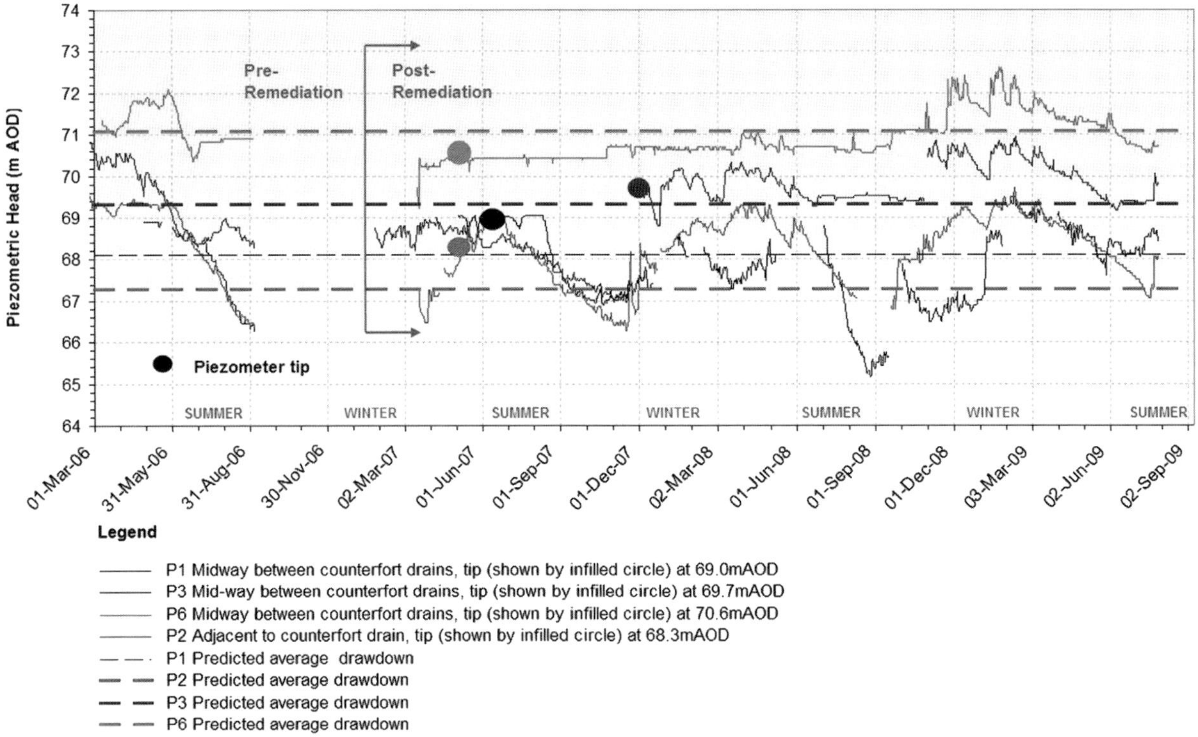

Fig. 12. M26 Sevenoaks cutting piezometric head *v.* time.

recharged the slope. Although this recharge should have been intercepted by the crest cut-off drain, it is possible that the Head Deposits were locally deeper and not fully intercepted by the cut-off drain. Localized recharge may have therefore been responsible for the more uniform pore water pressure response recorded near the crest of the cutting (P6) up to October 2008 (Fig. 11).

Discussion

The effectiveness of counterfort drains relies on the accuracy of the theoretical predictions and on the drains remaining efficient throughout the design life. Hutchinson (1977) noted that there are usually significant discrepancies between observed field data and predicted curves and that these discrepancies can be due to the fact that these solutions inherently adopt assumptions of saturated flow. When the soil desaturates suctions are measured and the transient response of the system will alter. Variations in rainfall and possible recharge at the crest of both cuttings may have influenced the effectiveness of the counterfort drains installed to remediate the highway slopes.

In both case studies the peak pore water pressures recorded in December 2008/January 2009 were higher than those during the whole of 2008 and this may be due to a trend of increasing annual rainfall. The regional data published by the Meteorological Office for Southern England for the period 2007 to date confirms that for the last 3 years the summer months in particular have been wetter that the average.

The piezometric readings for both case studies show a progressive trend of increasing pore water pressure after the initial suctions generated during construction, following which positive pore water pressures start to be measured. The loss of suction commences at the toe of the slope and progressively moves upslope with time suggesting that the counterfort drains are becoming less efficient with time and that they may actually be feeding water into rather than conveying it from the slope.

The measured pore water pressure data for both case studies show a seasonal variation with a gradual build-up of suction during the summer months followed by a rapid and sudden loss of suction as a wetting front passes down through the soil, possibly in response to rainfall. The general trend of a steady increase in pore water pressure with time occurs at both sites but is most noticeable in the M11 Chigwell cutting data.

The counterfort and cut off drains in both case studies were backfilled with acceptable Type A drainage material laid to a fall but with no carrier pipes or geotextile separator

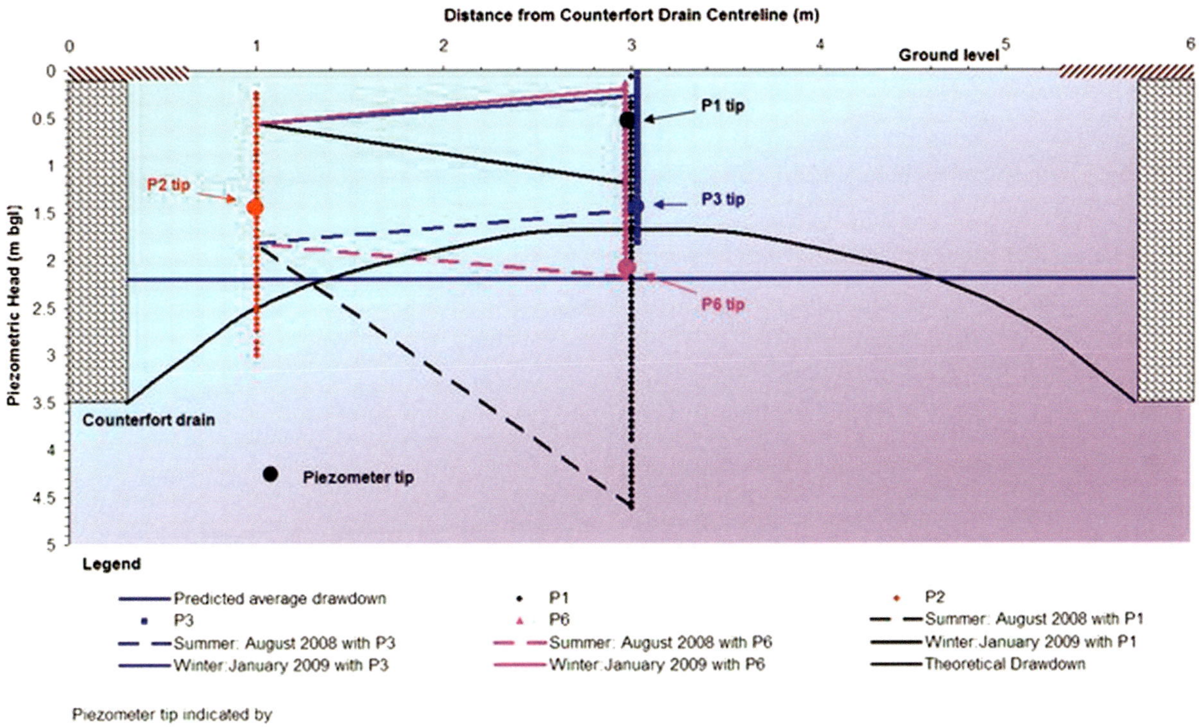

Fig. 13. M26 Sevenoaks cutting piezometric head variation: August 2008–January 2009.

membranes. The relatively shallow gradient of the counterfort drains and the lack of a geotextile separator may have permitted fines to enter the filter medium, possibly reducing the long term efficiency of the counterfort drains. However further investigations would be required to determine whether excess pore water pressures are being generated within the filter medium during periods of heavy rainfall.

It is possible that the counterfort drains may have become saturated during heavy rainfall resulting in the temporary increases in pore water pressures recorded in the piezometers for both cuttings.

Conclusions

This paper summarizes the findings of the research study undertaken by Mouchel (2010) which explored the current empirical approach to the design of counterfort drains in overconsolidated clays and compares the theoretical and predicted average drawdown with that achieved in practice in two motorway cuttings. The case studies comprised a 12 m deep cutting in London Clay on the M11 motorway at Chigwell, Essex and a 5 m deep cutting in Gault Clay on the M26 motorway near Sevenoaks, Kent.

In both case studies the installation of the counterfort drains arrested previous slope movements and therefore the remedial works appear to have been successful. However, the measured peak pore water pressures within the cuttings for both case studies exceed the theoretical design levels and are the equivalent in some instances to a ground water table close to ground level, indicating that the counterfort drains have been ineffective in limiting seasonally high peak pore water pressures. The success in remediating the cutting slopes was therefore more likely to be due to the reinforcing effect of the counterfort drains in the slope with some possible reductions in pore water pressure on the critical slip failure planes.

The study concluded that verification of the appropriateness of the current design methods based on charts such as those proposed by Hutchinson (1977) could not be established from the piezometer data recorded in each of the two case studies.

The study identified that the design charts predict that pore water pressures between the counterfort drains will be higher than the average water level and recommends that due consideration be given to designing to the predicted drawdown level at the mid-point between the drains rather than the average water level. An allowance for seasonal water level variations should also be accommodated in the design by anticipating peak water levels at least 1.5 m above the theoretical curve. This will most likely require a closer counterfort drain spacing as the height of the cutting will limit the drain depth.

The study also recommends installing carrier pipes in all counterfort and cut off drains to ensure that the drains effectively convey water from the slope. If there is concern over the migration of fines into the carrier drain then a suitable fabric should be wrapped around the pipe. In addition consideration should be given to using a coarser drainage material (Type B or Type C) in all drainage schemes, although such materials may require the installation of a separator membrane between the *in-situ* clay and the drainage material to prevent the migration of fines into the latter.

Where consideration is to be given to monitoring the effectiveness of counterfort drains the study recommends that piezometers be installed at depths below the design drawdown level such that positive pore water pressures are recorded in preference to recording suctions in shallow piezometers. Piezometers should also be installed within the counterfort drains to monitor whether or not excess pore water pressures are being generated within the filter medium during or following periods of heavy rainfall.

The authors would like to thank the Highways Agency, in particular the Project Research Project Sponsor (Jan Marsden) for the support and permission for publishing the information presented in this paper.

References

BROMHEAD, E. N. 1984. An analytical solution to the problem of seepage into counterfort drains. *Canadian Geotechnical Journal*, **21**, 657–662.

BSI. 2010. BS EN 13285:2010. *Unbound mixtures. Specifications.* British Standards Institution, London.

HUTCHINSON, J. N. 1977. Assessment of the effectiveness of correction measures in relation to geological conditions and types of slope movement. *Bulletin of the International Association of Engineering Geology*, **16**, 131–155.

HIGHWAYS AGENCY. 2009. *Manual of Contract Documents for Highway Works*, Volume 1, Specification for Highway Works, November 2009.

MOUCHEL. 2010. The use of deep counterfort drains as an effective method of stabilising cuttings constructed in overconsolidated clays. *Research Study for the Highways Agency*. Report 787303.

Stabilization of a railway embankment using electrokinetic geosynthetics

John Lamont-Black[1], Jean A. Hall[2*], Stephanie Glendinning[2], Chris P. White[1] & Colin J. F. P. Jones[1]

[1] Electrokinetic Limited, Drummond Building, Devonshire Terrace, Newcastle upon Tyne NE1 7RU
[2] School of Civil Engineering and Geosciences, Newcastle University, Drummond Building, Devonshire Terrace, Newcastle upon Tyne NE1 7RU, UK

**Corresponding author (e-mail: Jean.hall@ncl.ac.uk)*

Abstract: A trial was conducted on a Victorian embankment to evaluate the use of electrokinetic geosynthetics for embankment stabilization. The embankment had been constructed by end tipping London Clay fill onto unprepared ground and has a history of instability. Ground investigations indicated a core of firm material underlain by softened embankment fill and alluvium and overlain by softened material towards the ground surface. The slope was treated to a maximum of depth of 7 m using electrokinetic geosynthetics and electrokinetic soil nails. Over a period of 42 days, electrokinetic treatment forced water out of the ground by raising it >4 m above the initial groundwater level such that discharge from activated electrokinetic prefabricated vertical drains (ePVDs) was >25 times that of control drains. Cation exchange processes were active as part of the treatment and correlated with reductions in plasticity and shrinkage. Overall DC power consumption was approximately 11.5 kWh m^{-3} of soil treated. Post treatment boreholes demonstrated that the soft material beneath the embankment had become firm and exhibited apparent improvements in both undrained and drained strength parameters. Pull-out tests on electrokinetic soil nails demonstrated an improvement in the bond strength of the nails by an average of 263%. Inclinometer data showed a significant reduction of movement tending to zero after treatment.

Background

There are 20 000 km of earth structures (cuttings and embankments) on the UK highway and rail networks. Many earth structures on the rail network are old, and were not built to modern geotechnical engineering standards. Continuous maintenance of these slopes is essential and, according to Perry *et al.* (2001), cost £50 million in the year 1998/9 for embankments alone. The numbers of earthwork failures on the railway from 2003/04 to 2007/08 are shown in Table 1 and are increasing. (Network Rail Annual Return to the Office of Rail Regulation and Stakeholders, Network Rail 2008).

Clay embankments subjected to hot summer drying commonly require rail tracks to be re-ballasted or tamped to maintain consistent rail level and alignment. In the case of older forms of construction, failures were repaired (with historic repair methods) to keep the infrastructure available for use, but sometimes leaving the failure surface as an inherent zone of weakness. Older embankments therefore have a history of failure and repair where the underlying causes of current problems can be difficult to assess. The cost of hard remedial solutions (e.g. discrete pile stabilization or the construction of bored pile walls to support or stiffen the crest) is high, and there is an increasing interest amongst asset owners such as Network Rail in alternative approaches.

Toe weighting and/or slope regrading is a method commonly used to tackle the problem of the destabilizing loads by redistributing them from the crest to the toe, where they aid stability. These techniques are simple and require no

Table 1. *Earthwork failures on Network Rail*

2003–8 Year	*No. of earthwork failures*
2003/04	47
2004/05	54
2005/06	41
2006/07	90
2007/08	107

From: Radford T. A. (ed.) 2012. *Earthworks in Europe*. Geological Society, London, Engineering Geology Special Publications, **26**, 125–139, http://dx.doi.org/10.1144/EGSP26.15

specialist plant for installation. However, they do not address the problems of shrink–swell or pore water pressure changes and can act merely to delay failure rather than prevent it. As sustainability becomes increasingly important to asset owners, these methods, which can consume large quantities of primary aggregate and energy, are becoming less viable. Additionally, they may require additional land and access for relatively large plant.

Electrokinetic ground treatment using electrokinetic geosynthetics (EKG) is a novel slope treatment method which can tackle the causes of instability without the inherent use of natural resources or access difficulties. It can be deployed rapidly using small plant and has modest access requirements (Lamont-Black 2001).

Introduction to electrokinetics and electrokinetic geosynthetics

The water content of fine-grained soils is a fundamental physical property, and has a large impact on material strength. Hence it is vitally important to the stability of slopes created from fine-grained (clay-rich) materials. Controlling water in these soils is of great value to managing embankment stability but presents recurrent problems owing to their very low hydraulic permeability. Passive drainage systems can be ineffective as they become clogged, broken or have insufficient driving hydraulic head to cause water flow in low permeability materials. Electrokinetic (or EKG) 'active' drains can overcome these difficulties.

Electroosmotic flow is usually described in terms of the Helmholz Smoluchowski model (Smoluchowski 1914) and is defined in equation 1. This has a similar form to the hydraulic flow equation as defined by Darcy's Law.

$$Q_e = K_e \cdot V/L \cdot A \quad (1)$$

where:

Q_e = electroosmotic flow rate ($m^3\ s^{-1}$)
K_e = coefficient of electroosmotic permeability ($m^2\ sV^{-1}$)
V/L = applied potential gradient ($V\ m^{-1}$)
A = area across which flow is taking place (m^{-2}).

The flow driven by both hydraulic and electroosmotic mechanisms is governed by the permeability coefficients and the applied potential head (or gradient). For London Clay this can be up to 3 orders of magnitude greater for electrokinetically driven flow when compared to hydraulically driven flow (Pugh 1999*a*, *b*, *c*).

Despite the potential advantages of electroosmosis in geotechnical engineering, it has been historically under-utilized. The main reason for this has been ineffective control of the flow of electricity, water and gas and the use of ineffective electrode materials. Geosynthetics are materials which perform a range of functions including drainage, filtration, reinforcement, separation and containment (Pugh 2002). The concept of EKG is to add the functions of an electrode to these materials in order to resolve the historical problems of applying electrokinetics in practice. An overview of the development of the EKG concept is described in Jones *et al.* (2008). EKG materials have been developed for use in a variety applications including ground and lagoon consolidation (Jones 2002; Glendinning *et al.* 2005; Lamont-Black *et al.* 2005), slope stabilization (Lamont-Black *et al.* 2007), conditioning of organic materials (Lamont-Black *et al.* 2006*a*, *b*; White *el al.* 2008) and dewatering of wastes and sludges (Jones *et al.* 2005; Lamont-black *et al.* 2006*b*).

The EKG materials used in the field trial described herein are electrokinetic prefabricated vertical drains or ePVDs in combination with steel reinforcing rods.

Figure 1 shows the principal processes that are active in a section of clay rich soil under electrokinetic treatment. The application of a DC potential gradient creates electroosmotic flow from the anode to the cathode. By draining the cathode and preventing water ingress at the anode, a reduction in water content is accompanied by a major drop in pore water pressure, which starts initially at the anode and then spreads into the rest of the soil. The soil around the anode becomes cemented due to electrochemical changes associated with the anode; these can be enhanced or controlled by selecting the structure and composition of the anode together with conditioning fluids. Around the cathode, precipitation of hydroxide salts occurs due to the presence of OH^- ions in the soil. The effects of cementation and precipitation are to cause a reduction in the plasticity of the soil and an increase in cohesion in the soil. The bond strength between the anode and the soil is significantly increased, which is beneficial if, as in the case of the field trial, anodes are additionally employed as soil nails. The addition of lime as a conditioning fluid within the system can have several benefits including increasing electroosmotic flow rates and participating in cation exchange reactions to replicate the effects associated with lime stabilization. By incorporating the well established geosynthetic functions into the EKG electrode development, these EKG electrodes are able to be designed for a wide variety of applications.

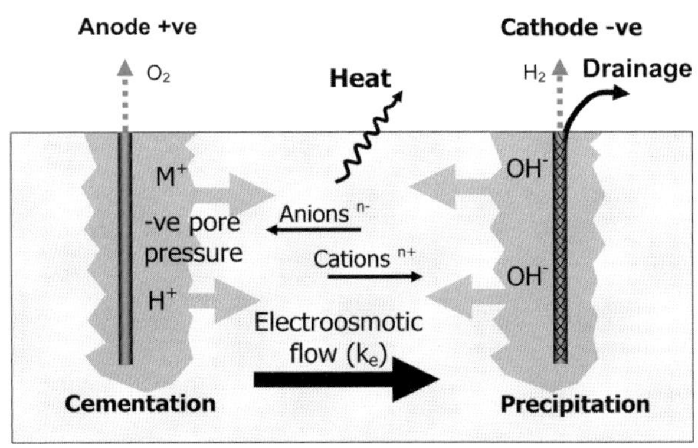

Fig. 1. Electrokinetic processes in fine-grained soils.

In general, the electrochemically induced changes in the soil can be considered permanent in fine grained soils. The reinforcement effects can likewise be considered permanent. The effects associated with changes in water content, pore pressure and consequent effects such as consolidation and reduction in void ratio may be permanent or non-permanent depending on the nature of the material being treated. The above effects and their duration are considered below in the context of railway embankments and with particular reference to material type and heterogeneity.

South Greenford trial

The railway line chosen for the trial was on a link line at South Greenford in the London Borough of Ealing, UK. The embankment is a Victorian structure approximately 9 m high with side slopes of approximately 22° (this included over-steepened sections towards the top of the slope, near the junction between the ballast/ash and the embankment fill and also towards the base of the slope). The embankment was constructed by end tipping fill onto unprepared ground; a construction method not uncommon on the UK rail network for the time.

Slope and track movements at numerous locations along the length of the embankment over many years have led to the remediation of sections of the slope (most notably using gabion baskets and slope re-grading) and the installation of slope monitoring instruments, principally inclinometers and piezometers. An assessment of the slope movements identified several sections of the slope as unstable, with some showing movement of over 6 mm per month (Arup 2006). Speed restrictions have been imposed on passenger and goods trains of 30 and 20 mph respectively.

Site conditions

The generalized site conditions are shown in Figure 2. The area of treatment was located so as to avoid interactions with a large tree on thc site. There were several areas of seepage or almost permanent dampness on the slope. The upper of these was coincident with the junction between the base of the ballast/ash and the top of the embankment fill and the lower area was coincident with the flattened zone at the base of the slope. Both areas of seepage were associated with zones of over-steepening immediately above. To maintain comparable analyses of results, the control array, of inactive anodes and cathodes, was located in a zone adjacent to the southern edge of the treated zone (non-hatched zone shown in Fig. 2).

Slope failure mechanism

Inclinometer readings indicated a distinct slip surface at approximately 2.5 m depth. However, prior to treatment it was only possible to determine approximate failure characteristics due to lack of geotechnical information. Data recovered during the installation of the electrodes allowed a more detailed ground model to be developed and thus a more accurate analysis. This post treatment analysis is described below.

In general terms, the embankment comprised a central core of firm to stiff heterogeneous fill composed dominantly of weathered London Clay with pockets of gravel or coarser material and occasional soft patches. This was bound above and below by softer material. Above the core material there was 0.5–1.5 m of soft clay fill and top soil. Beneath the core there was a mixture of soft to firm weathered London Clay fill and soft grey clay (alluvium). In the region of the toe of the embankment these two zones coalesced to create generalized softening in the toe zone.

The analysis of the slope using limit equilibrium methods showed a critical circular slip surface consistent with the inclinometer data, and having a factor of safety of 1.05 (Fig. 3). The piezometric levels were determined from the measurements taken in each of the individual cathode electrodes prior to active treatment. This generally coincided with the contact between the London Clay fill and the alluvium in the eastern area of the site bounded by 1, 14, 19 and 6 (Fig. 4) and is probably responsible for softening the base of the London Clay fill in the zone of water table fluctuation. The piezometric level measured in the region between C34 and C38 (furthest away from the large tree (Fig. 2) to minimize any potential influence of the tree on the groundwater conditions in the area) was generally lower and as a result the material forming the bulk of the slope remained firm to stiff in strength while the only the toe remained water softened.

Instrumentation

Additional instrumentation related specifically to the electrokinetic trial was installed in the embankment in June 2008. The location of these instruments is shown in Figure 3 and comprised an inclinometer at BH1 (mid slope) and three piezometers (of the vadose membrane tipped type) at BH1 (mid slope), BH2 (lower slope) and adjacent to existing inclinometer WS 17 (upper slope).

Electroosmotic design

The EKG treatment design comprised several elements including: electrode array and orientation; electrode spacing; and voltage and duration.

Electrode array

The electrode array is shown in Figure 4. It comprised tessellating hexagonal cells, with the hexagon being defined by anode stations and a cathode drain at the centre. The anodes and cathodes of three hexagonal cells, defined by cathode numbers 7, 20 and 33 near the top of the slope,

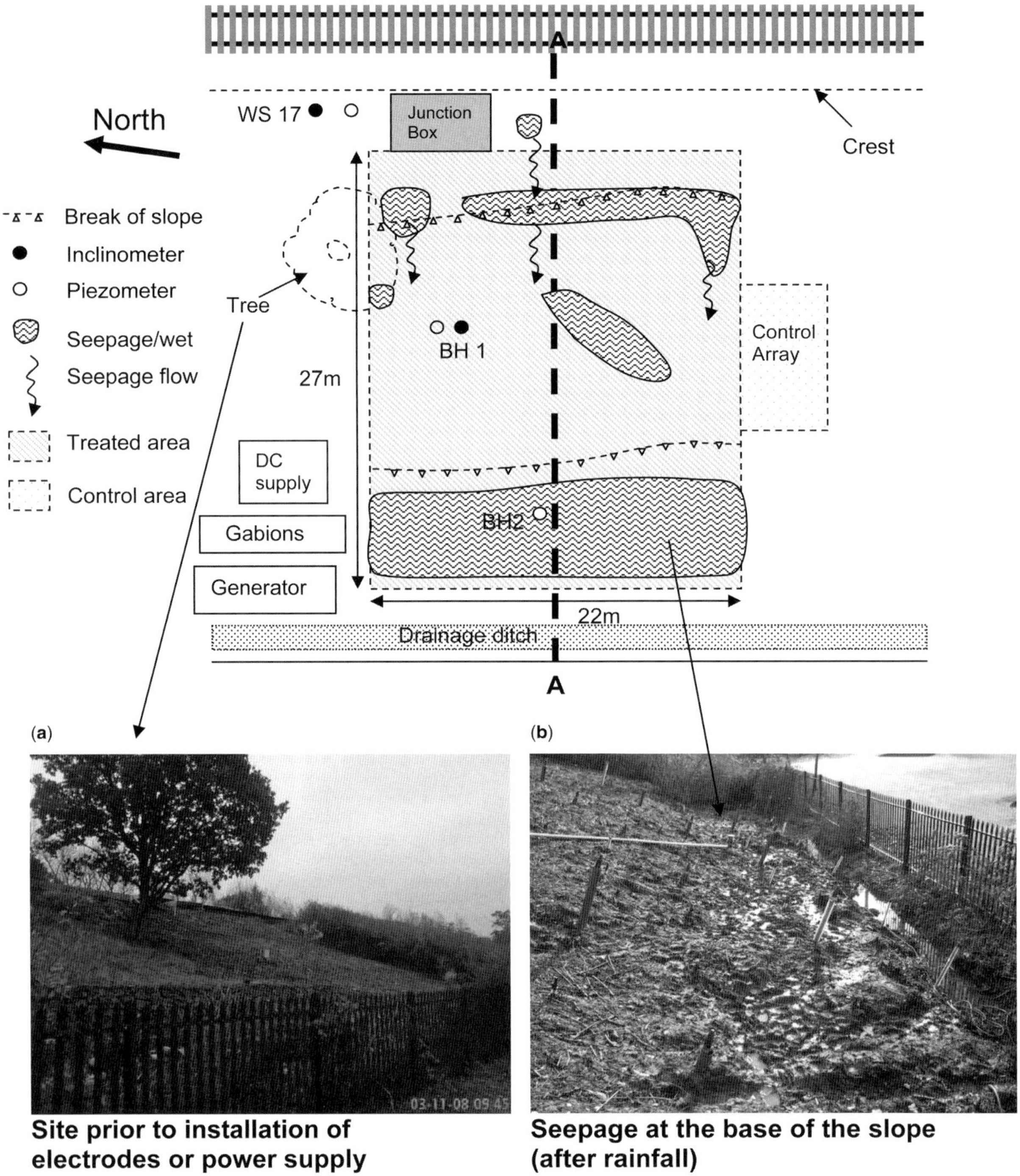

Fig. 2. General condition of the slope (section A–A shown in Fig. 3).

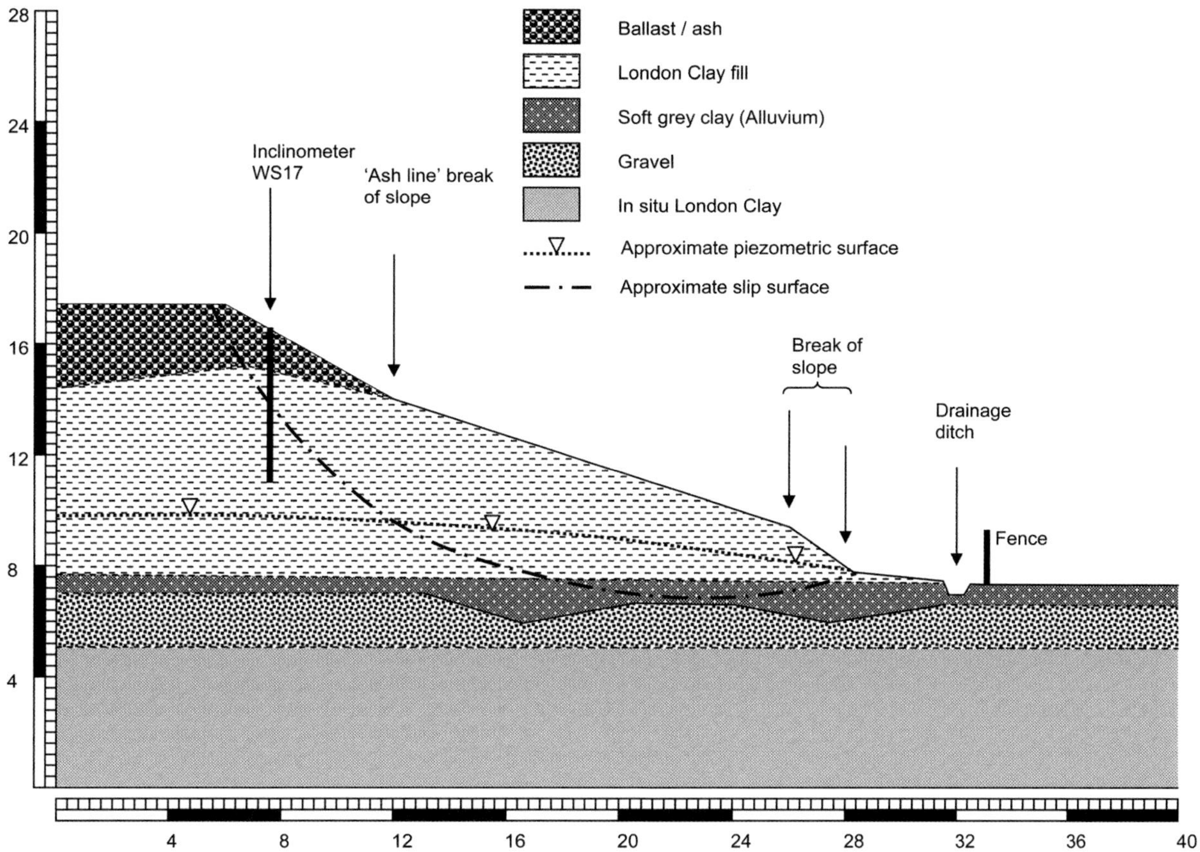

Fig. 3. Cross section of the embankment slope through Section A–A (Fig. 2). Vertical and horizontal scale bars in metres.

were installed at an angle 3° to the horizontal, dipping down slope. These cathodes were intended to drain the potentially problematic area associated with the ash/ballast. Additionally 2 control cathodes and 4 control anodes were installed as shown (Anodes 115 to 117 to the right of the treated area were the controls). All electrodes were installed at a spacing of 2 m.

Two cathodes were associated with anodes conditioned with a thin (10%) slurry of hydrated lime (cathodes 35 and 41, Fig. 4). This was included to assess the effects of lime dosing on electroosmotic flow and other factors such as current and anode bond strength. Locations of post-treatment boreholes and anode pull-out tests are also shown in Figure 4.

Setting up the trial

The anodes were installed first followed by the cathodes using a slope climbing drill rig equipped with a hydraulic hammer. The anodes were installed perpendicular to the slope to accommodate the angle of the embankment, the postulated position of the failure surface and the position of the gravel layer, as indicated in Figure 4. The reason for installing the anodes first was to enable the top of the gravel layer to be detected by a change in the percussive response of the drilling. With the exception of the sub horizontal array and the deep electrodes at the top of the perpendicular array (Cathodes 1, 8, 14, 21, 27, 34 and 40), all cathodes were installed to a depth of 250 mm above the level of the gravel as indicated by the anodes of the related cathode centered cell.

Running the trial

Following installation of the electrodes a DC voltage of 60–80 V was applied using a portable generator. The current was applied from 4 December 2008 to 19 January 2009. During this time several sources of data were recovered, including: a visual record of site conditions and weather; electric currents at 16 different locations in the

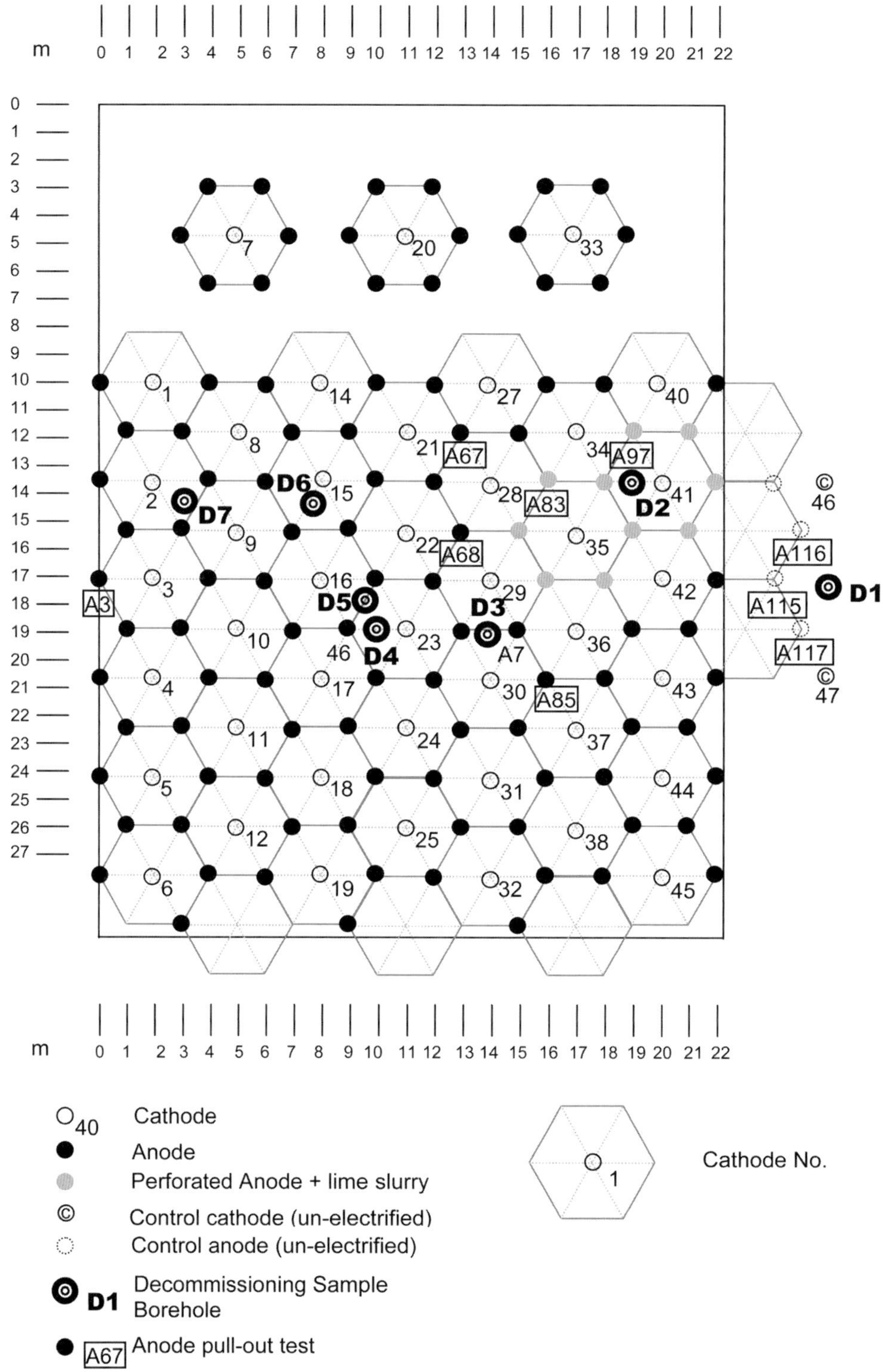

Fig. 4. Installed electrode array.

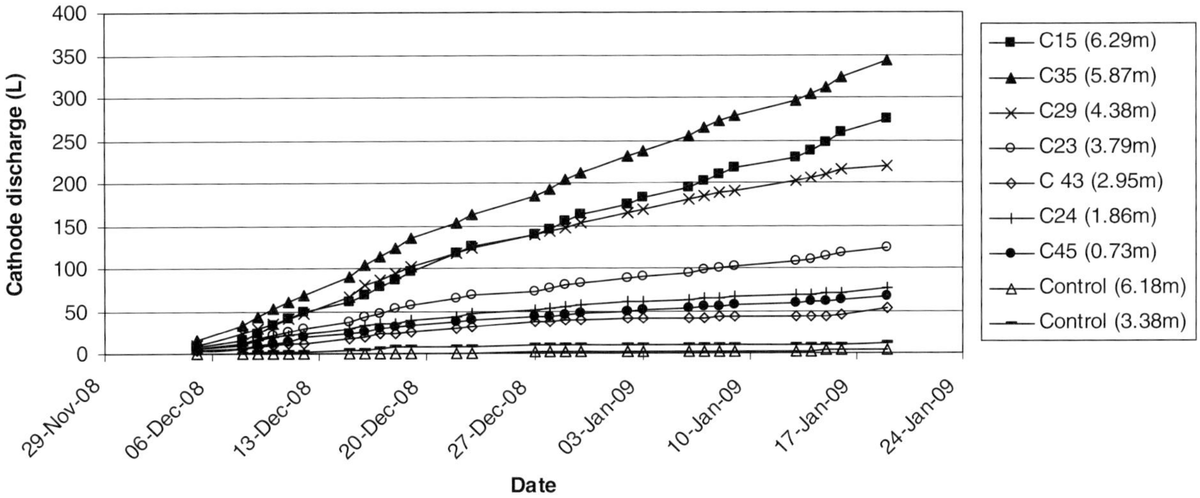

Fig. 5. Cumulative cathode discharge over time for a range of different cathode lengths.

array; piezometric levels in all cathodes (including controls); water discharge volumes from the cathodes; water quality and temperature and inclinometer readings.

Decommissioning took place from 19–23 January 2009 concentrating on direct ground investigations (Fig. 4). Seven inclined and vertical boreholes were drilled in the treated array and the control section. Within the boreholes data collection included Standard Penetration Tests (SPT) and *in situ* vane shear tests together with sampling of soils for subsequent laboratory testing. In addition to the boreholes drilled, nine anodes were selected for pull out tests (Fig. 4).

Results and analysis

This section presents the information extracted from the data collected during the treatment and the decommissioning. The results are presented in terms of water flow and quality, identified changes in soil properties and electrical energy.

Water flow and quality

In order to measure the volume of water produced at the cathodes, the cathodes were pumped out using a small submersible 12 V pump every two or three days during the course of the treatment period. It was noticed during a site visit on 27 December (four days after the previous cathode pumping procedure), that several cathodes were discharging water direct to the surface. The water recorded from the cathodes therefore represents a lower bound for the total volume of pumped water.

Figure 5 plots discharge over time for cathodes with a range of lengths. This indicates that the longer cathodes produced greater quantities of water and that the general indication is that the water production in most of the cathodes gradually decreased with time. Figure 5 also indicates that the control boreholes produced very little water in comparison to the main treatment cathodes. The total cathode discharge was approximately 6300 litres, representing a lower bound estimate.

The cathode discharge data and their analysis indicated the following trends:

(1) Cathodes were pumped out regularly and the rate of rebound gradually decreased throughout the course of the treatment indicating a gradual tailing off of flow from a peak of 250 L/day at the start of treatment to around 100 L/day at the end.
(2) The highest flow was achieved from cathode C35, (based on an uniform transmissivity assumption) which included conditioning of the anodes with a thin slurry of hydrated lime.
(3) Cathodes in the mid slope region yielded the lowest quantities of water per m length, those at the toe of the embankment yielded the highest and the cathodes near the top of the slope were intermediate producers. The above relationship was related to the lower slope cathodes being positioned in saturated soft clay and alluvium, which by its normally consolidated nature yields a greater quantity of water than stiffer over consolidated materials. This conclusion is based on the assumption that these normally consolidated materials are uniform in nature at this location and that electrokinetics particularly enhances flow through intact materials. It assumes there is no enhancement through

the fissures. Analysis by factoring the length of cathode to the discharge indicated that cathodes embedded into the soft materials at the toe of the embankment produced approximately 1.75× the average quantity per metre length. The upper slope cathodes produced more water than the mid slope cathodes owing to their proximity to the wet and soft materials associated with seepage from the water perched in the ballast/ash layer above the top of the London Clay embankment fill.

(4) Control or un-electrified cathodes were particularly poor producers of water, yielding on average 0.04 or 1/25th the quantity of the active cathodes.

Piezometer results

The piezometers were installed in three different locations in the embankment (Fig. 2). Unfortunately the data gathered were neither consistent nor conclusive. It is considered that the relative inclination of the metallic piezometer tubes with respect to the electrodes and the electric field distribution strongly influenced the results, by generating localized cathodic and anodic conditions. In future they should be installed parallel to the electric field and preferably using a non-metallic tube.

Groundwater quality

Cathode discharge was sampled and analysed for major ions. The data in Table 2 indicate a major difference between the water quality of the active cathodes and the controls.

Comparing the active cathodes to the controls revealed the following:

(1) A large reduction in calcium and magnesium ions in the active zones, which was interpreted to be as a result of precipitation in their hydroxide and/or carbonate forms adjacent to the cathode (Ewers *et al.* 2002).

(2) A significant rise in sodium and potassium, which was interpreted to be related to cation exchange where these ions are exchanged for ions of calcium (Ca^{2+}), magnesium (Mg^{2+}) and hydrogen (H^{+}). Sodium is the most easily replaceable cation and significantly the highest concentration was recorded adjacent to a cathode (C35) which had been conditioned with lime.

(3) The significant reduction in chloride ions indicated movement of these anions by ion migration towards the anode. The raised level in cathode C12 was probably related to the fact that this cathode was placed almost entirely within the soft clay/alluvium and close to the groundwater level, that is mineralogically dissimilar to the weathered London Clay comprising the bulk of the embankment.

(4) Reduced sulphate levels related to the movement of sulphate anions towards the anodes. cathode C12 had lower sulphate than the other active cathodes, which is probably related to its position in the soft clay/alluvium rather than the London Clay. As such this indicates a similar influence to that considered for chloride ions. The sulphate level in the control borehole was very high, close to the sort of levels found in gypsiferous springs (Lamont-Black *et al.* 1999) and is probably related to a localized zone of gypsum formation within the weathered London Clay used for the embankment fill.

These data and their interpretation indicate that chemical changes occurred in the soil principally as a result of cation exchange. This is an important effect because it is known to influence the plasticity of clays, which effects the relationships between water content, shear strength and shrink-swell behaviour. Atterberg limits (Table 3) determined as part of the investigation also indicate a significant rise in the plastic limit.

Table 2. *Groundwater quality analyses*

Test	*Units*	*Cathode identification*				
		C10	*C12*	*C16*	*C35*	*Control (upper)*
Calcium (Ca^{2+})	mg/l	37	110	75	58	290
Magnesium (Mg^{2+})	mg/l	bd	bd	bd	bd	150
Potassium (K^{+})	mg/l	960	1000	550	430	110
Sodium (Na^{+})	mg/l	640	410	630	750	300
Fluoride (F^{-})	mg/l	1.0	1.0	1.2	0.96	0.81
Chloride (Cl^{-})	mg/l	5.3	47	5.3	3.3	190
Nitrite (NO_3^{-})	mg/l	bd	bd	bd	bd	Bd
Bromide (Br^{-})	mg/l	bd	bd	bd	bd	Bd
Nitrate (NO_2^{-})	mg/l	0.54	0.44	1.3	1.1	1.3
Sulphate(SO_4^{2-})	mg/l	9.8	2.6	8.8	6.8	1700

bd, below detection.

Table 3. *Changes in soil properties*

Parameter	*Pre-treatment*		*Post-treatment*	
	Embankment fill	*Alluvium*	*Embankment fill*	*Alluvium*
c' peak	8	5	18	10
Φ' peak	25°	24°	32°	26°
Post peak c'	8	3	6	8
Φ' res	19	22	25	22
PI (%)	49	41	36	39
Liquid limit	75	66	78	90
Plastic limit	26	25	42	52

Changes in soil properties

Borehole investigations

Figure 4 shows the locations of the seven boreholes drilled during the decommissioning phase (and the location of anode pullout tests). These comprised seven percussion drilled boreholes five of which were vertical (D1, D2, D5, D6 and D7) and two shallow inclined boreholes (D3 and D4). Where possible *in situ* shear vane and SPT tests were undertaken and samples collected for laboratory testing. Engineering descriptions of the soils in the profiles were compiled from the detailed descriptions of the 44 cathode cores that were recovered during the installation stage.

These data are brought together in Figure 6 which shows a vertical cross section of the ground, perpendicular to slope contours and passing through cathode 14 (Fig. 4). Data from the decommissioning phase has been overlaid onto the pre-treatment profiles. The colour scales of the individual decommissioning holes show the recorded strengths of the material encountered along with the relevant *in situ* test type and result.

Engineering descriptions of materials in boreholes D5 and D6 indicate an improvement in the strength of the soft alluvium material beneath the embankment to a firm condition, this is confirmed in the high SPT values tested at the base of the holes

Soils testing on samples recovered from boreholes

The suite of soils tests conducted on the samples recovered from the boreholes were used to provide information for subsequent slope stability analyses. The test results are shown in Table 3. All testing was undertaken in accordance with BS1377 (BSI 1990).

When interpreting the information it must be borne in mind that the data set is limited and the embankment and underlying ground are heterogeneous in nature. Notwithstanding theses caveats, the test result suggests important changes in effective stress parameters and plasticity indices for the London Clay fill and the soft clay/alluvium.

The samples recovered from the decommissioning boreholes were described in accordance with BS 5930 1999:A2. These engineering descriptions, alongside the test results allowed for interpretative conclusions to be drawn.

From laboratory shear box testing, the untreated London Clay fill showed shear strength parameters in close agreement with values featured in published data (Potts *et al.* 1997). Electrokinetic treatment produced improvements in the peak values of effective cohesion, c' and the effective angle of friction, Φ'. This increase in c$'$ is attributed to the presence of sacrificial electrodes used during the treatment, corroding and migrating cations into the strata. These migrating ions had an apparent cementation effect within the soil. Changes in plasticity were also noted (Table 3). The implications of a rise in plastic limit for stability of the embankment are that it will reduce the plastic range of the soil, and critically by raising the plastic limit, the soil requires higher water contents to deform plastically. Indeed the natural water contents of many of the samples of the London Clay fill were around 28–30%. By raising the plastic limit to 33%, such material is unable to deform in a plastic manner.

The implications of a raised plastic limit for shrink swell behaviour are also positive, in terms of reducing the potential for seasonal volume changes in the soil which have been strongly linked with embankment and cutting instability (McGinnity 2009).

From Table 3 it can be seen that the main change detected from the treatment was that of changing the peak shear strength parameters of both the embankment fill and the soft clay and the increase in the plastic limit of the London Clay. This has important implications for both the stability of embankments and their shrink swell behaviour.

Anode pull-out tests

In order to assess the effectiveness of the treatment for improving the bond strength between anodes (acting as soil nails) and the soil, a total of nine (6 active and 3 control) anodes were pulled out using the slope climbing drill-rig fitted with a load cell. The results showed that there was a clear increase in pull-out resistance resulting from the electrokinetic treatment. The mean peak interface shear resistance mobilized on the solid walled unconditioned electrokinetic anodes was 40.8 kPa compared to 15.5 kPa for the control anodes, representing an increase in bond of 263%.

Electrical energy

Measurements of current using both logged and visual means showed an average current of 200 A, which rose to 230 A when the voltage was increased from 60 to 80 V. There was a weak correlation between current and discharge of water. The correlation was imperfect, because of inherent heterogeneity of the ground, and also because current can flow even when there may be little or no electroosmotic flow.

The trial was powered for 1010 h (42 days) with a DC energy consumption of 14 168 kWh and gave a unit volume power consumption of 11.5 kWh m^{-3} equating to

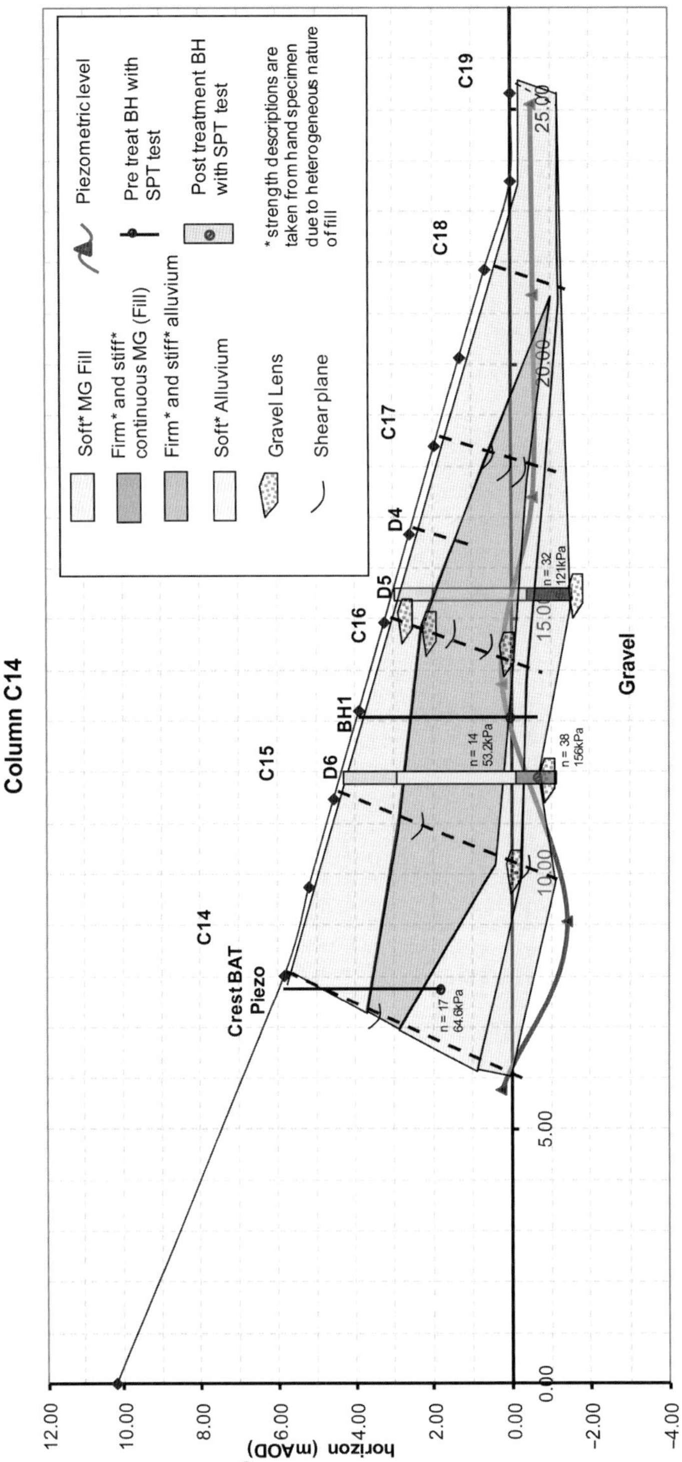

Fig. 6. Geotechnical cross sections of electrokinetic array passing through cathodes C1 (A) C14 (B) and C34 (C).

a power efficiency of 41% for the trial against the calculated theoretical maximum.

Effects on embankment stability

Embankment stability was assessed using the results from the inclinometer in BH1 (Fig. 2) and by the re-analysis of the slope using the improved soil strengths and installed EKG nails.

Inclinometer results

The location of the inclinometer is shown in Figure 2, immediately adjacent to cathode C9 (Fig. 3). Figure 7 shows the incremental movements of the inclinometer plotted with respect to the five project phases that is (i) pre-treatment; (ii) electrode installation; (iii) active treatment; (iv) decommissioning and (v) post-treatment.

In interpreting the inclinometer results the reference convention is that positive movement on the *A* axis indicates down slope movement and positive movement on the *B* axis indicates contour parallel movement. In general the deflections recorded in BH1 are modest, the most active zone being the top 1.0–1.5 m of the borehole. This represents the boundary between the softened London Clay fill and the stiffer material of the embankment fill (Fig. 6). The circular failure projected by the slope stability analyses would pass through BH1 at a depth of approximately 4.0–4.5 m. The data from BH1 indicate a slight increase in down slope movement during the treatment phase, which then flattened off during the post treatment phase. This is viewed as an encouraging indication that the treatment has successfully reduced or stopped slope movement.

The movements resolved in the *B* axis can be similarly correlated to the phases of the trial. Significant movements in the upper part of the soil profile in the up chainage direction occurred post installation during August, September and October. This movement was reversed by the installation of the electrodes and the magnitude and rate of reversal increased during the active treatment. After decommissioning and into the post treatment phase, all levels in the borehole indicated much reduced or effectively zero movement in the *B* axis.

Slope stability analyses

Analysis of the pre-treatment slope condition indicated a circular failure, passing through the softened clay/alluvium horizon lying above the gravels with a factor of safety of approximately 1.

Following the electrokinetic treatment of the slope, a series of stability analyses were undertaken to determine the change in the stability of the slope. The analyses assumed a Bishop circular failure plane with the electrodes taken to act as soil nails if appropriate. A range of different conditions were analysed to take into account the very best and worst cases of long-term pore water pressure, soil strengths and soil–nail bond characteristics (Table 4).

For those materials which were not affected by EKG treatment that is the ballast, gravel and *in situ* London Clay, published peak values of c' and Φ' and bulk density were used for input into the model.

The embankment fill and the underlying soft clay were the subjects of the EKG treatment. The peak c' and Φ' values of these units used for the pre and post treatment analytical conditions were based on laboratory testing. Pre-treatment parameter values for the London Clay fill were measured and matched, quite closely, published values for London Clay. The post treatment shear strength values for the London Clay fill and soft grey clay/alluvium are shown in Table 3. The bond strength of the EKG nails used in the model was determined from the pull-out test results.

Overall, electrokinetic treatment raised the factor of safety (FoS) without the inclusion of reinforcement and with the inclusion of reinforcement by 153% and 178% respectively. With a long-term extreme condition of the water table rising to the base of the ballast (fully saturated embankment), the analysis showed that without EKG treatment this condition would result in failure, as indicated by a FoS of 0.57. However, post EKG treatment, the embankment would be stable with a raised water table, particularly when the reinforcement action of the electrodes is included.

In the long term, the strength of the embankment could reduce as the EK treatment may be permanent with only part of the embankment material (as discussed below). This has been modelled and even with a raised water table the embankment would remain stable.

Discussion

The EKG treatment comprises several distinct effects including:

- electroosmotic flow and pore pressure change;
- soil nailing/reinforcement;
- enhancement of soil nail bond strength;
- passive drainage;
- changes in soil strength parameters;
- changes in soil plasticity characteristics.

The following sections consider these in terms of embankment stabilization and their contribution to improved short and long-term stability.

Electroosmotic flow and pore pressure change

The electrokinetic treatment was applied to only two material types, that is the embankment fill materials and the underlying soft clay/alluvium on which it was originally placed. This section will consider the nature and variability of these materials and how they are affected by electrokinetic treatment.

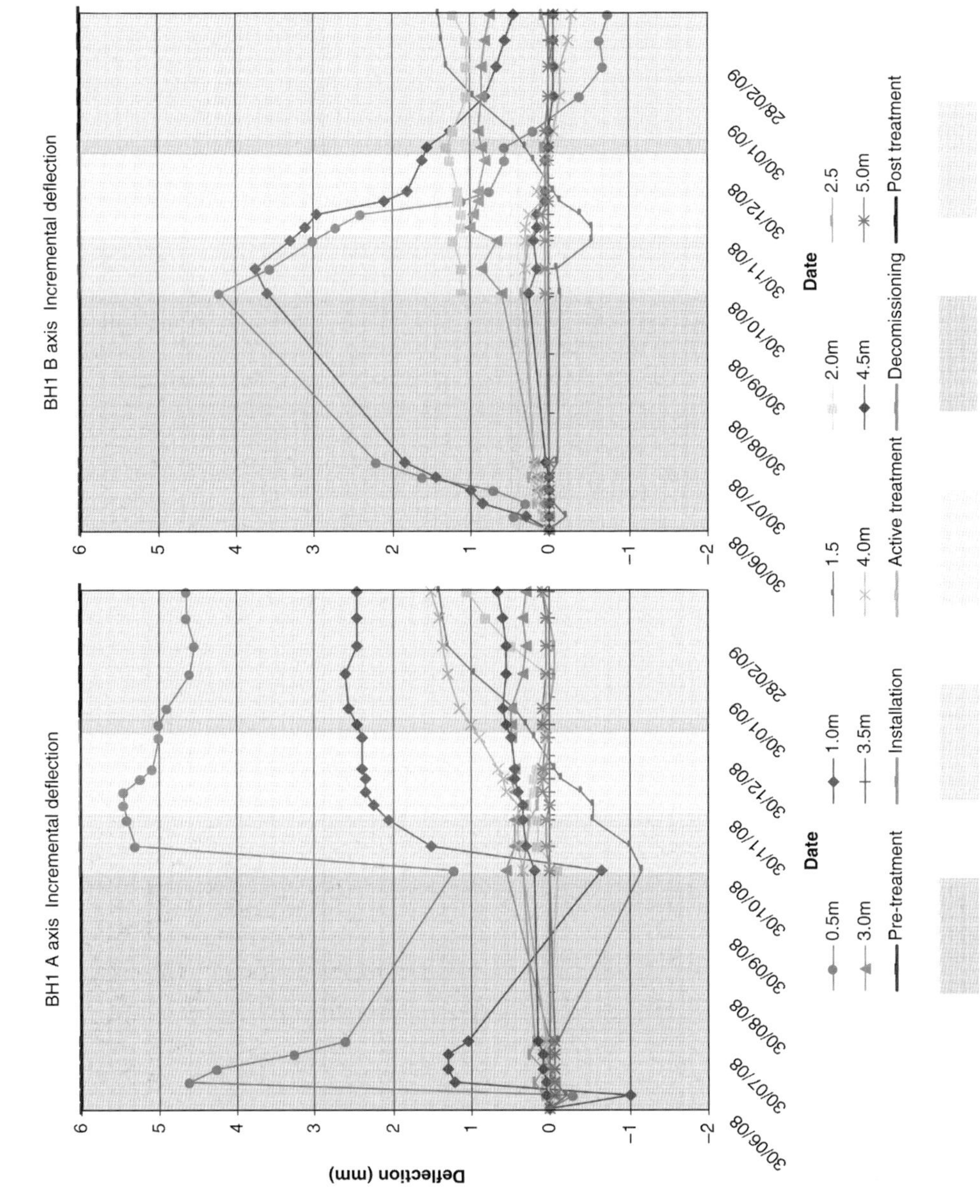

Fig. 7. Borehole 1 incremental deflection measurements in the *A* axis and *B* axis for the period 30 June 2008 to 25 March 2009.

Table 4. *Results of slope stability analyses*

Analysis	*Reinforcement*	*FoS (ULS)*	*Increase of FoS*
Observed conditions (using measured water table data)			
A1 Pre treatment (TGP data)	No	0.81	–
A2 Pre EK treatment	No	0.96	Base
A3 Post EK treatment	No	1.47	153%
A4 Post EK treatment	Yes	1.71	178%
Hypothetical conditions (assuming a raised water table)			
A5 Initial condition	No	0.57	Base
A6 Post EK treatment	No	1.00	175%
A7 Post EK + nails	Yes	1.27	223%
A8 Post EK softened + nails	Yes	1.24	218%

FoS, factor of safety; ULS, ultimate limit state.

Embankment fill material

The structure of the embankment as indicated from the investigations at South Greenford and many other railway embankments in and around London (e.g. McGinnity 2009) is characterized in Figure 8.

The 'core' of the embankment fill material was highly variable ranging from stiff to very soft. The London Clay fill consisted of a 'clod and matrix' structure created by poor compaction of clods of heavily over consolidated London Clay. Thus it contained lumps of over consolidated material with soft interconnecting patches that may have been washed into place over time. There were extended zones of softened embankment fill at the base of the embankment in the zone of water table fluctuation and an extended zone of mixed softened/stiff material in the zone of surface weathering, particularly on the embankment shoulders. There was highly softened material beneath the ballast. This is likely to have been caused by water collecting at the base of the ballast. Over time the soft material can be pumped or piped upwards into the ballast resulting in a reduction in the effective depth of the ballast. This structure is not uncommon, particularly in the London region of the rail network.

Electroosmosis affects the soil across which the voltage gradient is sustained in the following ways:

(1) Electroosmotic flow leading to negative pore water pressures in stiff materials (clod) or consolidation in soft materials (matrix). Electroosmotic flow alone will rapidly improve strength in the short term in all (soft–stiff) fine grained materials. Soft, normally consolidated zones within the embankment fill will selectively consolidate under the increased effective stress, leading to a permanent increase in peak drained shear strength. The effectiveness and benefits are increased if the treatment is conducted under saturated conditions;

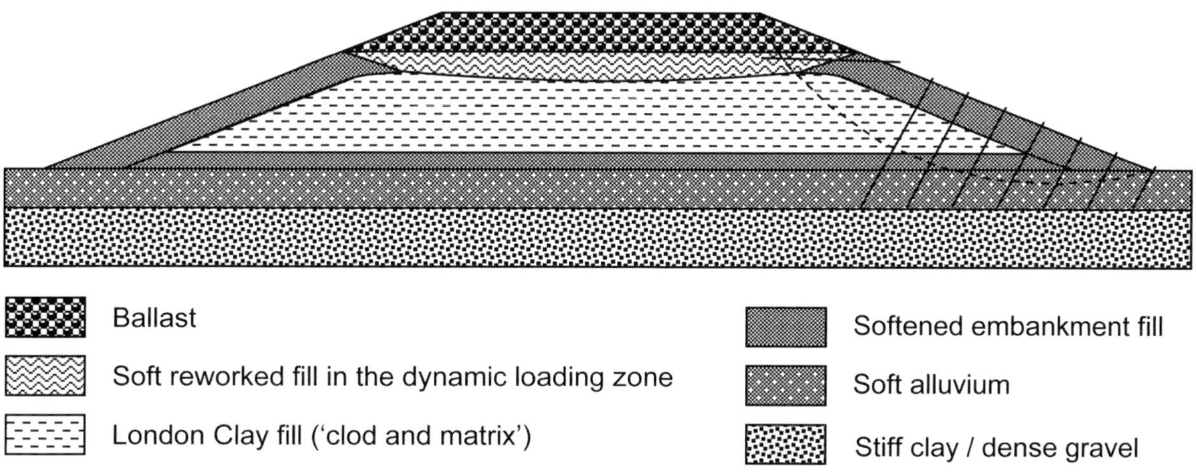

Fig. 8. Characterization of the types of material present in an embankment (based on South Greenford) including the disposition of electrodes.

the improvement under partially saturated conditions may not be so great.

(2) The 'clod and matrix' structure (representing the bulk material behaviour) will approximate to a normally consolidated material on a macro level (i.e. clods within a continuous matrix of softened material) and will consolidate under this increased effective stress, thus increasing peak values of c' and Φ' of the bulk material. The overconsolidated and intact lumps or clods will not themselves, consolidate but will respond to electroosmotic flow and develop a reduction in pore water pressure. The suction effect will be lost over time.

(3) The amount of consolidation will be proportional to the initial void ratio (assuming saturated conditions) and therefore the wettest zones will be improved preferentially over the drier zones.

(4) The overall amount of consolidation achieved (and hence increase in peak strengths) will be proportional to the ratio of matrix material to intact or overconsolidated material.

The assumption of saturated conditions is made in order to reflect the wet times of year when the slope movements are noticed and it is of significance that electroosmosis can create pore water suctions under fully saturated conditions.

In-situ *material*

At South Greenford, the material immediately beneath the embankment fill was dominantly soft grey highly plastic clay (alluvium) which was overlain in a few places by a thin band of topsoil (buried during the original placement of the fill). A reduction in pore water pressure will result in an increase in undrained shear strength (c_u) whilst the reduced pore pressure is maintained. Such an increase in suction would be maximized under saturated conditions (when electroosmosis is most effective). Consolidation under such conditions will yield a reduction in void ratio which will not be lost fully upon wetting up and will increase peak drained shear strengths.

In all cases, the changes in strength caused by consolidation only are represented by a return to peak values. Given sufficient strain these can revert to pre-treatment post-peak values. The likelihood of such movements is however reduced using EKG treatment by utilizing its other functions of reinforcement, drainage and plasticity changes.

Soil nailing/reinforcement

The objective of soil nailing is for the nails to act in tension and anchor the material in the (active zone) identified as at risk of failure, to a more stable material in the (resistant zone). As such the nail needs to bond well the soil in both zones.

By using electrokinetic anodes as soil nails an enhanced bond between the soil and the anode nail is achieved through migration of ions from a sacrificial electrode. This results in additional chemical changes to the soil such as cementation. There are well known examples from the literature which attest to the longevity of such effects (Soderman & Milligan 1961; Milligan 1994; Ewers *et al.* 2002; Micic *et al.* 2003; Shang *et al.* 2004).

The electrokinetic treatment resulted in a 263% increase in the soil/nail bond. Provided that attention is given to the wall thickness of the nail and the nature of the ground into which it is installed (with respect to long term corrosion), then this component of the treatment will be permanent.

Changes in soil strength parameters and longevity implications

The investigations at South Greenford indicated improvements in c' and Φ'. The values measured were in excess of those recorded for the London Clay in the literature, particularly in the zone of the anode. Both of these changes reflect changes in cementation and friction in the soil. As such these proffer changes in drained shear strength in addition to the improvements caused by consolidation. The changes are caused by a range of factors relating to cation exchange, cementation of new minerals and boundary flocculation effects.

Unlike Milligan (1994) much of the more recent work (Lo *et al.* 1991; Shang *et al.* 2004) has not investigated the longevity of electrokinetic treatment, but showed that the production of insoluble soil precipitates and cementing agents is unlikely to be reversed especially in low permeability soils such as clays.

As discussed, most effects will be permanent, an exception to which is the generation of pore water suction. The other exception relates to increases in drained shear strength resulting from consolidation. If significant strains should occur in the future (possibly causing a serviceability limit state failure) the angle of friction of the improved material would reduce. However, improvements in c' and Φ' resulting from the EKG treatment would prevent large strains developing in the first place. If strains were to occur then these would quickly invoke a contribution from the reinforcement to resist further movement.

Conclusions

The trial of EKG for embankment stabilization at South Greenford is considered to be a success because it has treated the slope such that measurable improvements in the slope behaviour and geotechnical characteristics of the soil have been confirmed. These improvements have lead to significant improvements in the calculated factor of safety for the slope, both in the short and long term. The results have also allowed significant improvements in the understanding of the longevity of EKG treatment of clay soils, and in particular clay fill materials. This was defined as being a combination of:

(1) consolidation of soft materials, causing an increase in shear strength (at least at small strains);

(2) increase in bond strength between the soil and the anodes through the formation of cementitious materials (which resist stresses and therefore prevent large strains occurring), and
(3) relief of pore water pressures via the inclusion of permanent drains. EKG treatment has the potential to offer an effective alternative to more traditional means of slope stabilization.

The authors wish to acknowledge the funding from Network Rail and the Technology Strategy Board in support of this trial. They also wish to thank Dean and Dyball Rail Ltd. (now Birse Rail Ltd) for Project managing the works and Geotechnical Engineering for installing the electrodes. Together with Tony Gee and Partners who provided technical support.

References

ARUP. 2006. *South Greenford Monitoring WEL 7 m 40c to 8 m 14c Up and Down*. Unpublished report by Arup Ltd. for Alfred McAlpine Project Services. Doc ref REP/316/0602.

BRITISH STANDARDS INSTITUTE. 1990. *British Standard Methods of Tests for Soils for Civil Engineering Purposes*: BS 1377. BSI, 2 Park Street, London WA1 2BS.

EWERS, B., SMITH, D. & FITYUS, S. 2002. *Electro-Osmotic Testing of Ballina Clay*. Environmental Geomechanics, Monte Verita, Switzerland, 295–301.

GLENDINNING, S., JONES, C. J. F. P. & LAMONT-BLACK, J. 2005. The use of electrokinetic geosynthetics (EKG) to improve soft soils. *In*: INDRARATNA, B. & JIAN CHU, (eds) *Ground Improvement Case Histories*. Elsevier, London, 1001–1049.

JONES, C. J. F. P., GLENDINNING, S. & SHIM, G. S. G. 2002. Soil Consolidation using electrically conductive geosynthetics. *In*: DELMAS, P., GOURC, J. P. & GIRARD, H. (eds) *Proceedings of the 7th International Conference on Geosynthetics*. Nice, France, 22–27 September 2002, 1039–1042.

JONES, C. J. F. P., LAMONT-BLACK, J., GLENDINNING, S. & PUGH, R. C. 2005. New applications for smart geosynthetics. *In*: RATHJE, E. M. (ed.) *Geo-Frontiers 2005*. ASCE Special Publications and GRI-18, Austin, Texas, 130–142.

JONES, C. J. F. P., LAMONT-BLACK, J. ET AL. 2008. *Recent research and applications in the use of electrokinetic geosynthetics*. Keynote Paper, EuroGeo 4, 6–8 September 2008, Edinburgh (Proceedings on CD).

LAMONT-BLACK, J. 2001. EKG: the next generation of geosynthetics. *Ground Engineering*, October 2001, 22–23.

LAMONT-BLACK, J., YOUNGER, P. L., FORTH, R. A., COOPER, A. H. & BONNIFACE, J. P. 1999. Hydrogeological monitoring strategies for investigating subsidence problems potentially attributable to gypsum karstification. *In*: BECK, B. F., PETTIT, A. J. & HERRING, J. G. (eds) *Hydrogeology and Engineering Geology of Sinkholes and Karst*. Balkema, Rotterdam, 141–147.

LAMONT-BLACK, J., GLENDINNING, S., JONES, C. J. F. P., HUNTLEY, D. T. & SMITH, R. 2005. The development of in situ dewatering of lagooned sewage sludge using electrokinetic geosynthetics (EKG). *In*: LOWE, P. & HORAN, N. (eds) *Proceedings of the 10th European Biosolids and Biowastes Conference*, November 2005. Wakefield, UK. Aqua Enviro Technology Transfer, Wakefield. Paper 21, 8.

LAMONT-BLACK, J., HUNTLEY, D., GLENDINNING, S. & JONES, C. J. F. P. 2006*a*. The use of Electrokinetic geosynthetics (EKG) in enhanced performance of sports turf. *In*: KUWANAO, J. & KOSEKI, J. (eds) *Proceedings of the 8th International Conference on Geosynthetics*. Yokohama, Japan, 18–22 September 2006. Millpress, Rotterdam, 453–456.

LAMONT-BLACK, J., HUNTLEY, D. T., GLENDINNING, S. & JONES, C. J. F. P. 2006*b*. Case history: the use of electrokinetic geosynthetics (EKG) in belt press dewatering. *In*: KUWANAO, J. & KOSEKI, J. (eds) *Proceedings of the 8th International Conference on Geosynthetics*. Yokohama, Japan, 18–22 September 2006. Millpress, Rotterdam, 535–538.

LAMONT-BLACK, J., JONES, C. J. F. P., GLENDINNING, S., HUNTLEY, D. T. & FOURIE, A. B. 2007. Laboratory evaluation of the potential for electrokinetic belt filter press of Kimberlite slimes. *In*: FOURIE, A. & JEWELL, R. (eds) *Proceedings 10th International Seminar on Paste and Thickened Tailings*. Perth, Australia, 13–15 March 2007, Australian Centre for Geomechanics.

LO, K. Y., INCULET, I. I. & HO, K. S. 1991. Electro osmotic strengthening of soft sensitive clays. *Canadian Goetechnical Journal*, **28**, 62–73.

MCGINNITY, B. 2009. *Geotechnics in London Underground – Opportunities and Challenges*. Presentation to Equipe Training, Banbury, U.K., 26 March, 2009.

MICIC, S., SHANG, J. Q. & LO, K. Y. 2003. Improvement of of the load-carrying capacity enhancement of skirted foundation element by electrokinetics. *Canadian Geotechnical Journal*, **40**, 949–301.

MILLIGAN, V. 1994. First application of electro-osmosis to improve friction pile capacity-three decades later. *Proceedings of the 13th ISSMGE*, **5**, 1–5, New Delhi, India, Balkema.

NETWORK RAIL 2008. Independent Reporter. Annual Return Audit 2008. Final Report. Halcrow Group Ltd. http://www.rail-reg.gov.uk/upload/pdf/reporters-ara08-halcrow.pdf

PERRY, J., PEDLEY, M. & REID, M. 2001. *Infrastructure Embankments – Condition Appraisal and Remedial Treatment*. CIRIA, London, Publication C550.

POTTS, D. M., KOVACEVIC, K. & VAUGHAN, P. R. 1997. Delayed collapse of cut slopes in stiff clay. *Geotechnique*, **47**, 953–982.

PUGH, R. C. 1998*a*. *Electroosmotic testing of London Clay*. University of Newcastle unpublished internal Report No. GR/L66090/SC1, May 1998.

PUGH, R. C. 1998*b*. *Further electroosmotic testing of London Clay*. University of Newcastle unpublished internal Report No. GR/L66090/SC2, May 1998.

PUGH, R. C. 1998*c*. *Testing of London Clay from Wimbledon Southfields Embankment*. University of Newcastle unpublished internal Report No. GR/L66090/SC3, May 1998.

PUGH, R. C. 2002. *The application of electrokinetic geosynthetic materials to uses in the construction industry*. PhD Thesis, Newcastle University, 277.

SHANG, J. Q., MOHAMEDELHASSAN, E. & ISMAIL, M. 2004. Electro chemical cementation of offshore calcareous soil. *Canadian Geotechnical Journal*, **41**, 877–893.

SMOLUCHOWSKI, M. 1914. *In*: GRAETZ, L. (ed.) *Handbuch der Elektrizitat und Magnetisums*. **2**, J A Barth, Leipzig, Germany.

SODERMAN, L. G. & MILLIGAN, L. 1961. capacity of friction piles in varved clay increased by electro-osmosis, *Proceedings of the 5th ICSMFE*, Paris, France, **2**, 143–147.

WHITE, C., GLENDINNING, S., LAMONT-BLACK, J., JONES, C. J. F. P. & TAYLOR, T. 2008. The application of electrokinetic geosynthetics in the advanced conditioning and composting process of sewage sludge. *EuroGeo*, **4**, 6–8, September 2008, Edinburgh.

The use of soil moisture deficit as a trigger mechanism for targeted monitoring of earthworks slopes

G. J. Macdonald, A. R. Vooght* & S. Parkin

Mouchel Ground Engineering, Mouchel, Export House, Cawsey Way, Woking, Surrey, GU21 6QX, UK
*Corresponding author (e-mail: alan.vooght@mouchel.com)

Abstract: Seasonal effective stress changes can result in progressive failure of clay embankments. These seasonal effective stress changes are reflected in the soil moisture deficit (SMD) values. This paper explores the relationship between slope failures and SMD for both embankment and cuttings and proposes an SMD trigger level at which more cost effective targeted monitoring of earthwork slopes, considered to be at risk of failure, should be undertaken. The paper presents a methodology for obtaining and reviewing SMD data and discusses the use of rainfall and temperature forecasts as a predictor of future short-term rapid changes in SMD.

It has been shown that there is a relationship between slope failures such as those shown in Figures 1 and 2 and soil moisture deficit (SMD) (Kovacevic *et al.* 2001) and (Ridley *et al.* 2004*b*).

Earthworks failures are commonly associated with periods when the SMD for trees is low or at zero with serviceability problems (settlement) being reported when the SMD for trees is high due to shrinkage. Earthworks failures are also associated with prolonged dry periods followed by a sometimes short but intense wet period. This suggests that instability may be associated with a rapid change in SMD.

Climate change is expected to bring more extreme weather patterns to the UK with wetter winters, hotter summers and more extreme rainfall events which may arrive in short heavy rain storms. Rapid changes in SMD could therefore

Fig. 1. M25 junction 28 Franks Farm embankment failure.

Fig. 2. M11 junction 4–5 Chigwell cutting failure.

From: Radford T. A. (ed.) 2012. *Earthworks in Europe*. Geological Society, London, Engineering Geology Special Publications, **26**, 141–149, http://dx.doi.org/10.1144/EGSP26.16

be expected to become ever more frequent possibly leading to more slope failures.

This paper explores the relationship between SMD and slope failure and presents the conclusions from a research study (Mouchel 2010) into the use of SMD as a trigger mechanism for targeted, more cost effective monitoring of earthwork slopes. The research study, undertaken by Mouchel for the Highways Agency, correlated instrumentation data relating to slope movements recorded at seven earthworks (six on embankments and one in cutting) in SE England with SMD data for each. SMD for both grass and trees and precipitation data have been obtained from the Meteorological Office and correlated with the results of the slope instrumentation data recovered at each earthwork over a number of years. Mouchel (2010) shows there is a strong relationship between SMD for trees and slope movement and that the SMD for trees could be used as a trigger mechanism for identifying periods when targeted slope monitoring would be most cost effective.

Soil moisture deficit

SMD is defined as the height of water (mm) per unit surface area required to return a partially saturated soil to its field capacity water content. Field capacity is defined as the quantity of water contained within the soil profile when the storage capacity of the profile has been reached and further precipitation does not pond on the surface. Low SMD values indicate that the soil is at or close to its field capacity and is therefore wet with a high moisture content. A high SMD value indicates that the soil is dry with a low moisture content.

SMD is influenced primarily by precipitation, temperature, the soil type and type of vegetation cover. The latter is particularly important as vegetation influences the depth of the profile from which moisture is extracted and therefore the quantity of precipitation required to restore the field capacity. Other factors which may influence SMD include evaporation, wind speed, wind direction and humidity.

SMD data is provided by the Meteorological Office on a weekly or monthly basis through MORECS (Meteorological Office Rainfall and Evaporation Calculation System). Although MORECS was developed for agricultural purposes and is available on a 40 × 40 km grid covering mainland UK and Northern Ireland, the program can derive site specific SMD data.

SMD is calculated for different vegetation cover (trees or grass) and for differing soil types. SMD for trees increases to a higher maximum compared to that for grass cover and often does not reach zero in the winter whereas the SMD values for grass cover nearly always falls close to zero during the winter months. The SMD for trees is considered to be a good indicator of slope stability. If there is insufficient rainfall during the winter months to replenish the moisture abstracted by the trees during the summer months, then the SMD for trees will remain positive during the winter period and stability will be enhanced.

Slope failure and soil moisture deficit

Contemporaneous research on the effects of clay shrinkage on embankment movement was reported by various contributors to a British Geotechnical Association (BGA) workshop on the '*Maintenance of ageing earthworks*' held at the Institution of Civil Engineers in January 2008. The studies, which are ongoing, are reported to show embankment heave in the winter associated with high moisture content, positive pore water pressures and low SMD with settlement and negative pore water pressures (suctions) and high SMD in the summer. The presence of vegetation was shown to increase the amount of soil suction during the summer when plant activity was greatest. Remnant suctions were maintained at depth during the winter.

The role of permeability was highlighted with many historic rail embankments being constructed of loose tipped clay fill resulting in 'clods' of stiffer low permeability clay in a higher permeability matrix of softer material and potential voids. It was suggested that the permeability of these railway embankments might be at least one order of magnitude greater than that of the clods alone.

The use of SMD as a tool to predict instability was discussed at the BGA workshop. It was concluded that the SMD data from the MORECs grid was not sufficiently accurate and that more local data was needed to provide more robust predictions. This was highlighted by the findings from the BIONICS (The Biological and Engineering Impacts of Climate Change on Slopes Trial Embankment Project (led by Newcastle University) site which observed that SMD values varied significantly on opposing sides of an embankment.

Kovacevic *et al.* (2001) showed that embankments constructed of plastic clays may suffer delayed progressive failure if subject to seasonal effective stress changes. Movements occur during shrinkage and swelling which cause cumulative strain and the progressive development of a rupture surface at reduced strength. Kovacevic *et al.* (2001) modelled the effects of cycles of pore pressure changes caused by tree roots in an embankment of dumped London Clay fill. The finite element analyses indicate that seasonal movements result with the crest settling and the slopes moving outwards during the summer (shrinkage) and the crest heaving with no measurable outwards movement (swelling) during the winter. The net result is a small permanent outward movement per pore pressure cycle. As this occurs a shear surface develops in the base of the fill with the shear strength going from peak towards residual (progressive failure) and eventually a gravity slide develops. The number of swell–shrink cycles required for a slide to occur depends on how quickly the shear strength drops

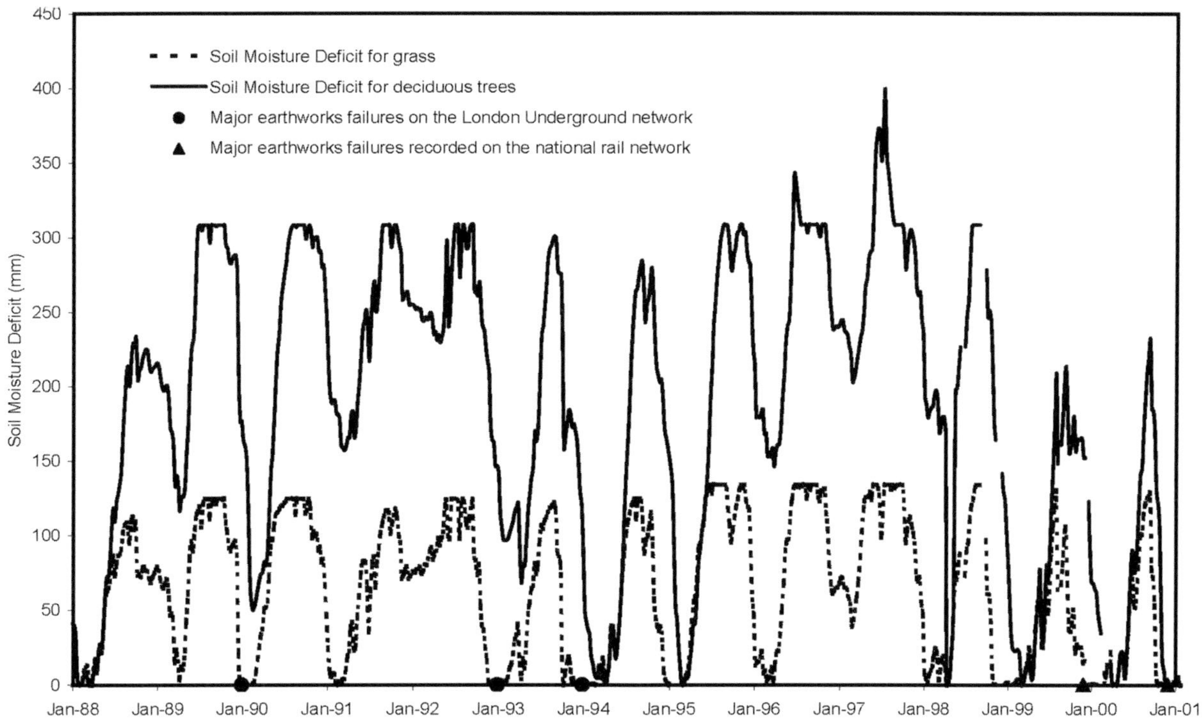

Fig. 3. Variation of SMD for London area with major earthworks failures indicated (after Ridley *et al.* 2004*b*).

from peak to residual and the magnitude of the pore pressures at the end of winter.

A reduction in strength with swell–shrink cycling allows the failure to occur at lower pore water pressures. Stiffness of the fill has little influence. Displacement is due primarily to plastic yield. The number of cycles to cause collapse increases approximately linearly with the reduction in pore water pressure. Swelling and shrinkage can therefore result in progressive failure of a slope. Strain softening at the toe results in deformation which regresses back into the slope resulting in an eventual shallow or potential deep seated failure.

Kovacevic *et al.* (2001) found that this type of embankment movement could be correlated with SMD for grass. Slope failures tend to coincide with winters with long wet periods where the SMD falls to zero. On grassed slopes the zone of pore pressure change is shallower and results in shallow (<2 m deep) slides. Ridley *et al.* (2004*b*) plotted the relationship between SMD and major earthworks failures since 1988. They noted that failures are frequently associated with periods when prolonged dry periods are followed by an intense wet period, where the SMD falls to zero. Problems of serviceability can also occur when the SMD for trees is high due to shrinkage. Figure 3 shows the SMD for both deciduous trees and grass annotated when major earthworks failure occurred on the rail network.

A second mechanism of failure was also identified during the BGA 2008 workshop described as a ratcheting movement produced by opening and closing of desiccation cracks in the down-slope direction. This was thought to be an important mechanism on slopes without trees. The workshop concluded that there were many variables that were not adequately understood and therefore any research carries uncertainty.

Pore pressures and vegetation

Seepage pressures in slopes are mainly controlled by infiltration and evapotranspiration, run-off and the permeability of the ground. The seasonal fluctuation in superficial pore pressure is complex as it is controlled by root growth and cracking as well as by the consolidation properties of the soil. Vaughan *et al.* (2002) noted that plant roots require oxygen as well as water for growth and that in saturated clay the roots must crack the soil to obtain this oxygen supply. When the slope wets up pore pressures reflect flow parallel to the slope surface in this cracked zone. The cracked zone also controls the swelling process and the lag in pore pressure response at depth. The rate of root growth probably controls the shrinkage process. The presence of

Fig. 4. M25 motorway junction 5–6 Rooks Nest Farm cutting failure.

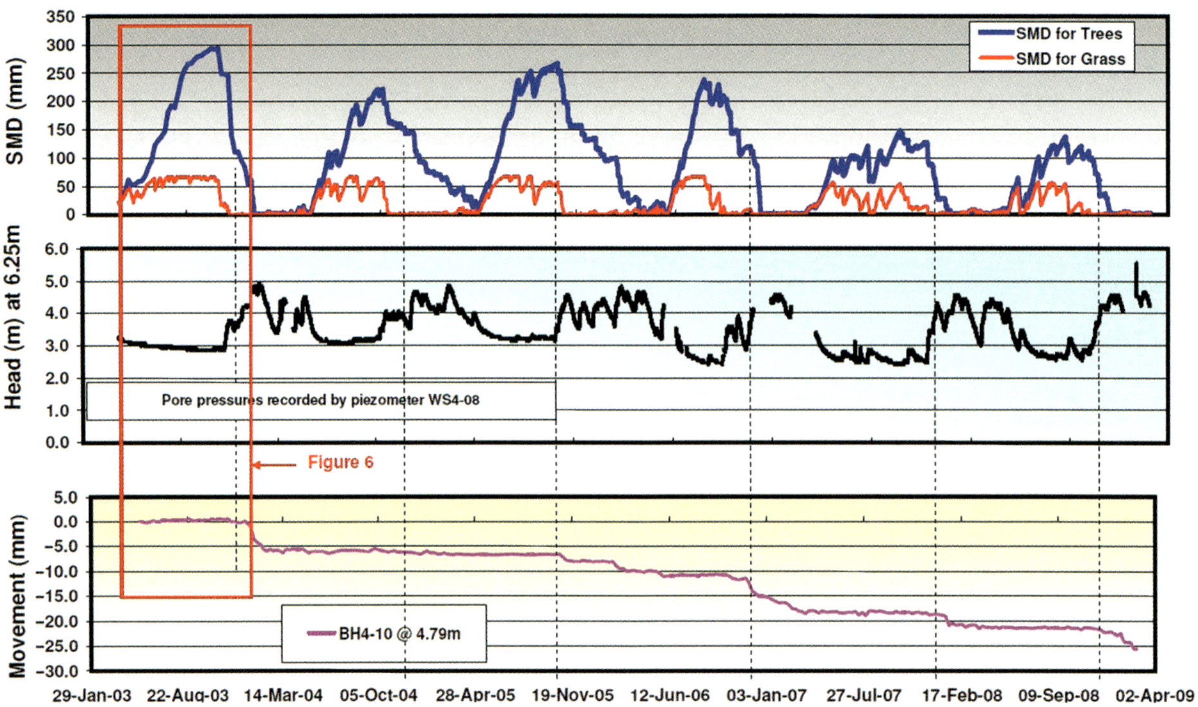

Fig. 5. Plot of slope movement, pore water pressure and SMD with time for Rooks Nest Farm.

vegetation on the surface will also tend to lower the zero pressure line.

The maximum suction usually occurs in the lower mid-slope with the lowest seasonal variation being encountered at the bottom of the slope. Suction can be lost very rapidly when the ground begins to wet, probably due to rainfall penetrating desiccation cracks (Ridley *et al.* 2004*a*).

The size, maturity and type of vegetation was shown to be important with slopes covered by larger, high water demand mature trees which generate much higher suction than grassed slopes (Ridley *et al.* 2004*b*). Consequently pore water pressures in grassed slopes tend to return to hydrostatic conditions during wet winters whereas slopes vegetated with mature trees commonly retain a residual suction. Ridley *et al.* (2004*b*) reports that measurements taken after the exceptionally wet winter of 2000/2001 showed that pore pressures came close to the worst credible conditions with a hydrostatic pressure profile from a zero pressure line at the surface suggesting that lowering of the zero pressure line by the presence of deciduous trees cannot be relied on through a period of prolonged wet weather.

The presence of trees therefore increases the soil suction generated in the summer and the range in pore pressure during a seasonal cycle. Where the summer suctions are such that a remnant suction is preserved during the winter then the number of cycles to slope failure increases. The presence and type of vegetation is therefore important in increasing the operational life of an embankment.

Case studies

M25 Rooks Nest Farm

Rooks Nest Farm is located between Junctions 5 and Junction 6 of the M25 motorway, near Godstone, Surrey on the shallow south-facing slope of the North Downs escarpment. The M25 motorway at this location has been cut into the slope with fill placed on the downslope side of the motorway. Rooks Nest Farm occupies the area downslope of the motorway. The slope is underlain by Made Ground and Head deposits overlying the Gault Clay.

During early 2001 the ground adjacent to the motorway displayed ongoing continual movement producing surface depressions and cracking in the hard shoulder and running lane 1. This creep movement was caused by reactivation of two relict slip failure planes some 5–9 m below ground level along the interface between the weathered and unweathered Gault Clay, downslope of the motorway.

Monitoring of the slope between March 2003 and April 2005 identified that periods of increased movements coincided with low SMD values and high pore water pressures.

To control the pore water pressures in the slope a series of siphon wells were installed parallel to the motorway between August 2005 and April 2006, together with additional instrumentation (Fig. 4). The correlation between SMD, pore water pressure and slope movement for this slope is shown for the period between 2003 and 2009 on Figure 5 and in detail for 2003 in Figure 6.

The following detailed observations are taken from the plot of SMD, pore water pressure and slope movement for 2003 (Fig. 6).

- The summer of 2003 was hotter than average and this is reflected in the SMD for trees which reached a maximum of approximately 300 mm in October of that year. The shallow movement between 29/9/03 and 7/10/03 occurred when the SMD value for trees was close to the summer maximum and is considered to be related to shrinkage cracking.
- The reduction in SMD for trees on 3/11/03 is associated with 55 mm of rainfall resulting in a minor upslope movement and is likely due to swelling of the clay.
- Larger downslope movements commenced around 17/11/03 when the SMD for trees fell from 245 mm on 18/11/03 to 155 mm on 25/11/03, that is a reduction of 90 mm over 1 week. Although not shown on Figure 6, two peaks in weekly rainfall occurred during week ending 3/11/03 (55 mm) and 24/11/03 (82 mm of which 57 mm occurred over 2 days, 22/11/03 and 23/11/03), that is just prior to and during the period when the slope movements commenced. The SMD for trees continued to fall towards 60 mm by the end of December 2003 by which time the cumulative downslope movement had reached approximately 5 mm.
- At the time of the commencement of slope movement on 17/11/03, pore pressures started to increase from a summer head of 2.85 m towards a maximum of 4.8 m at the end of December 2003.
- Slope movements appear to commence when the weekly change in SMD for trees exceeded 25 mm/week.

A similar pattern of slope movement, increase in pore water pressure and change in SMD was observed for the period 2004–2009 and is discussed in detail in Mouchel's (2010) research study.

M26 J2A to J1, Sevenoaks, Kent

The site is located on the westbound carriageway of the M26 motorway between Junction 2A and Junction 1 near Sevenoaks, Kent.

The earthwork comprises a 5 m deep cutting in Gault Clay with an average slope angle of 1 in 6 (10°), planted with a mixture of mature trees and shrubs (Fig. 7). The failure comprises a shallow translational movement along a 65 m length of the batter slope with multiple backscarps, up to 600 mm in height and a 1 m high toe bulge.

The motorway cutting was formed through Head deposits and into the underlying weathered Gault Clay.

The remedial works comprised reprofiling the slipped area and the installation of 3.5 m deep counterfort filter drains

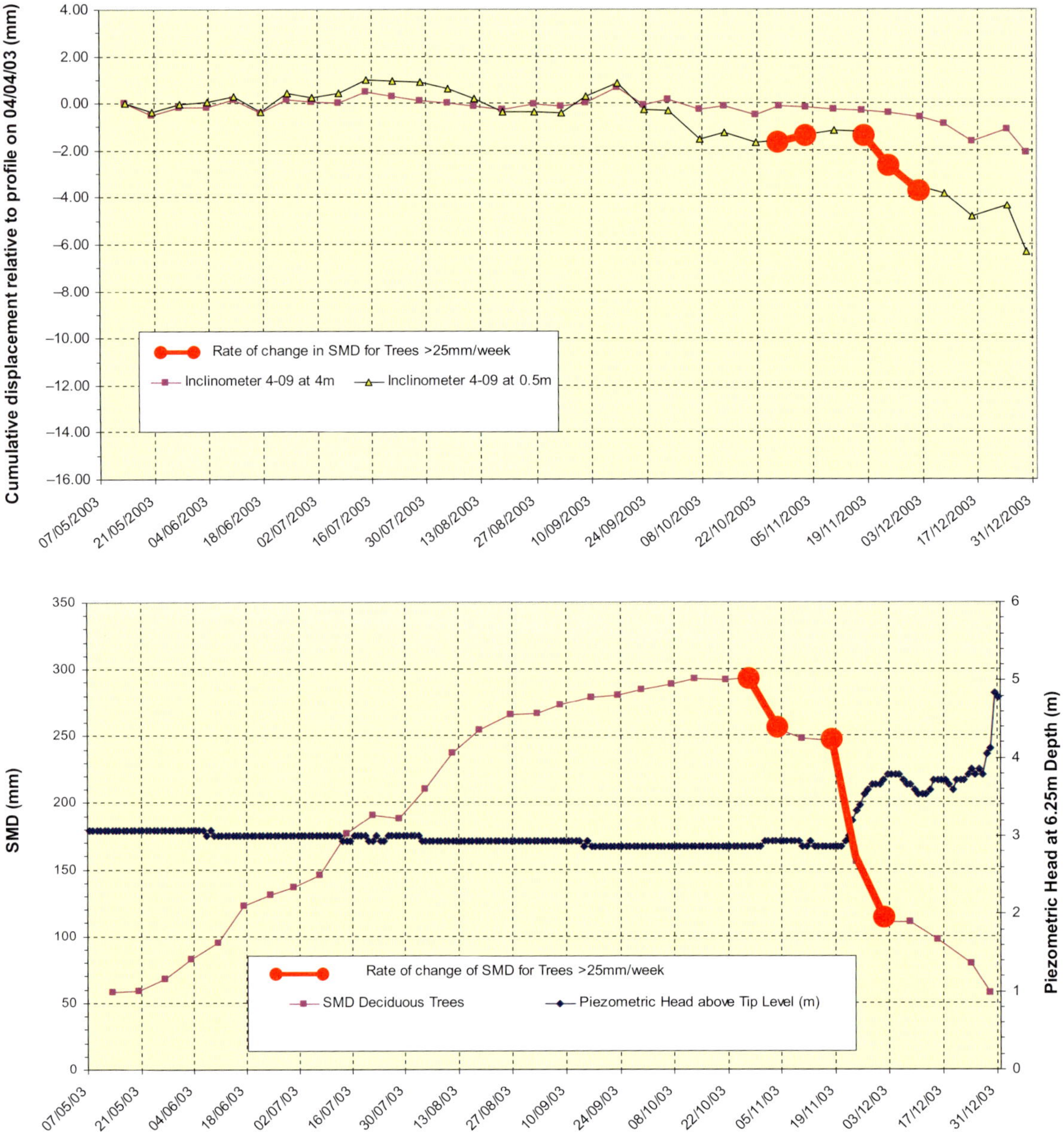

Fig. 6. Composite plot of deflections, SMD, pore water pressure and weekly rate of change in SMD for Rooks Nest Farm, 2003.

0.6 m wide at 6 m centres with a toe drain and a cut off drain at the crest of the slope. The drains were installed between October 2006 and December 2006.

The following detailed observations are taken from the plot of SMD, pore water pressure and slope movement between 2006 and 2009 (Fig. 8).

- The SMD for trees reduced from a peak of 220 mm on 30/09/05 to zero on 14/02/06, remaining at or close to zero until 30/5/06 when SMD values began to rise during the summer months.
- Inclinometer readings were taken monthly between January 2006 and September 2006 and therefore weekly

Fig. 7. M26 junction 2A to J1 Sevenoaks cutting failure.

variations have not been recorded. Minor creep movements occurred from February 2006 to May 2006 with the SMD for trees at or close to zero.

- Pore water pressures at the toe (P2) and mid slope (P3) were falling during the period February 2006–May 2006 with a slight lag between the mid slope and toe response, which may be indicative of recharge from the overlying head deposits near the crest of the cutting.
- Significant downslope movements commence around 11/07/06 when the SMD for trees was rising at a rate of >25 mm/week towards the peak summer value of 217 mm on 15/08/06. Rainfall varied from 8.4 mm in June 2006 to 11 mm in July 2006 before increasing to over 100 mm in August 2006. The downslope movement that commenced on 11/07/06 may therefore be due to shrinkage of the clay with the late July/August 2006 downslope movement being precipitated by the heavy rainfall in August 2006 which in turn is reflected in the decrease in SMD from the summer maximum.
- Downslope movements within the cutting continue until 10/01/07 with up 34 mm of movement recorded at shallow depth. However, inclinometer readings were suspended during the installation of counterfort drains between October 2006 and December 2006 and therefore detailed movements during this period are unknown.
- The rate of change in SMD exceeds 25 mm per week as shown on Figure 8.
- The piezometers at the toe and mid slope were installed in the unsaturated zone at shallow depth (1.45 m) above the water table and may therefore be susceptible to rapid changes in pore water pressure and in particular to a rapid loss of suction.

Conclusions

The research study (Mouchel 2009) identified that, for the seven earthwork slopes studied, significant slope movements occur when the SMD for trees reduces or increases at a rate of >25 mm per week.

Slope movements continue when the SMD for trees reduce to and are maintained at or close to zero during the winter months. This relationship appears to apply to both embankments and cuttings, although with the latter groundwater conditions tend to be more complex and influence the relationship.

The case studies have shown that there is a close correlation between rainfall, pore water pressure and the SMD for both trees and grass and slope movement.

Two types of slope movements have been identified:-

- Where the slope is vegetated with grass, shrubs and or sparse trees, the study has shown that slope movements are generally precipitated by an increase in pore water pressure and therefore occur when the SMD for trees is falling. This behaviour is seen in most of the case studies.
- Where the slope is heavily vegetated with mature trees and or there are mature trees at the toe of the slope, then slope movements are generally precipitated by shrinkage and desiccation of the clay and occur when SMD for trees are rising. This type of movement is often seasonal and cyclic with a downslope movement on a rising SMD in summer due to shrinkage/desiccation followed by an upslope recovery in winter on a reducing SMD as the clay swells.
- A variation in the latter behaviour was observed where upslope movement at a shallow depth occurs on a rising SMD and is likely to be associated with rainfall entering desiccation cracks, resulting in local swelling of the clay.

In order for the rate of SMD for trees to be effectively used as a trigger mechanism for targeted monitoring of slopes, the likely failure mechanism must be determined and SMD data must be obtained and reviewed in an appropriate timescale. A preliminary assessment of the likely failure mechanism may be made by categorizing the slope into one of the above based on vegetation cover and inspection data.

Site specific SMD data can be provided by the Meteorological Office on a daily or weekly basis. It is recommended that for each earthwork asset that is considered to be at risk, monthly historical SMD data for trees is obtained for a minimum of two years in order that yearly trends may be established. The trends will identify the months during which SMD values for trees are likely to fall from the summer maxima or rise from winter minima.

For the months following the summer maxima and winter minima, and for one month before (to allow for seasonal variations), weekly SMD for trees should be obtained for the particular earthwork slope and the data examined to determine the rate of change in SMD. Where the rate of change exceeds 25 mm per month either on a falling SMD or a rising SMD for trees, the monitoring of slopes considered to be at risk should commence.

Although SMD can be updated at sub-daily intervals, there is no means of forecasting changes in SMD in advance. However, the use of surrogates for SMD in the

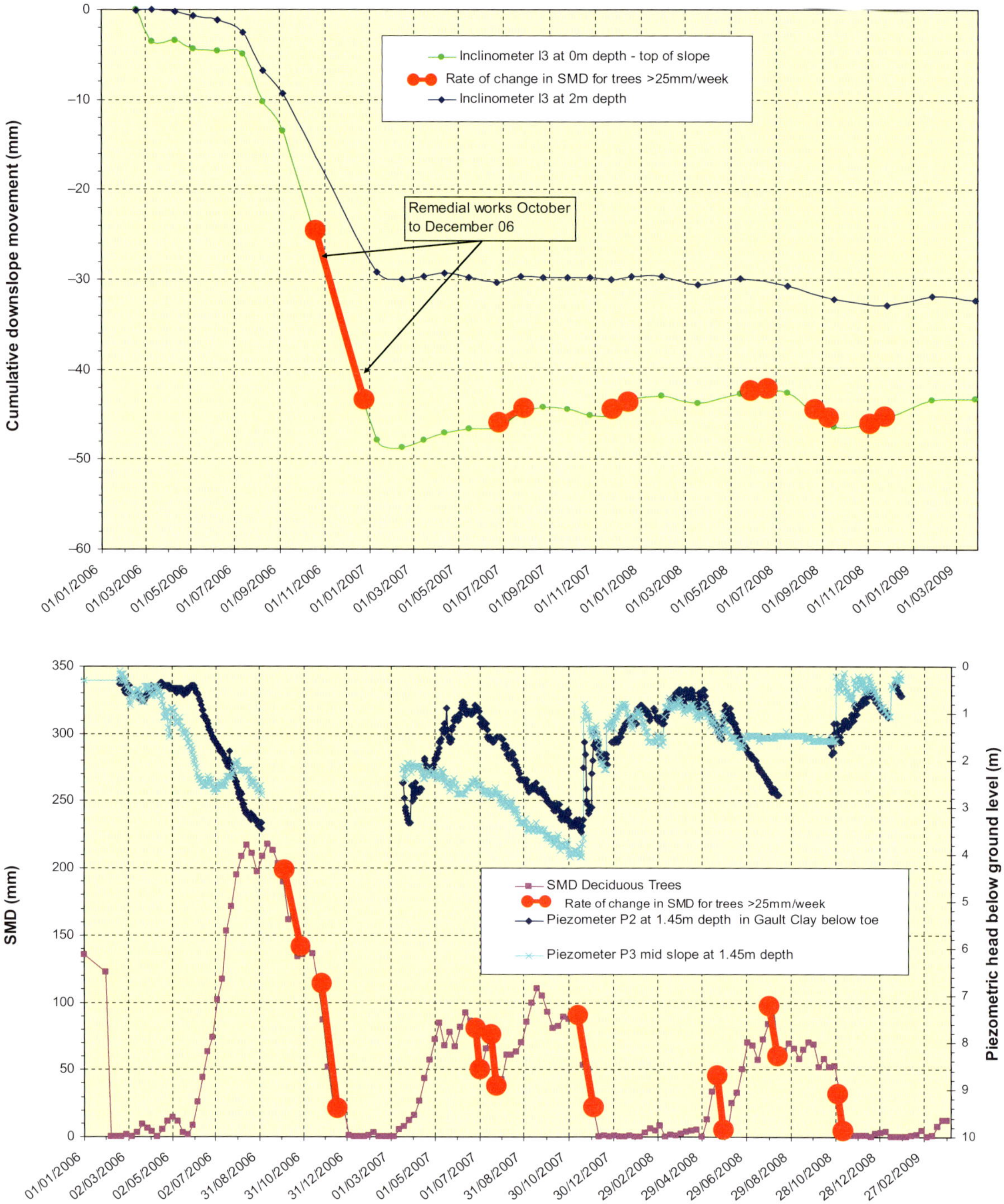

Fig. 8. Composite plot of deflections, SMD, pore water pressure and weekly rate of change in SMD for M26 Sevenoaks cutting.

form of rainfall and temperature forecasts could be useful in anticipating short-term rapid changes in SMD. Both of these variables are used by the Meteorological Office in the estimation of SMD.

The research study (Mouchel 2010) suggested that the recommended forecast and alert aspects should be further investigated to include:

- incidence of slips associated with heavy rainfall events;
- the use of an antecedent precipitation index or intensity-duration indicator as an adjunct to SMD monitoring;
- Meteorological Office warnings and forecasts associated with the above;
- incidence of slips associated with extended dry periods; and
- drought indicators, for example degree-days, which might provide useful triggers.

The research study (Mouchel 2010) also identifies the requirement for further research:

- to develop a similar rate of change of SMD for grass cover;
- to review to determine the relationship between swell shrink movements and slope deformation in more detail utilising combined inclinometer and extensometer plots.

Full details of the correlation between rainfall, pore water pressure and SMD for each of the case studies are included in the research report (Mouchel 2010).

The authors would like to thank the Highways Agency, in particular the Project Research Sponsor (Jan Marsden) for the support and permission for publishing the information presented in this paper.

References

BRITISH GEOTECHNICAL ASSOCIATION WORKSHOP. 2008. Maintenance of ageing earthworks. *Ground Engineering*, August 2008, 12–14.

KOVACEVIC, N., POTTS, D. M. & VAUGHAN, P. R. 2001. Progressive failure in clay embankments due to seasonal climate changes. *Proceedings of the 15th International Conference on Soil Mechanics & Geotechnical Engineering*, Istanbul, Turkey. A. A. Balkema, **3**, 2127–2130.

MOUCHEL. 2010. *Soil Moisture Deficit and Slope Movement*. Research Study for the Highways Agency. November 2010, Report 787303, 1–58.

RIDLEY, A. M., VAUGHAN, P. R., MCGINNITY, B. & BRADY, K. 2004*a*. Pore pressure measurements in infrastructure embankments. *In*: JARDINE, R. J., POTTS, D. M. & HIGGINS, K. G. (eds) *Advances in Geotechnical Engineering. Proceedings of the Skempton Conference*. Thomas Telford, **2**, 922–932.

RIDLEY, A. M., VAUGHAN, P. R. & MCGINNITY, B. 2004*b*. The role of climate and pore water pressure in assessment of infrastructure embankments. *Proceedings of the Institution of Civil Engineers, Geotechnical Engineering*, **157**, 193–198.

VAUGHAN, P. R., KOVACEVIC, N. & RIDLEY, A. M. 2002. The influence of climate and climate change on the stability of embankment dam slopes. *Reservoirs in a Changing World. Proceedings of the 12th International Conference*. The British Dam Society. Thomas Telford, London, 353–366.

Investigating the impacts of climate change on slopes: field measurements

D. G. Toll[1]*, J. Mendes[1], D. Gallipoli[2], S. Glendinning[3] & P. N. Hughes[3]

[1] School of Engineering and Computing Sciences, Durham University, Durham DH1 3LE, UK
[2] School of Civil Engineering, University of Glasgow, Oakfield Avenue, Glasgow G12 8LT, UK
[3] School of Civil Engineering and Geosciences, Newcastle University, Newcastle upon Tyne NE1 7RU, UK

*Corresponding author (e-mail: d.g.toll@durham.ac.uk)

Abstract: Climate change has the potential to have significant effects on the stability and serviceability of earthworks slopes, impacting on the performance of transport infrastructures. This paper describes how a unique facility for engineering and biological research was established in NE England through the BIONICS Project (Biological and Engineering Impacts of Climate Change on Slopes). It describes the building and monitoring of a full-scale embankment representative of road/rail embankments in the UK. The paper presents the results of monitoring of pore water pressure carried out between 2007 and 2009. The pore water pressures within the upper 3 m of the embankment have been largely positive, in some case approaching hydrostatic conditions after wetter periods. This is largely due to the wet nature of the summers in both 2007 and 2008 since monitoring began, as well as the application of artificial inundation using a climate control system. Negative pore water pressures (suctions) of the order of −30 kPa were recorded at greater depths below 3 m. The experimental facility provides essential field measurements that can be used to calibrate numerical models of soil responses to climatic changes and gain a better understanding of the response of engineered UK fills to climate events.

A major concern that engineers face with assessing earth structures is whether structures can deal with the current climate patterns and if they will maintain their serviceability when subjected to different climate patterns. Models of climate change predict that the UK, certainly in SE England, could face more extreme rainfall events during winter periods (involving more intense rainfall) and also longer dry spells during summer seasons. Climate change therefore has the potential to have significant effects on the stability and serviceability of earthworks slopes (Kilsby *et al.* 2009).

The BIONICS project (Biological and Engineering Impacts of Climate Change on Slopes: http://www.ncl.ac.uk/bionics) is a large cooperative research project involving academic partners and industrial stakeholders focusing on the effects of the expected climate change on infrastructure embankments (Hughes *et al.* 2009). BIONICS is a £1.1 m project, largely funded by EPSRC, looking at the potential impacts of climate change on UK infrastructure slopes. It is a collaboration between the Universities of Bristol, Dundee, Durham, Loughborough, Nottingham-Trent coordinated by the University of Newcastle upon Tyne.

The BIONICS project has involved constructing an instrumented embankment to investigate the response to changing climatic conditions. The embankment has been equipped with a system of sprinklers and covers to control climatic conditions. It has been instrumented with piezometers, tensiometers, water content and temperature sensors, inclinometers and extensometers. The experimental facility provides essential field measurements that can be used to calibrate numerical models of soil responses to climatic changes. This paper presents the results of monitoring of pore water pressure carried out September 2007 and March 2009.

The BIONICS embankment

The BIONICS embankment was built at Nafferton Farm, near Newcastle upon Tyne, and construction was completed in November 2005 (Hughes *et al.* 2009). The embankment dimensions are 90 m long by 6 m high (Fig. 1) with side slopes of 1 in 2 (V:H) and a 5 m width crest. The end slopes of the embankment were constructed with reinforced soil with an inclination of 45°.

The embankment was divided into six different panels (not including the end slopes) as shown in Figure 1. The four central panels were 18 m long and were constructed to represent two different types of construction methods ('well compacted' and 'poorly compacted' as will be

From: Radford T. A. (ed.) 2012. *Earthworks in Europe*. Geological Society, London,
Engineering Geology Special Publications, **26**, 151–161, http://dx.doi.org/10.1144/EGSP26.17

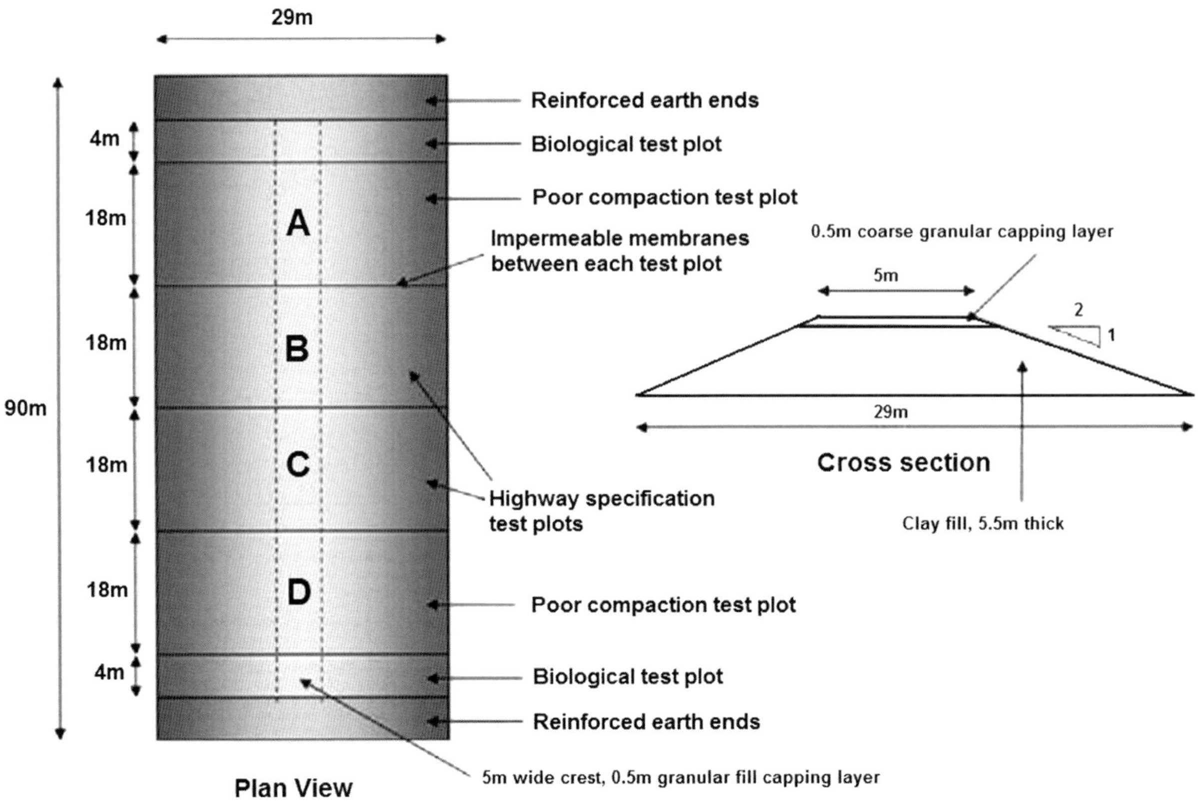

Fig. 1. BIONICS embankment. Plan view and cross section.

defined later). The outermost panels (each 4 m in length) were for biological studies.

The 'poorly compacted' panels (panels (A) and (D) in Fig. 1) were constructed using limited compactive effort to represent railway embankments built over 100 years ago in Victorian times. These panels were placed in 1 m lifts and were simply compacted using a tracked hydraulic excavator; plant movement on the panel was kept to a minimum to limit compaction as much as possible. The 'well compacted' panels (panels (B) and (C) Fig. 1) were constructed to represent modern highway embankments using current highways specifications for compaction. Layers were compacted according to Method 3 as set out in the Highways Agency Specification for Highway Works (Highways Agency 1998). The fill was placed in 300 mm layers and was subjected to nine passes by a self propelled vibrating smooth drum roller.

Vertical impermeable membranes were placed between panels and end slopes to ensure engineering and hydraulic isolation. At the top of the embankment a 0.5 m thick layer of coarse fill (crushed basalt) was built. The coarse, free draining material was placed in order to simplify boundary conditions by preventing surface cracking on the crest of the embankment but still to allow water access into the embankment.

A 200 mm layer of top soil was placed on the side slopes of the embankment to facilitate vegetation growth and simulate current infrastructure embankments. In July 2006, a seed mixture consisting of 30 grasses and wildflowers was sown on the embankment. A survey of plant species growing on the slopes of the embankment undertaken in July 2007 (Glendinning *et al.* 2009) identified 35 species. Eleven species were very common: *Lolium perenne* (Perennial Ryegrass), *Cynosurus cristatus* (Crested Dog's-tail), *Poa trivialis* (Rough Meadow-grass), *Agrostis capillaries* (Common Bent), *Holcus lanatus* (Yorkshire Fog), *Ranunculus repens* (Creeping Buttercup), *Trifolium repens* (White Clover), *Lotus corniculatus (*Bird's-foot Trefoil), *Vicia sativa* (Common Vetch), *Medicago lupulina* (Black Medick) and *Trifolium pratense* (Red Clover). Twelve species had established from the seed mixture sown and a number of species had established from the seed bank or colonized from adjacent habitats; a number of these were arable weeds, including *Cirsium arvense* (Creeping Thistle). Species richness was higher on the south-facing slopes than the north-facing slopes (10.9 species per m^2 compared with 9.1 species per m^2).

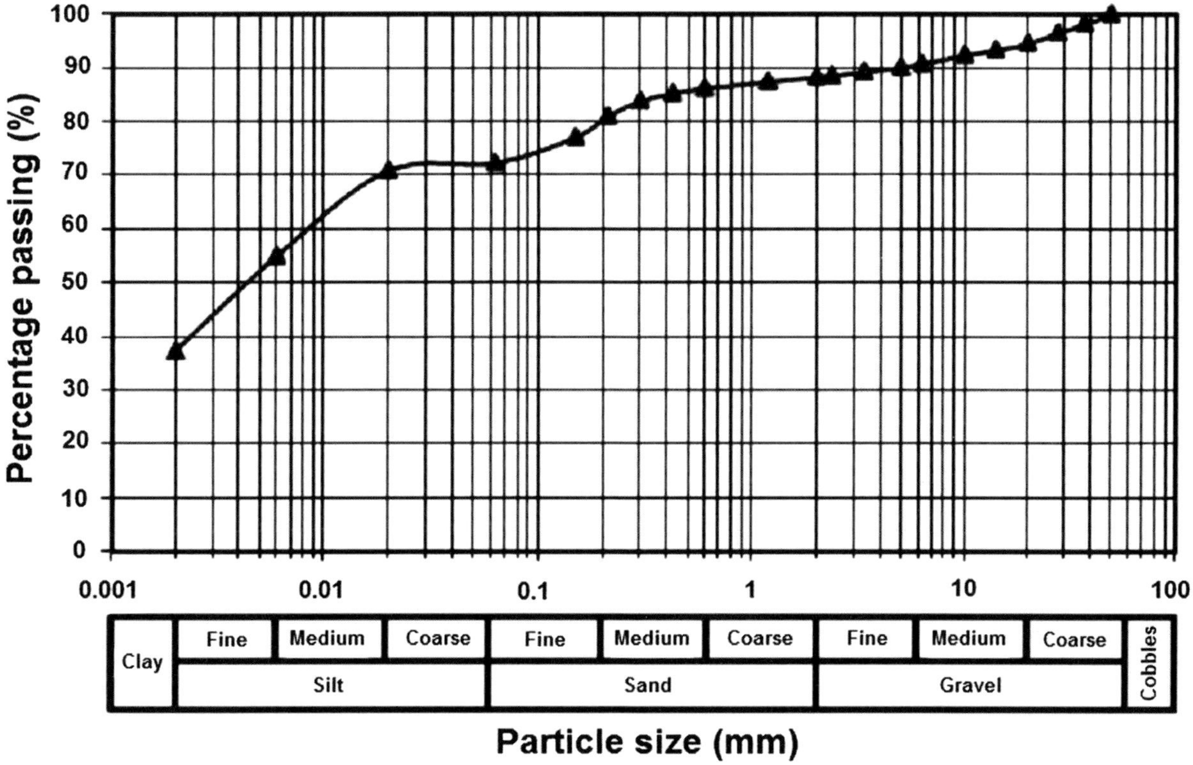

Fig. 2. Particle size distribution for the BIONICS fill material.

The fill material

The fill material used to build the BIONICS embankment was a glacial till (Durham Lower Boulder Clay), a common fill material in NE England and hence representative of earthwork construction. The material was a single sourced material from a stock pile in County Durham with the particle size distribution curve shown in Figure 2. The Atterberg limits (determined using BS 1377 1990: Tests 4.3 and 5.3) are shown in Table 1. The fill material can be classified as a sandy clay of intermediate plasticity.

Laboratory compaction curves for the BIONICS soil using the BS light 2.5 kg hammer (Proctor) compaction test and the BS heavy 4.5 kg hammer (Modified Proctor) compaction test (BS1377 1990: Tests 3.3 and 3.5) are shown in Figure 3. For light compaction the maximum dry density (MDD) was 1.71 Mg m^{-3} at an optimum water content (w_{opt}) of 15.5%. For the modified test a maximum dry density of 1.80 Mg m^{-3} was achieved at w_{opt} of 13%.

During construction, density measurements were taken to assess the field compaction level (Hughes *et al.* 2007).

Table 1. *Atterberg limits for the BIONICS fill material*

Natural water content w_n (%)	*Liquid limit w_l (%)*	*Plastic limit w_p (%)*	*Plasticity index PI*	*Liquidity index LI*
22.6	43.3	23.7	19.6	−0.05

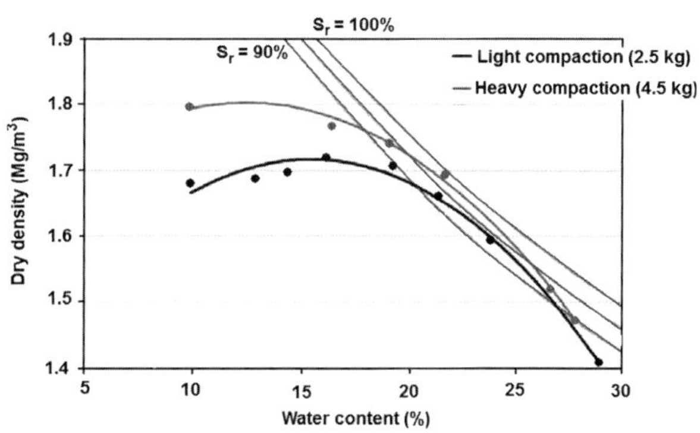

Fig. 3. Laboratory compaction curves for the BIONICS fill material.

The construction was performed wet of optimum, with water contents ranging from 16–24%. Most of the measurements in the well compacted panels showed densities of 94–101% of MDD from laboratory light compaction (average 99%). However, a small number of measurements on layers compacted nearer to optimum (with water content below 17%) showed densities near to MDD for heavy compaction (equivalent to 106% of MDD from light compaction). The densities achieved for the poorly compacted panels were 90–99% of MDD from light compaction (average 94%).

A laboratory permeability test on fill material from a well compacted panel showed a saturated permeability value, $k_s = 8.77 \times 10^{-11}$ m s^{-1} while a laboratory measurement on the material from a poorly compacted panel gave a result of $k_s = 1.6 \times 10^{-10}$ m s^{-1}.

Instrumentation

As an experimental embankment the BIONICS embankment was intended to be used as a platform for new testing methodologies that can help the investigation and determination of soil behaviour. The instrumentation at the BIONICS embankment can be divided in two major groups as shown in Table 2. The first group includes general instrumentation commonly found in instrumented embankments. The second

Table 2. *Summary of the embankment instrumentation (after Hughes et al. 2009)*

	Parameter measured	*Typical usage*	*Primary information on failure mode(s)*
General instrumentation			
Standpipe piezometers	Piezometric surface, point measurement	Geotechnical monitoring	Shrink–swell, re-wetting
In-place inclinometers	Slope movement, continuous monitoring	Geotechnical monitoring	Progressive failure
Probe inclinometers	Slope movement, point measurement	Geotechnical monitoring	Progressive failure
Magnetic extensometers	Settlement/heave, point measurement	Geotechnical monitoring	Shrink–swell, surface creep
Specialized instrumentation			
Durham tensiometers	High values of positive and negative pore water pressure continuous measurement	Experimental, under development by Durham University	Shrink–swell, re-wetting
Geotechnical Observations (GEO) tensiometers	High values of negative pore water pressure, point measurement	Geotechnical monitoring	Shrink–swell, re-wetting
Flushable piezometers	Positive and negative pore water pressure, continuous measurement	Geotechnical monitoring	Shrink–swell, re-wetting
Acoustic waveguides	Slope movement, continuous measurement	Experimental, under development by Loughborough University	Progressive failure
Theta probes	Volumetric soil water content, continuous measurement	Agriculture irrigation control and geotechnical monitoring.	Shrink–swell, re-wetting
Water content profile probes	Volumetric water content, point measurement	Agriculture irrigation control	Shrink–swell, re-wetting
Temperature probes	Soil temperature, continuous measurement	Agriculture, biology, hydrology, meteorology	Shrink–swell, re-wetting
Weather stations	Rainfall, wind speed/direction, pressure, humidity, air temperature, sunlight (inputs to potential evapo-transpiration calculation; indirect, requires calculation)	Meteorology, agriculture, hydrology, biology	Shrink–swell, re-wetting
Remote methods: Lidar, terrestrial laser scanner, GPS	Movement, condition of vegetation	Remote ground surface monitoring	Progressive failure, surface creep

group comprises newly developed technologies. Both groups of instrumentation are widely distributed through the embankment in all types of panel, as can be observed in the instrumentation plan for the well compacted panel (b) in Figure 4.

The general instrumentation comprised extensometers (at the centre of the crest in each panel) inclinometers in both the north and south slopes (a total of six per panel) and standpipe piezometers spread out through the embankment to measure pore water pressure.

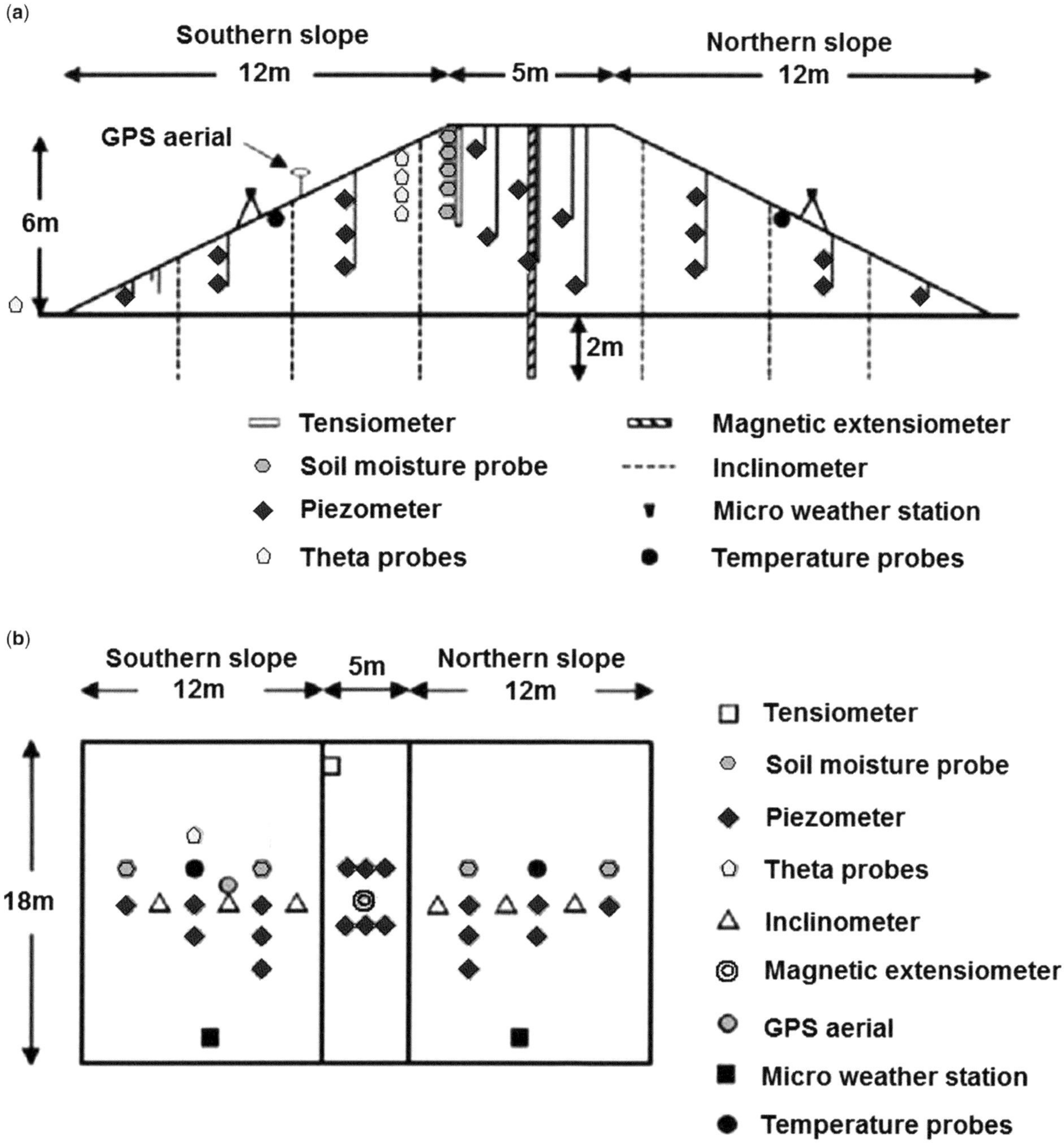

Fig. 4. Instrumentation plan for panel b: (**a**) cross section and (**b**) plan view (Hughes *et al.* 2009).

The specialized equipment included tensiometers and flushable piezometers for measurement of negative pore water pressure (suction). This paper describes in detail the installation of high capacity tensiometers (suction probes) developed by Durham University. In addition, flushable piezometers from Geotechnical Observations (GeO) capable of measuring negative pore water pressures down to −100 kPa were installed. High capacity suction probes (Ridley & Burland 1993) from GeO were also available for measurements in dry vertical boreholes. Further instrumentation included Theta probes and moisture profile gauges to measure volumetric water content. Collection points were also used to determine the surface runoff.

Three different weather stations were used to monitor the climate at the embankment. One of the weather stations was located 300 m from the embankment and monitored the annual weather pattern (i.e. rainfall, air speed, etc.); this station complies with Meteorological Society Guidelines (Overton 2007). The other two weather stations (micro weather stations) were installed on top of the north and south slopes of the embankment; these did not fully comply with Meteorological Office guidelines but were deliberately positioned closer to ground level in order to capture the differences in climate between different faces of the embankment. The latter two weather stations were used to monitor the local climate effects, most particularly when the climate control system was in operation.

Climate control system

A climate control system was developed for the embankment to be able to impose different scenarios predicted for climate change. The system employed at the BIONICS embankment can recreate extreme scenarios; designed to reproduce intense rainfall and dry events. To reproduce extreme rainfall events, rainfall sprinklers mounted on poles were installed on panels A and B. The location of the sprinklers across the embankment ensured a spatially uniform rainfall at the ground surface with appropriate droplet sizes.

To reproduce dry events, segments of the embankment were covered with a retractable transparent flexible cover system. With this design, precipitation from natural climate events could be prevented from reaching the embankment. Due to its proximity to the surface (1 m above the ground) temperature loss was also greatly reduced. By reducing infiltration and by inducing a small increase in temperature, higher rates of evaporation could be generated to represent drying conditions.

High capacity tensiometers

A novel experimental set-up for field monitoring of soil suctions was developed for the BIONICS project (Mendes *et al.* 2008). The experimental set-up used Durham University-Wykeham Farrance high capacity tensiometers (Lourenço *et al.* 2006) to provide real-time continuous measurements of suction inside the embankment. The wide measuring range of the tensiometers (suction of 2 MPa) allows the monitoring system to be used in most natural and constructed earth structures.

The tensiometers were installed with the help of a probe locator (Fig. 5), which consisted of a 3 m long PVC cylinder buried in the ground. The probe locator had an outer diameter of 90 mm and an inner diameter of 70 mm. Four small openings existed on the sides of the probe at depths of 0.5 m, 1 m, 1.5 m and 2 m (Fig. 5b) with one further opening on the bottom cross-section at a depth of 3 m (Fig. 5c). Each of these openings was connected to the surface by means of a flexible hose with an inner diameter of 19 mm. The flexible hoses ran inside the probe locator and served as guides for inserting or removing tensiometers (Fig. 5a). These guide hoses were fitted at the bottom extremity with a cylindrical clasp made of aluminium, which tapered the inner diameter from 19 mm down to 14 mm and fitted tightly around the tensiometer preventing movement of soil inside the guide. The inner space at the top of the probe locator was injected with silicone and foam to seal the gap around the guide hoses, to avoid any infiltration of water or other kind of material from the surface.

The tensiometers were fitted with a nylon tube (over the electrical cable) which was glued to the back of the tensiometer. This nylon tube had a twofold function, to facilitate insertion and removal of each tensiometer individually and to protect the electrical connection. In case of cavitation or fault, tensiometers could be easily removed from the ground by pulling the nylon tube to which they were strongly attached. After re-saturation or repair, it was also possible to install the tensiometer back in the ground by pushing the nylon tube until the tensiometer was firmly inserted in the cylindrical clasp at the end of the guide hose. The nylon tube, which has outer and inner diameters of 8 mm and 6 mm respectively, was sufficiently stiff to allow a considerable force to be applied during installation without buckling. The tensiometer casing was smoothed to remove sharp edges to allow easy removal and insertion.

Electrical cables from each tensiometer ran through the guide tubes inside the probe locator and were connected to a logging unit sited on the embankment crest. The logging unit was in turn connected to a personal computer, hosted in a nearby hut, which read the suction data and transmitted it by means of an internet broadband connection. This allowed remote real time records of the suction profile to be obtained with time as well as with depth.

Two probe locators were installed in February 2007 in two sections of the BIONICS embankment, one in a well compacted section and one in a poorly compacted section. It was initially intended to grout the probe locators in place. However, measurements of suction across specimens of grout (Mendes 2011; Toll *et al.* 2011) showed that the grout was unable to transmit suctions in excess of 600 kPa when placed in contact with soil. This means that suction

Fig. 5. A probe locator on the BIONICS embankment before installation. (**a**) Detail of the guide tubes emanating from the top; (**b**) detail of the side suction station; and (**c**) detail of the base suction station.

values measured through such a grout could give a false reading at these higher suction levels and could underestimate the true suction in the soil.

Given that it was initially uncertain whether suction levels in the embankment would exceed this limit, it was cautiously decided not to seal the gap by using the cement–bentonite grout but rather to rely on the natural closure of the borehole around the probe locator. Adequate sealing around the probe locator was later confirmed by pumping out water from the location of one tensiometer. There was no response shown by adjacent tensiometer positions, confirming there was no preferential connection path along the sides of the borehole locator.

Results of field monitoring

Field monitoring was performed using the Durham University-Wykeham Farrance tensiometers at the BIONICS embankment. The system to measure pore water pressure was installed in April 2007. Figure 6 shows continuous measurements for the period of October 2007 to March 2009 for a well compacted panel. Although high suctions (>150 kPa) were measured on the embankment during construction in 2005 (Hughes *et al.* 2007) the values since November 2007 were small (less than 20 kPa). Such low values of suction might be expected as the period May–July 2007 was the wettest on record for 250 years and caused extensive flooding in the UK. September 2008 was also a very wet month, with 158 mm of rainfall and this followed a period May–July 2008 when artificial inundation was applied with the climate control system, totalling 456 mm. The responses do demonstrate that the tensiometers were able to record positive pore water pressures as well as suctions.

The profile of pore water pressures with depth for a well compacted panel is shown in Figure 7. For reference, a hydrostatic pore water pressure line is shown in the figure, referenced to zero pore water pressure at the top of the clay embankment, that is assuming no pore water pressure would build up in the 0.5 m thick coarse ballast layer. The initial pore water pressures show generally positive values in the upper 3 m of the embankment although suctions exist at 4.5 m depth (measured by flushable piezometers). These lower values could represent construction induced suctions, but are also likely to be affected by under-drainage of the embankment by field drains. The initial values are lower than hydrostatic conditions, even in the upper 3 m.

By December 2007, the pore water pressure profile had increased somewhat, but still falls below the hydrostatic

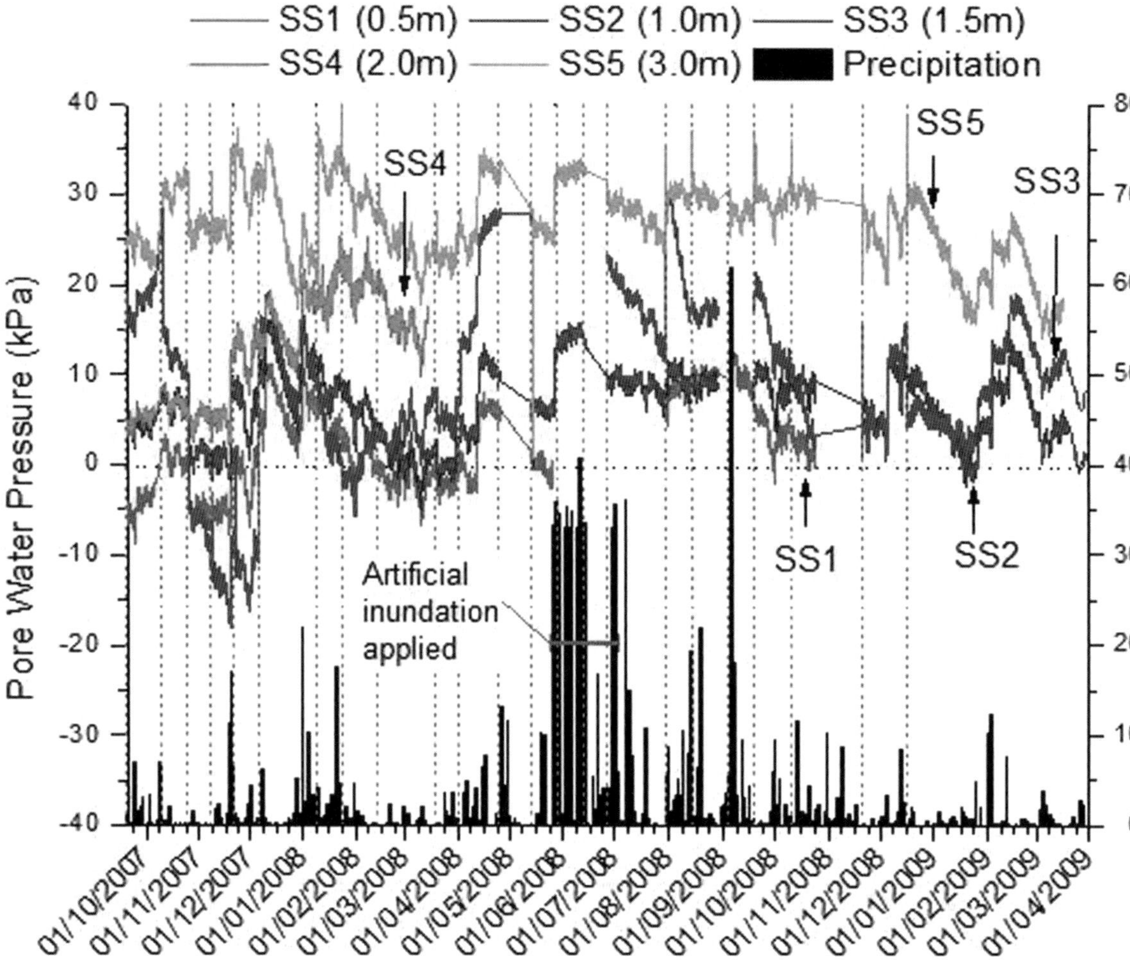

Fig. 6. Pore water pressure records for a well compacted panel (SS indicates suction station at different depths). Vertical spikes show daily precipitation.

line. However, by January 2008, after a period of heavy precipitation, pore water pressures near the surface (0.5 and 1 m depth) rose above the expected hydrostatic condition. They approach a hydrostatic profile defined by a water table above the top of the clay, suggesting that the coarse (ballast) layer had become flooded and water pressures had built up within this layer.

By March 2008, pore water pressures had dropped back to similar values to December 2007. Then in June 2008, after artificial inundation using the climate control system, pore water pressures rose again to approach the upper hydrostatic profile with the coarse ballast layer flooded. By December 2008, after 6 months of only natural precipitation, pore water pressures had dropped back again to similar values to December 2007. By March 2009, values had dropped back further to some of the lowest values since the initial installation. This was the result of a dry winter period with little rain from mid December 2008 until the end of January 2009.

These results show the cycles of pore water pressure that might be expected in an embankment, with higher pore water pressures (near to hydrostatic) during winter months (typically January) falling back to lower values in March. Lower values might normally be expected during summer months, but the climate during the summers of 2007 and 2008 were 'untypical' with high levels of summer rainfall (as noted earlier, the period May–July 2007 was the wettest summer period on record in the UK for 250 years). Also, artificial inundation was applied to the embankment during the period May–July 2008.

Monitoring of settlement of the embankment was carried out using magnetic extensometers installed at depths

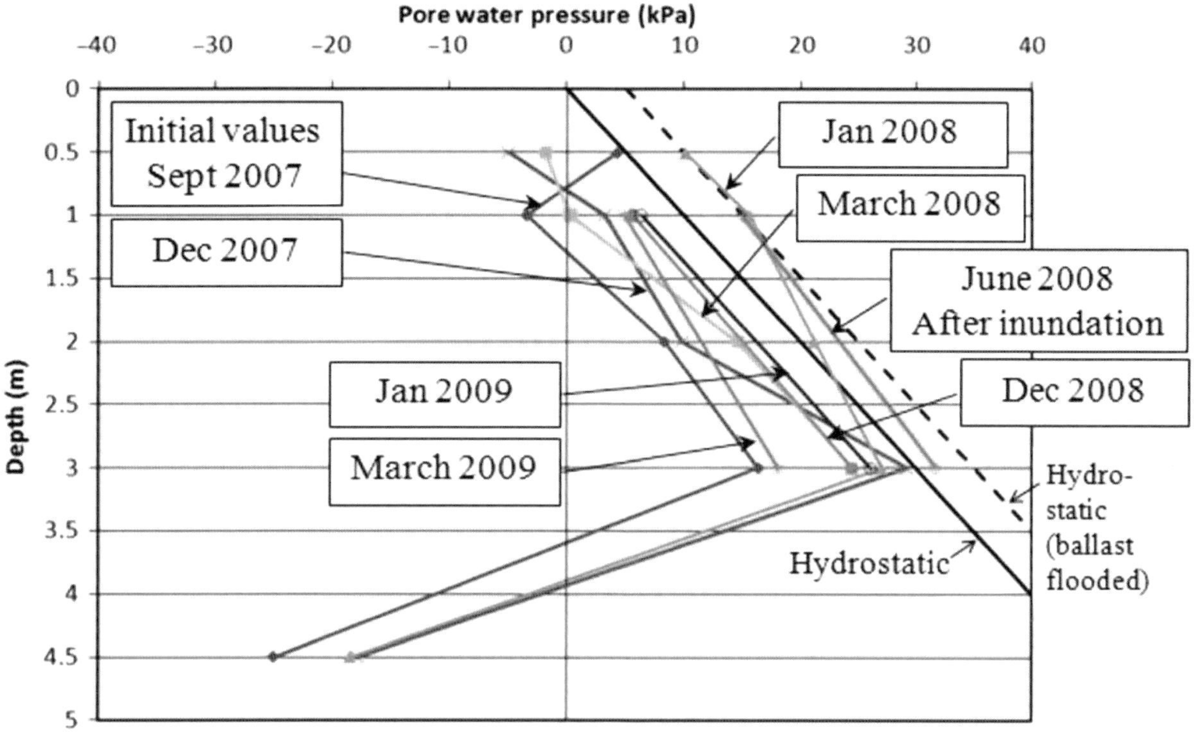

Fig. 7. Pore water pressure profiles for a well compacted panel.

ranging from 1.13–8.19 m (below the base of the embankment (Hughes *et al.* 2009). Results of settlements within the upper 3 m of the embankment in a well compacted panel are shown in Figure 8, for comparison with the pore water pressure readings in Figure 6. Following the larger rainfall events in December 2007 and January 2008 that a small surface heave (reduction in settlement) of the order of 5 mm was recorded. This is consistent with an increase in pore water pressure during this period (Fig. 6). From January to April 2008, pore water pressures dropped and resulted in a downward movement of the embankment (settlement). During the artificial inundation in June/July 2008 the pore water pressures showed an increase which produced a heave of 2–3 mm. Therefore, pore water pressure cycles cause shrink–swell cycles in the fill material of the embankment.

Further monitoring over coming years will now be required to understand the pore water pressure responses to rainfall over the longer term. A dry summer would be particularly welcome to give an indication of the generation of suctions during a dry period.

The vegetation on the BIONICS embankment comprises grasses and wildflowers and will therefore only affect the hydrology and mechanical properties of the near-surface soil. It is unlikely that transpiration/interception produced by grasses and wildflowers will be sufficient to exacerbate seasonal shrink–swell cycles to any significant depth (although will have an effect at the near surface), as observed by Loveridge & Anderson (2007) but changes to run-off and infiltration properties are likely to be affected (Glendinning *et al.* 2009).

Plans for numerical modelling

The monitoring described in this paper will be used to calibrate a numerical model that couples the unsaturated flow with the geo-mechanical behaviour of the fill material. The model has been described by Davies *et al.* (2008) and Rouainia *et al.* (2009) and makes use of a combined hydrological model (Shetran) and a geotechnical model (FLAC). Laboratory data on the soil water retention behaviour and constant water content triaxial tests carried out on unsaturated specimens will be used to identify the geo-mechanical properties (Mendes 2011). The triaxial tests were carried out on specimens that have been subjected to wetting and drying after compaction.

One of the main uncertainties in modelling infiltration and runoff on clay slopes is the permeability and how this changes with water content and with time. The unsaturated permeability to water can drop by 4–5 orders of magnitude

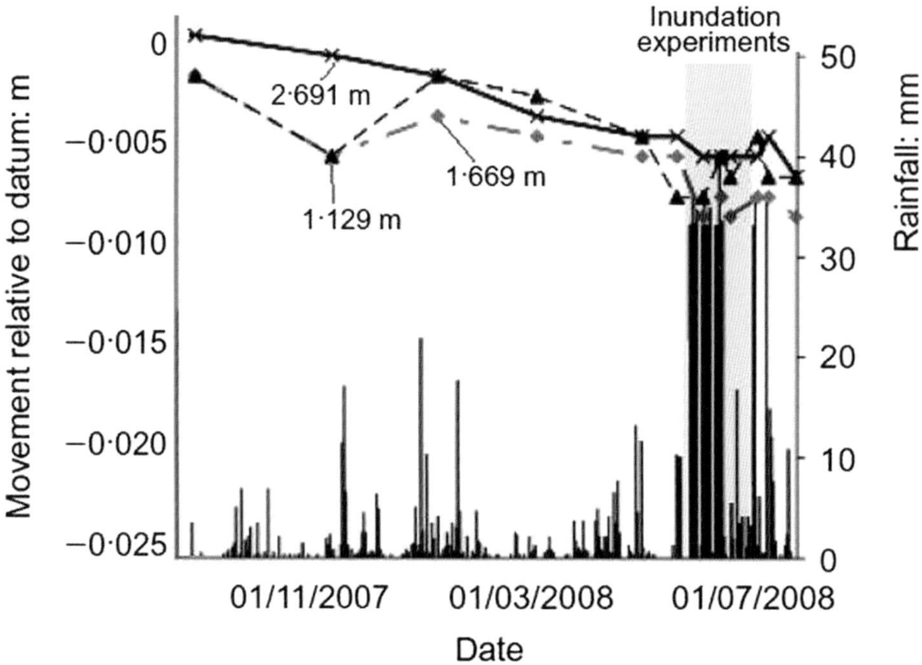

Fig. 8. Settlement monitoring for a well compacted panel (Labels indicate the depth of installation of the magnetic extensometers).

relative to the saturated permeability as the soil desaturates. The near surface permeability also changes temporally as desiccation cracks form and as vegetation develops and creates root passages. Tsaparas & Toll (2002) found that near surface permeability needed to correctly predict infiltration rates in numerical models were two orders of magnitude higher than values measured in the field below the surface desiccated zone.

These uncertainties make it difficult to undertake numerical prediction of the soil responses based only on laboratory data. Therefore, the numerical model has to be calibrated against field monitoring to ensure that the soil water retention properties and permeability functions adopted are able to correctly model the observed responses.

Once the model has been calibrated against current climate conditions, it can then be used to predict changes that might occur in the future. The climatic conditions for modelling future scenarios will be generated using the weather generator described by Kilsby *et al.* (2007). The use of this generator for modelling the impacts of climate change on slopes is described by Davies *et al.* (2008).

Conclusions

The BIONICS project is a large cooperative research project involving academic partners and industrial stakeholders focusing on the effects of the expected climate change on infrastructure embankments. Part of the project has involved constructing an instrumented embankment equipped with a system of sprinklers and covers to control climatic conditions. It has been instrumented with piezometers, tensiometers, water content and temperature sensors, inclinometers and extensometers.

A novel experimental set-up for field monitoring of soil suctions has been developed for the BIONICS project. The wide measuring range of the tensiometers (suction of 2 MPa) allows the monitoring system to be used in most natural and constructed earth structures. The system allows multiple tensiometers to be used within a single borehole and can provide continuous measurements with time and at different depths.

The results of monitoring from September 2007 and March 2009 show the cycles of pore water pressure that might be expected in an embankment. Higher pore water pressures (near to hydrostatic) were observed during winter months (typically January) falling back to lower values in March. Lower values might normally be expected during summer months, but the climate during the summers of 2007 and 2008 was ‘untypical’ with high levels of summer rainfall, as well as being subject to artificial inundation. Further monitoring is required to gain an indication of the suctions that can be developed during a dry summer. The embankment has demonstrated shrink/swell cycles responding to the changes in pore water pressure, with surface heaves of the order of 5 mm being observed when pore water pressures increase.

References

BS 1377. 1990. *Methods of Test for Soils for Civil Engineering Purposes*. Milton Keynes: British Standards Institution, Milton Keynes.

Davies, O., Rouainia, M. & Glendinning, S. 2008. Numerical predictions of seasonal pore pressure fluctuations using FLAC tp flow. *In*: Toll, D. G., Augarde, C. E., Gallipoli, D. & Wheeler, S. J. (eds) *Unsaturated Soils: Advances in Geo-Engineering, Proceedings of the 1st European Conference on Unsaturated Soils*, Leiden. CRC Press/Balkema, **1**, 817–822.

Glendinning, S, Loveridge, F., Starr-Keddle, R. E., Bransby, M. F. & Hughes, P. N. 2009. *Role of Vegetation in Sustainability of Infrastructure Slopes*, Special Issue of Engineering Sustainability, Proceedings of the Institution of Civil Engineers, 162, No. ES2, 101–110.

HIGHWAYS AGENCY. 1998. *Specification for Highway Works*. Manual of Contract Documents for Highway Works, Volume 1. The Stationary Office, London.

Hughes, P., Glendinning, S. & Mendes, J. 2007. Construction Testing and Instrumentation of an Infrastructure Testing Embankment. *In*: Liong, S. Y., Phoon, K. K. & Toll, D. G. (eds) *Proceedings Expert Symposium on Climate Change: Modelling, Impacts & Adaptations*. Singapore, Research Publishing Services, 159–166.

Hughes, P. N., Glendinning, S., Mendes, J., Parkin, G., Toll, D. G., Gallipoli, D. & Miller, P. 2009. *Full-Scale Testing to Assess Climate Effects on Embankments*. Special Issue of Engineering Sustainability, Proceedings of the Institution of Civil Engineers, **162**, No. ES2, 67–79.

Kilsby, C. G., Jones, P. D. et al. 2007. A daily weather generator for use in climate change studies. *Environmental Modelling and Software*, **22**, 1705–1719.

Kilsby, C., Glendinning, S., Hughes, P. N., Parkin, G. & Bransby, M. F. 2009. *Climate-Change Impacts on Long-Term Performance of Slopes*. Special Issue of Engineering Sustainability, Proceedings of the Institution of Civil Engineers, 162, No. ES2, 59–66.

Lourenço, S. D. N., Gallipoli, D., Toll, D. G. & Evans, F. D. 2006. *Development of a commercial tensiometer for triaxial testing of unsaturated soils*. Geotechnical Special Publication, **147**. American Society of Civil Engineers, Reston, **2**, 1875–1886.

Loveridge, F. & Anderson, D. 2007. What to do with a vegetated clay embankment? *EMAP Slope Engineering Conference 2007, London*. Thomas Telford, London.

Mendes, J. 2011. *Assessing the Impact of Climate Change on an Instrumented Embankment: An unsaturated soil mechanics approach*, PhD thesis, Durham University (http://etheses.dur.ac.uk/612/).

Mendes, J., Toll, D. G., Augarde, C. E. & Gallipoli, D. 2008. A system for field measurement of suction using high capacity tensiometers. *In*: Toll, D. G., Augarde, C. E., Gallipoli, D. & Wheeler, S. J. (eds) *Unsaturated Soils: Advances in Geo-Engineering, Proceedings of the 1st European Conference on Unsaturated Soils*, Leiden. CRC Press/Balkema, 219–225.

Overton, A. K. 2007. *A Guide to the Siting, Exposure and Calibration of Automatic Weather Stations for Synoptic and Climatological Observations*. Royal Meteorological Society, Reading.

Ridley, A. M. & Burland, W. K. 1993. A new instrument for the measurement of soil moisture suction. *Géotechnique*, **432**, 321–324.

Rouainia, M., Davies, O., O'Brien, T. & Glendinning, S. 2009. *Numerical Modelling of Climate Effects on Slope Stability*. Special Issue of Engineering Sustainability, Proceedings of the Institution of Civil Engineers, 162, No. ES2, 81–89.

Tsaparas, I. & Toll, D. G. 2002. Numerical analysis of infiltration into unsaturated residual soil slopes. *Proceedings of the 3rd International Conference on Unsaturated Soils*, Recife, Brazil. Swets & Zeitlinger, Lisse, **2**, 755–762.

Toll, D. G., Lourenço, S. D. N. et al.. 2011. Soil suction monitoring for landslides and slopes. *Quarterly Journal of Engineering Geology and Hydrogeology*, **44**, 23–33.

Measuring and mitigating the environmental impact of earthworks and other geotechnical processes

Nick O'Riordan[1] & Alan Phear[2]*

[1] Arup Group Ltd, 560 Mission Street, San Francisco, USA
[2] Arup Group Ltd, The Arup Campus, Blythe Gate, Blythe Valley Park, Solihull, UK

**Corresponding author (e-mail: alan.phear@arup.com)*

Abstract: Considerable progress has been made in the assessment, management and maintenance of existing earthworks and in the selection of methods of construction of new earthworks. Techniques involving multi-criteria analysis, carbon-footprinting and life cycle assessment of competing schemes provide transparent methodologies for use in environmental impact assessment.

This review paper provides examples of use of the various techniques, and in particular concentrates on areas of innovation including:

- reuse of excavated tunnel spoil
- methods of ground treatment to minimise earthwork impact (e.g. *in-situ* soil strengthening, green walls etc).
- combining performance-based design, under environmental load cases such as earthquake and flood, with carbon footprint and cost analyses

Earthworks construction and other geotechnical processes are required to remove, replace, stiffen, strengthen or stabilize the ground in response to the requirements of the developing world. The environmental impact of such construction processes is regulated throughout Europe, although the emphasis on mitigation methods varies from country to country. In the past, mitigation was partly informed by custom and habit: there have been many situations where the consequences of the mitigation works in terms of associated carbon footprint have added enormously to the carbon budget of individual projects. Equally the logic of and benefit from certain activities in terms of energy efficiency and carbon footprint has often been untested by those carrying out environmental impact assessment. It is perhaps a paradox that current environmental legislation for infrastructure projects does not require a detailed examination of benefit or 'payback period' in terms of carbon footprint and energy efficiency in construction, maintenance and end-of-life replacement.

There is an enormous variety in geotechnical processes in terms of plant, materials, cost, effectiveness and risk. In recent years, there has been a realization that inappropriate processes can lead to early failure, excessive initial cost, high maintenance, remedial works and in extreme cases, complete replacement. Increasing urbanization is placing considerable demands on energy production globally. We can anticipate that energy efficiency in design, construction, maintenance and replacement of infrastructure will become even more important over time. The capturing of information to feed into whole life evaluation activities is challenging, and often there is no single, correct answer.

Cost alone is found to be too crude a variable on which to base a course of action. Nevertheless the activities of economists in the field of environmental accounting and sustainability have resulted in workable connections across the cost-energy-emissions-compliance-societal benefit-time polygon.

Benchmarking of 'what we do now' in terms of cost, emissions, energy usage, supply chains etc. is entirely possible. By understanding current processes we can devise feedback systems that enable the proper exercise of judgment at any stage of the stewardship of that finite resource which is the ground.

Strategic guidance

A recent Sustainable Development Commission report entitled *Prosperity without Growth* www.sd-commission.org.uk states that the world currently produces 770 g of carbon dioxide per $ of income and that this figure must reduce to 6 g per $ (!) by 2050, if all 9 billion of us are to enjoy European income levels in 2050 and simultaneously stabilize atmospheric carbon dioxide at 450 ppm. Construction activity is generally agreed to contribute about 25% to global carbon emissions, and so it will be particularly

From: Radford T. A. (ed.) 2012. *Earthworks in Europe*. Geological Society, London, Engineering Geology Special Publications, **26**, 163–173, http://dx.doi.org/10.1144/EGSP26.18

essential to reduce emissions in construction, in order to pursue sustainable development.

Economic growth relies on the confidence of society in all its forms and the predictability of outcome of a certain course of action.

The EU Sustainable Development Strategy directive, renewed in 2006, has a three-part approach, setting the following objectives by 2020:

- To reduce greenhouse gas emissions by 20%
- To reduce primary energy usage by 20%
- To increase use of renewable forms of energy to 20% of the energy mix

There is thus a strong driver to put ground engineering activities into a context which can show improvements in these indicators of sustainable development. The relevant standard is ISO 14064 (International Organization for Standardization 2006).

In the UK, the Environment Agency (2007) has produced a Carbon Calculator for construction activities http://www.environment-agency.gov.uk/static/documents/carbon_calculator_2_1883909.xls and this has been extended specifically for highway-related projects by the Highways Agency http://www.ha-partnernet.org.uk/minisite/PNET_UI/Features/ClimateChangePage/Documents/CCT-Tool-MajorProjects-v4b.xls (Parsons Brinckerhoff/WSP 2008 and Highways Agency 2008). These documents provide a basic framework from which an inventory of emissions in terms of carbon dioxide equivalents can be prepared for significant movements of earthworks and other materials. Such calculation systems enable broad decisions to be made but are not sufficient to allow different methodologies to be compared.

As experience in the use of carbon calculators grows, so such calculations can be incorporated into design, optioneering and value engineering activities. Infrastructure clients, designers, and contractors are all finding carbon accounting to be a useful tool, but in different ways. Designers are increasingly using this method in combination with traditional cost comparisons to decide between different schemes. Contractors are increasingly applying carbon accounting to their construction management systems: This is because it has been realized that carbon reduction is a useful way to combine environmental management and 'lean construction' methods. There is great legal and ethical pressure on contractors to manage and minimize the environmental impact of their work. There is also pressure to increase operational efficiency, reduce costs and make best use of resources. Both topics are concerned with controlling waste. Environmental impact (i.e. emissions) from construction plant is closely linked to the efficient use, or otherwise, of that plant and carbon accounting is a good way to show this.

Good construction practice will also reduce both waste and environmental impact. Examples of this are coordinated planning of the different workstreams, e.g. by workspace booking, and by making good quality workmanship a key objective: Poor workmanship increases waste, costs and the carbon footprint.

The impact of environmental decisions on earthworks

As a result of extensive environmental protests in the UK in the 1990s against new transport infrastructure projects (e.g. M3 Twyford Down, Winchester, A34 Newbury by-pass, Manchester Airport second runway, etc.), there is increasing pressure in the UK to widen motorways within the existing land-take. In addition on some highway widening schemes, there is great environmental pressure to minimize the impact of the required earthworks on the existing vegetation. This is to reduce ecological fragmentation, as well as to maintain trees which screen the road from neighbouring properties. However it is debatable whether the substantial cost and effort of carrying out widening works in this way justifies the retention of the vegetation.

On a recent highway widening scheme within the existing land take that the authors were involved with, there was an environmental undertaking that not more than 3 m of vegetation could be removed behind the back of the widened verge. This requirement resulted in extensive use of soil-nailed slopes in cuttings and steepening up of embankments using high-quality imported granular fill. The constrained working space resulted in frequent construction delays along the site during installation. It is likely that a less costly solution could have been developed without the 3 m environmental limit. If greater space had been made available, more efficient construction methods could have been adopted and these would probably have reduced the carbon emissions during the works. The authors suggest that embodied energy and associated carbon footprint calculations can enable construction impacts to be balanced against traffic flow consequences. This is developed further below.

Mitigation of environmental impact

Most of the environmental impacts of earthworks can be minimized or substantially mitigated by considering and planning for them as early in the design process as practicable. An environmental impact assessment is an integral and essential part of the planning of any infrastructure, or earthworks, project. Examples of other activities that should be carried out at project planning or feasibility stage are ecological and environmental baseline surveys, transport studies, geotechnical and geoenvironmental site investigations, and baseline groundwater monitoring. Access routes for construction traffic, working hours, etc. also need to be considered and negotiated at this stage; and neighbours, third party stakeholders and the wider public should be consulted. Likewise preliminary suitability assessments for earthworks material excavated from the site should be carried out so that their re-use can be optimized and waste minimized. Residual and short-term environmental impacts (such as noise, dust, etc.) should be managed with a site environmental management plan (e.g. see CIRIA C650; Charles & Connolly 2005). Qualitative scoring systems can be useful, such as the CEEQUAL system promoted in the UK by the ICE and CIRIA organizations (Campbell-Lendrum & Feris 2007).

Across Europe the regulatory framework (by using instruments similar to the UKs Landfill Tax and the Aggregates Tax) is seeking to ensure that good environmental practice is increasingly also good commercial practice. Thus minimization of waste through beneficial re-use and/or recycling of site-won materials are increasingly important, particularly with regard to earthworks. Tunnel spoil management, management of earthworks materials in landscaping and material exchanges will increasingly be used, particularly in projects of regional or national importance where a planned approach can be made between the promoter, regulator, stakeholders and the designers and constructors (e.g. Warren *et al.* 2003; O'Riordan 2012).

Earthworks surfaces are vulnerable to erosion by surface water run-off until they are topsoiled and seeded. Silt-laden surface water run-off can cause considerable pollution to nearby watercourses if not properly managed. CIRIA C648 (Murnane *et al.* 2006) contains excellent guidance on control of water from linear construction projects. This is an environmental impact of earthworks that is frequently poorly managed by contractors, but can readily be mitigated by a good erosion management plan.

The need to manage topsoil properly and sustainably is increasingly recognized. It is a finite resource that has taken thousands of years to develop. However there is a general lack of awareness and understanding within the construction industry of the need to handle and treat topsoil carefully to minimize construction impacts on its properties. Good practice includes storage in low stockpiles, working of the soil in appropriate (i.e. dry) weather and minimizing damage to the structure of the soil by avoiding excessive compaction or trafficking. Site-won topsoil resources should be safeguarded and reused wherever possible. The Department for the Environment Food and Rural Affairs (DEFRA) has published a code of practice on the sustainable use of soils on construction sites (DEFRA 2009).

Aesthetic considerations alongside good engineering are essential to sustainable urban development, and this has long been recognized in parts of Europe (Head 2004). Some infrastructure clients, such as the Highways Agency in the UK are very aware of the need for earthworks solutions to have an attractive appearance (see Highways Agency 2003*b*). One example of this is the use of local stone to face gabion walls: where the M40 motorway passes through the Chiltern Hills, a range of chalk hills north-west of London, local flint stone (which occurs within the chalk) has been used to face the gabion walls. Guidance on this is given in the Highways Agency Advice Note HA 85/01, and there are many examples of similar practice in the Czech Republic, France, Italy, Germany, Spain and Switzerland (Vanicek & Vanicek 2008).

Ecological initiatives such as relocating fauna such as great crested newts or dormice, translocating ancient woodland, or seeding of earthworks slopes with wildflowers are topics that the public can relate to. However there is a need to balance the ecological impact of earthworks with impacts that are less newsworthy such as waste reduction or carbon minimization.

Embodied energy and other environmental indicators

Embodied energy (EE) is defined as the total energy (in Joules) that can be attributed to the use of an item or component. For the construction industry, embodied energy includes the energy used in extraction of the raw materials from the earth; the processing of that raw material into finished products; transportation to suppliers and then on to site; the construction process; the demolition and recycling; and the construction and maintenance of any associated temporary works.

Quantifying embodied energy is important because it encompasses associated environmental impacts such as resource depletion and greenhouse gas emission. Research into the relationship between embodied energy and carbon dioxide shows a high correlation; 1 GJ of embodied energy produces 0.098 tonnes carbon dioxide (INUI *et al.* 2011; O'Riordan 2012). Therefore, although there are no direct environmental impacts associated with embodied energy, this link to carbon dioxide gives a context for interpreting embodied energy data. Fundamentally, there is little absolute knowledge of the energy footprint of current ground engineering operations. Reid & Clark (2000) developed a whole life cost (WLC) model for earthworks slopes based upon available failure information. The implications of this work were that slopes as shallow as 1 V:5H could be justified on economic grounds. Different conclusions could be reached using multi-criteria analysis techniques in which emissions and/or energy consumption parameters are introduced into the assessment.

More recently Workman & Soga (2004) evaluated the embodied energy associated with the tunnel construction between the Stratford and the St Pancras terminus of the High Speed 1 (HS1) project. This work has been extended into activities such as retaining wall design and construction, railway trackbed design and maintenance, and embankment construction on soft ground (Chau *et al.* 2007; O'Riordan 2012).

Examination of current practice can produce counter-intuitive results. For example Harbottle *et al.* (2007) examine the performance of five remediation projects and conclude that on an emissions basis, the 'best' overall solution is to provide a cover system, rather than use geotechnical processes such as soil stabilization or washing, bioremediation or transport to landfill.

There is thus an emerging knowledge base that can be used and extended more generally into ground engineering activities.

By establishing energy usage and associated greenhouse gas equivalents for geotechnical processes, a common vocabulary can emerge that will facilitate dialogue across disciplines. For example, the carbon dioxide emissions consequences of choosing to provide a new road, on a particular route, to provide better mobility around a congested town can be compared with the construction and maintenance emissions associated with the road design and construction (O'Riordan 2012). The consequences in terms of emissions of shorter or longer construction times, and the associated processes to achieve them, can be explicitly provided. Similarly maintenance decisions and processes can be examined.

Using multi-criteria analysis techniques in which quantifiable parameters are used explicitly, geotechnical processes can be put into the context of sustainable design, construction and maintenance practice that until now has concentrated on the end-user's behaviour inside buildings.

Performance based design and carbon calculations

Performance based specifications are defined by the United States Transportation Research Board (TRB) (1995, 2002) as:

> Specifications that describe how the finished product should perform over time. For highways, performance is typically described in terms of changes in physical condition of the surface and its response to load, or in terms of the cumulative traffic required to bring the pavement to a condition defined as 'failure'. Specifications containing warranty/guarantee clauses are a form of performance specifications.

The perceived advantages of performance-based specifications include provision of a contractual environment that encourages innovation, use of designer's or contractor's experience, focusing the client on performance rather than historical empirical relationships, and appropriate apportioning of the risk between client and contractor (Pidwerbesky & Alabaster 2004). The potential limitations include engaging a contractor who turns out to be either unwilling or unable to construct the road to the specified quality and paying for a product that does not meet the long term performance expectations.

The risks can be mitigated by mutual understanding between the client and its contractors of the elements of the performance-based specifications, mandatory quality assurance requirements and good prediction of long-term behaviour, often using probabilistic methods in combination with historic performance data.

The effects of performance-based specifications are wide reaching and the roles of the different parties to the contract can change significantly (Highways Agency 2003*a*). The contractor needs to possess or out-source skills such as highways design and advanced materials testing. The highway authority's role changes from designer/supervisor to that of quality auditor.

The use of probabilistic methods to establish the adequacy of a design enables more efficient and economic solutions to be developed. Embankment and cutting stability can conveniently be treated in this way (see for example Reid & Clarke 2000). Chau & Soga compiled an unpublished report in 2007 (*Technical Environmental Consideration Chapter for the International Familiarisation of ISO Code for Geotechnical Earthquake Design*) on a Japanese study in which various ground treatment methods were considered for highway embankments in the Tokyo area. These embankments have to accommodate high seismic forces. The parameters studied are shown in Figure 1 and the outcome of embodied energy calculations shown in Figure 2. By combining these figures with the probability of failure of the selected treatment system under the design earthquake Chau and Soga report the relationship in Figure 3, from which can be seen that the geotextile reinforced embankment provides the most energy efficient solution: little is to be gained in strengthening the soil to resist the design earthquake using soil mixing, in this case.

Combining construction embodied energy/carbon footprint with environmental benefits

O'Riordan (2007) combined embodied energy calculations with a range of ground treatment methods for building new roads on soft clays. The selection of design parameters for very soft clays, and the search for the dominant failure mode and consolidation characteristics for a given problem, is extremely challenging. O'Riordan & Seaman (1994) addressed the practical problem of selecting an appropriately robust method for the delivery of design, and identified a wide range of methods available at that time that can be used singly and in combination.

The constraints of time, cost and reliability impose considerable demands on the designer of earthworks and structures on soft clay. Proprietary solutions developed in one country to resolve a particular problem may be inappropriate in another. Supply chains can become extended, and the 'carbon footprint' for a given solution can become very large indeed.

Table 1 shows embodied energy intensity (EEI) values, excluding the energy expended in delivery to site, given separately, evaluated for some soft clay construction components that can be used to evaluate options. It follows that a dominant factor in almost all alternative construction processes is whether a design option contains steel and/or ordinary Portland cement. These figures apply to sites in the UK mainland, and would need to be modified to reflect the supply chain in other countries.

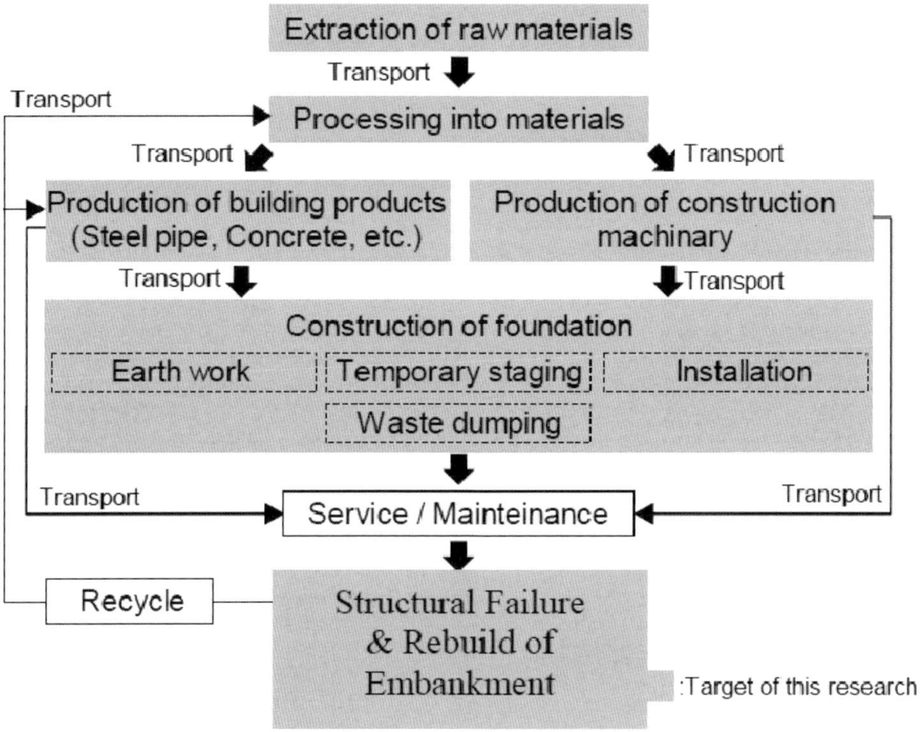

Fig. 1. Flow chart of the process boundary (Chau & Soga pers. comm.).

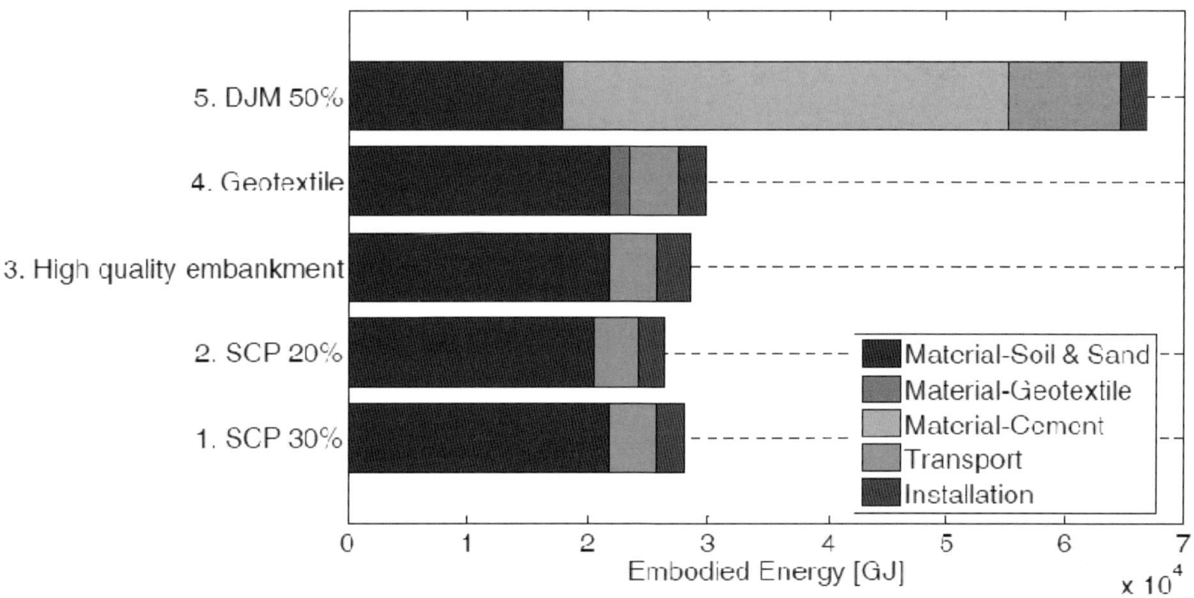

Fig. 2. Construction embodied energy of the embankment designs. SCP are sand column piles and DJM is a form of deep soil mixing using the 'wet' process (CEN 2005) (Chau & Soga pers. comm.).

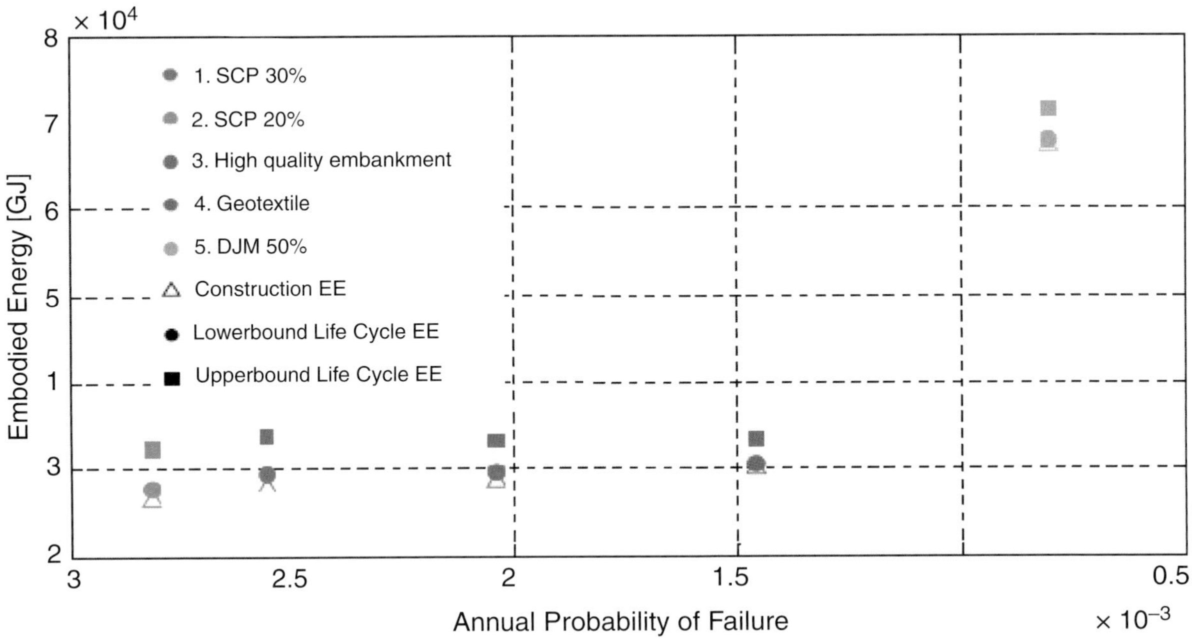

Fig. 3. Life cycle embodied energy for different aseismic ground treatment designs (Chau & Soga pers. comm.).

If we consider a hypothetical 9 m thick embankment which undergoes 2 m settlement using vertical drains to control stability, the embodied energy associated with construction, excluding transport to site for the ground treatment components is approximately:

Embankment fill	12600 MJ m^{-2}
Vertical drains	1.5 MJ m^{-2}
Geogrid	40 MJ m^{-2}

If the 2 m settlement, and the associated time for consolidation can be avoided using bridge approach support piling (BASP) on the basis of Table 1, the comparable EE becomes

Embankment fill	9800 MJ m^{-2}
Driven pile	6000 MJ m^{-2}
Tensar geogrid	40 MJ m^{-2}

For comparison, Workman & Soga (2004) report that a 100 m^2 masonry-built house in the UK uses a little over 600 GJ in construction (or 6000 MJ m^{-2}), and the figure for a large 52 storey office building in Australia is reportedly 2590 TJ (or approximately 25900 MJ m^{-2}).

From these simple examples it can be seen that for deep dry soil mixing to approach either BASP or stage constructed options, either reduced OPC products would need to be developed, or a reduced coverage, around 50% or below would be needed.

If the environmental benefits for early provision of, for example, a new road by-pass on such an embankment, can be expressed in reduced end-user energy emissions, then a direct comparison of environmental benefit for these various forms of construction on soft ground can be made. Furthermore, the effect of length of supply chain for

Table 1. *Some embodied energy intensity values for evaluation of soft clay engineering options*

Component	*EEI value*
Driven 300 mm PC pile, 10 m long, 2 tonne capacity	6 GJ per pile
Deep dry soil mixing (DDSM) @ 100 kg m^{-3} Ordinary Portland Cement (OPC) from gas-fired power station energy source, 10 m deep, 90% coverage	10 GJ m^{-2}
Vertical, 100 mm wide Prefabricated drain, 10 m deep @1 m c/c	1.5 MJ m^{-2}
Geogrid such as Tensar SS40 @ 0.53 kg m^{-2}	40 MJ m^{-2}
11 tonne truck, average daily running speed = 50 km h^{-1}	2 MJ km^{-1} (pro-rata for lower daily running speeds)

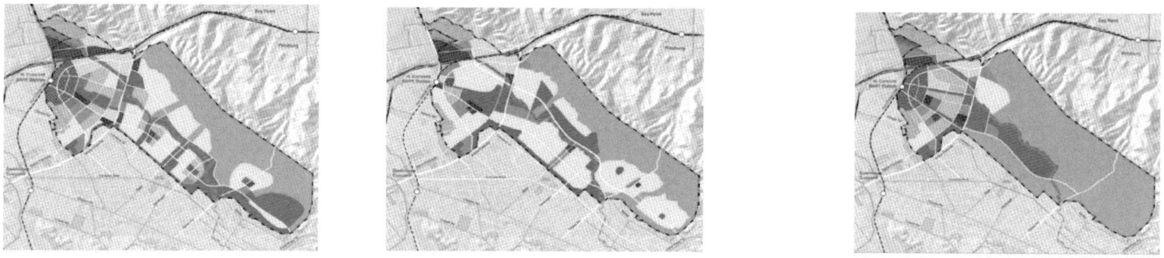

Alternative 1 Alternative 2 Alternative 3

	Alternative 1 Business-as-Usual	**Alternative 2** Maximum build-out of site	**Alternative 3** Concentrated, transit-oriented development
Regional mobile emissions over 'No Project'/'do nothing'	95,208	145,766	52,446
Stationary emissions (TCO2e)	400,470	457,074	350,028
Total gross emissions (TCO2e)	**495,678**	**602,840**	**402,474**
Service population (residents + jobs)	39,200	59,600	45,800
Greenhouse gas efficiency rate (TCO2e/person)	**12.6**	**10.1**	**8.8**

Fig. 4. Total carbon footprint of alternative development scheme layouts, in terms of transport provision.

individual components can be evaluated, and decisions taken, to balance a remote, but energy efficient design solution, with a nearby component of lower efficiency.

Figure 4 presents a comparison between 'business as usual' for a development on the west coast of the USA, with two alternative development and transport provision strategies. As the construction requirements for each scheme will vary, it follows that the 'payback' period in terms of carbon expenditure during construction and savings due to lower emissions as a result of that construction can be calculated.

It can be seen from Figure 4 that there are substantially different outcomes after construction when the schemes are in use. These outcomes can be readily associated with the carbon dioxide footprint associated with the construction of the alternative schemes. In a study of a major new motorway comprising 24 km of dual three lane highway with two large interchanges and 29 structures (Pantelidou *et al.* 2012), the effect of reduced congestion after the construction was explicitly calculated, together with the carbon footprint of the works.

Figure 5 shows the division of the carbon emissions calculated for the primary components. The average figure can be compared with other benchmarked projects. For example Workman & Soga (2004) showed an embodied energy of the CTRL rail tunnel construction from Stratford to St Pancras of 128 GJ m^{-1}, or approximately 12.8 tonne carbon dioxide per metre. To put the total carbon dioxide construction footprint into context, Figure 6 shows the effect in reduction of congestion over the *status quo* ('do nothing') after the new motorway is built. The work was carried out using VISSIM software and standard vehicle emission data.

The calculated improvement was approximately 6000 tonne carbon dioxide per year, as shown in Figure 6. The payback period, with reference to Figure 5, amounts to 80 years. The dominance of high carbon footprint for the non-earthworks components in this example therefore has a significant effect on environmental payback period. This outcome enables the following decisions to be identified, in order to reduce the environmental payback period:

- is the motorway the best solution for dealing with congestion: could a shorter payback period arise from provision of other transport modes, such as rail?
- can earthwork components be increased and structures reduced?
- can pavement materials with a lower carbon footprint be used?

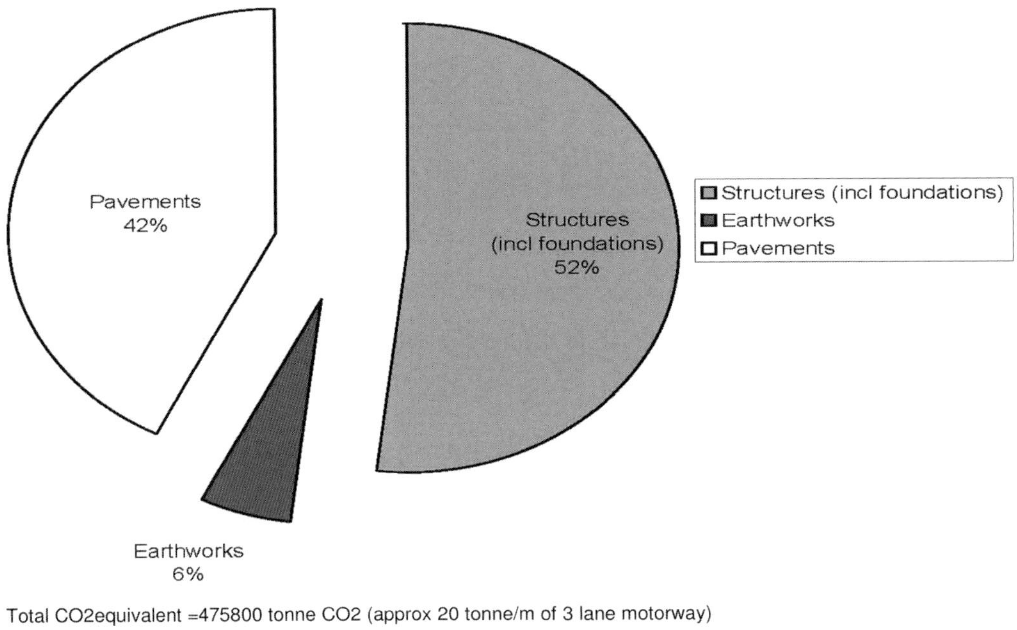

Fig. 5. Motorway construction carbon dixoide impact assessment (Pantelidou *et al.* 2012).

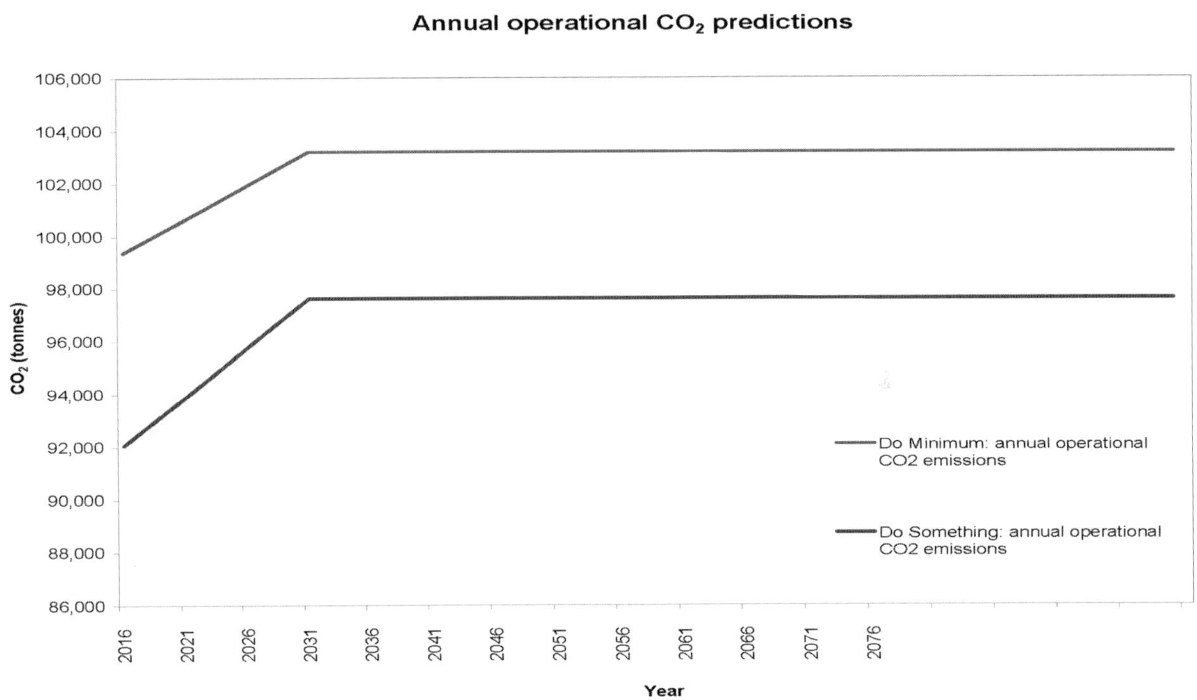

Fig. 6. Predicted change in traffic carbon emissions as a result of provision of new 24 km motorway (Pantelidou *et al.* 2012).

In this way, innovative, low carbon footprint methods of design and construction can be introduced to infrastructure projects.

Tunnel spoil and earthworks

Tunnel spoil arisings from tunnels can pose particular challenges. The primary aim of the tunneller is to create the tunnel as quickly and safely as possible. The spoil produced, its quality and re-usability are secondary factors to the tunnel designer. Excavations using traditional mining techniques, roadheaders etc. can produce material that can be used without further treatment. Excavations using tunnel boring machines produce material that is governed by the method of advancement of the face, and the muck-away process. Without positive intervention in the design of the muck-away process, the final product can be a slurry of very high water content. Prediction of the quality of the arisings can be very difficult at the planning stage and consequently a ready market for the material outside the project is impossible to find, and a frequent outcome is that spoil is taken as a quasi-waste to landfill. Two benchmark examples of re-use adjacent to the main construction works from the High Speed 1 project in the UK are described below (after Phear & O'Riordan 2004).

Stratford landraise

The majority of the length of the twin 7.2 m internal diameter London tunnels for HS1 was driven westwards and eastwards from Stratford to enable the most of the spoil (2.5 million m^3) to exit there. The Stratford site covers about 50 hectares and was used for nearly 150 years for railway yards, foundries and coal and oil stores. The contamination resulting from these uses was remediated and a 7 m high development platform is now being constructed over the whole footprint of the site using the tunnel arisings. The tunnels were constructed using Earth Pressure Balance tunnel boring machines. The material which was excavated was primarily silty sand, and later London Clay. The designer and project manager for HS1, Rail Link Engineering, undertook extensive laboratory and other studies prior to construction to investigate the

Fig. 7. Stratford, five years after completion of rail tunnels and station box (left centre of picture).

likely water content and strength of the spoil on exit from the tunnels. The aim of these studies was to estimate how much treatment would be required to make the spoil trafficable and thus enable placement in the landraise. The effects of the foam and lubricants which are added at the face to assist tunnelling have significant effects on the moisture content, as does the dewatering of the tunnel face. Some of the spoil was therefore treated with quicklime to dry it out sufficiently to allow satisfactory placement in the landraise. The amount of treatment was optimized as the works have progressed.

The rest of the London tunnels were driven westwards from the eastern portal at Barking. The spoil arising there (0.5 million m^3 of mainly London Clay) was placed on local derelict land to create a platform for a housing development. Figure 7 shows the current Spring 2009 state of the landraise in spring 2009: it has become the base for Stratford City, adjacent to the site for the 2012 Olympic Games. In Workman & Soga (2004), the benefits of placing the tunnel spoil in the landraise were found to be only slightly more energy efficient than taking the same material to landfill 100 km north of the site by rail. This shows, as in the 24 km motorway example above, that benefits beyond pure carbon accounting have to be taken into account in decision-making.

Re-use of spoil from Thames tunnels on HS1

The HS1 Thames tunnel passes under the River Thames in twin 2.5 km long tunnels of 7.2 m internal diameter. These were dug through 10–12 m of alluvium, up to 6 m of terrace gravels and Upper Chalk (which comprises most of the tunnel length). The total unbulked spoil volume was 0.25 million m^3. It was decided to use the arisings from this tunnel to infill an old chalk quarry near the tunnel and thus create a development platform. Again the designers undertook extensive laboratory and other studies prior to construction to investigate how much treatment of the chalk spoil would be needed on exit from the tunnels. These studies included visits to other tunnels through chalk geology and laboratory trials at Imperial College, in London. The contractor also carried out several trials after contract award. These studies and trials are described in more detail by Warren *et al.* (2003).

The chalk contained up to 15–20% flints, some of which are as large as footballs and are highly abrasive. The tunnels were dug using slurry tunnel boring machines (TBM). The chalk here is generally a soft low density limestone and the action of the TBM broke it down to a slurry with a size range of from 150–0.002 mm. A crusher near the TBM face broke down the flints to smaller than 150 mm size and the combined flint and chalk slurry was then pumped to the surface along a pipeline. The tunnel arisings were passed through a primary separation plant comprising screens and hydrocyclones to remove as much as possible of the coarser fraction (greater than about 60 μm size). The remaining slurry was then passed through a bank of centrifuges to separate out the silt sized 'paste' from the transport water. This paste was then treated with cement and then re-combined with the coarser fractions and transported to the fill area on a covered conveyor, where it was successfully placed with a maximum thickness of 10 m.

Conclusions

Sustainable practice in earthworks design and construction can play a significant part in reducing energy use and carbon emissions. Tools for the calculation of carbon and embodied energy are readily accessible and these enable design and construction choices to be made. In turn, this can change customs and habit, to the benefit of all.

Performance-based design methods, in which probability of failure and increased maintenance during the design life can be established, are well suited to carbon and energy accounting methods presented herein. Choices can be made transparent and readily communicated across the multiple disciplines involved, and the public at large. Construction emissions can be compared with the environmental benefits of the changes from 'business-as-usual' baselines, and environmental payback periods can be compared.

References

Campbell-Lendrum, E. & Feris, J. 2007. Trialling Ceequal on a London railway embankment. *Proceedings of the ICE, Engineering Sustainability*, **16**, 71–76.

CEN. 2005. EN 14679. *Execution of Special Geotechnical Works: Deep Mixing*. CEN (European Committee for Standardization), Brussels.

Charles, P. & Connolly, S. 2005. *Environmental Good Practice on Site*. CIRIA Report C650. 2nd edn.

Chau, C., Nicholson, D., O'Riordan, N. & Soga, K. 2007. Embodied energy as an environmental impact indicator for geotechnical infrastructures. *Proceedings of the Geocongress*, New Orleans November 2007. ASCE.

DEPARTMENT FOR THE ENVIRONMENT FOOD AND RURAL AFFAIRS. 2009. *Construction Code of Practice for the Sustainable Use of Soils on Construction Sites* (available from www.defra.gov.uk).

ENVIRONMENT AGENCY. 2007. *Carbon Calculator for Construction Activities* (available from www.environment-agency.gov.uk/).

Harbottle, M. J., Al-Tabbaa, A. & Evans, C. W. 2007. Sustainability of land remediation part 1: overall analysis, ICE Proceedings. *Geotechnical Engineering*, **161**, 75–92.

Head, P. 2004. *A practical framework for the development of construction innovation through sustainable development: 20 years of progress, B4E Conference*, Maastricht, October.

HIGHWAYS AGENCY. 2001. *Road improvement within limited land take*, Highways Agency Advice Note HA 85/01, Design Manual for Roads and Bridges 10.2.1.

HIGHWAYS AGENCY. 2003*a*. Developing performance specifications for Earthworks: Phase 1 Scoping Study. Consultation Document, April.

HIGHWAYS AGENCY. 2003*b*. *Building better roads: towards sustainable construction.*

HIGHWAYS AGENCY. 2008. *Highways Agency carbon calculation and reporting requirements*, Interim Advice Note 114/08.

INTERNATIONAL ORGANIZATION FOR STANDARDAIZATION. 2006. ISO 14064-1. *Greenhouse Gases Part 1: SPECIFICATION with Guidance at the Organisation Level for Quantification and Reporting of Greenhouse Gas Emissions and Removals.* http://www.iso.org/iso/catalogue_detail?csnumber=38381

MURNANE, E., HEAP, A. & SWAIN, A. 2006. *Control of Water Pollution from Linear Construction Projects – Technical Guidance*. Construction Industry Research and Information Association, London CIRIA Report C648, (available from www.ciria.org).

O'RIORDAN, N. 2007. *Technical Sustainability of Construction on Very Soft Ground.* Keynote paper, Conference on Soft Ground Engineering, Portlaoise, Ireland. Engineers Ireland.

O'RIORDAN, N. J. 2012. Sustainable and resilient ground engineering. *Ground Engineering in a Changing World. Proceedings of the 11th Australia–New Zealand Conference on Geomechanics.* Melbourne, 15–18 July 2012.

O'RIORDAN, N. J. & SEAMAN, J. W. 1994. *Highway Embankments over Soft Deposits: Guidelines for Design and Construction.* UK Transport Research Laboratory, TRL Contractor Report 341. DTp/Highways Agency.

PARSONS BRINCKERHOFF LTD/WSP. 2008. Highways Agency Carbon Accounting Tool – Explanatory Report. Report prepared for Highways Agency (available from www.ha-partnernet.org.uk/sustainability)

PHEAR, A. G. & O'RIORDAN, N. J . 2004. Earthworks for the Channel Tunnel Rail Link High Speed Railway, UK. *Proceedings of the ASCE Conference on Geotechnics for Transportation*, Los Angeles, USA.

PIDWERBESKY, B. D. & ALABASTER, D. 2004. New Zealand's experience with performance specifications for pavement design and construction. Proceedings of the 83rd Annual Conference of Transportation Research Board, Washington, DC.

REID, J. M. & CLARK, G. T. 2000. Whole life cost model for earthworks slopes. Transport Research Laboratory, Report TRL 430, UK.

TRANSPORTATION RESEARCH BOARD. 1995. *Performance-Related Specifications for Highway Construction and Rehabilitation.* NCHRP Synthesis 212, Washington, DC.

TRANSPORTATION RESEARCH BOARD. 2002. Glossary of Highway Quality Assurance Terms. Transportation Research Circular E-C037, Washington, DC, April.

VANICEK, I. & VANICEK, M. 2008. Earth structures in Transport, Water and Environmental Engineering. Springer.

WARREN, C. D., PHEAR, A. G, SCHULTEIS, T. & GREGG, I. 2003. Treatment of chalk spoil from CTRL Thames Tunnel. Proceedings of the Underground Construction Conference 2003, ExCeL, London, September.

WORKMAN, R. & SOGA, K. 2004. *The Embodied Energy of Tunnel Construction*. Technical Report, Cambridge University Engineering Department.

Earthworks emissions in construction of a highway

H. Pantelidou[1]*, D. Nicholson[1], L. Hughes[2], A. Jukes[3] & L. Wellappili[3]

[1] Associate, Arup Geotechnics, 13 Fitzroy Street, London W1T 4BQ
[2] University of Cambridge, Department of Engineering, Trumpington Street, CB2 1PZ, UK
[3] Geotechnics, Highways Agency, 5 Broadway, Broad Street, Birmingham B15 1BL

**Corresponding author (e-mail: heleni.pantelidou@arup.com)*

Abstract: Accounting for carbon emissions and energy consumption for the production of materials, transportation to and from site and construction plant operation provides an indicator of infrastructure's environmental impact and could assist in engineering option selection.

A comprehensive review of the embodied energy/embodied carbon calculation was carried out for the Highways Agency, producing a simple spreadsheet calculator for geotechnical structures. With a bill of quantities format, the calculator aims to present embodied energy/embodied carbon in concert with cost information and serves as a decision support tool for option selection. Information on this research work can be found on the Highways Agency's Research Compendium website.

Most of the emissions/energy consumption are locked in the construction materials used (production of concrete, steel, asphalt etc from raw materials) with small contributions from the transportation to and from site and all the associated construction work on site.

The carbon emissions related to the construction of a new highway in the UK are discussed, with particular emphasis on the contribution of the associated earthworks and how they compare with the overall construction and operational emissions. Earthworks emissions are calculated to be less than these from the structures and pavement construction. Capitalizing on use of earthworks could help in the effort to reduce the carbon footprint of infrastructure.

Sustainability is the biggest challenge for engineering in the 21st century. The construction industry is a major energy user and a significant contributor of carbon emissions into our environment. The detrimental effect of the built environment on the natural resources, environmental pollution, climate change potential, as well as its social and economic cost is large, but not yet properly quantified and understood. In an effort to measure its environmental impact and what needs targeting to reduce the harmful effects on the environment, the construction industry urgently needs to agree a set of indicators for monitoring and regulating its impact, ultimately aiming to achieve a sustainable construction. Two of the most relevant such indicators are the Embodied Energy (EE) and Embodied Carbon Emissions (ECO_2). These two indicators are closely linked and can be calculated using the same principles and methodology.

This paper briefly summarizes the background of the EE/ECO_2 calculation and then discusses the results from an example calculation for a new UK motorway construction.

Review of embodied energy/carbon emissions

Embodied energy is the sum total of the energy used – from raw material extraction, transport, manufacturing, assembly, installation plus the capital, environmental and other costs – to produce a service or product from the start, through its disassembly, to deconstruction and/or decomposition. (Green Building Glossary website: http://www.oikos.com/library/green_building_glossary.html)

Embodied carbon expresses the same process as EE but in terms of the total carbon dioxide (CO_2) emitted. CO_2 emissions are highly correlated with the energy consumed in manufacturing construction materials. On average, 0.098 tonnes of CO_2 are produced per gigajoule of embodied energy (CSIRO 2007: http://www.csiro.au/org/CMSE). In the following, the methodology for calculating ECO_2 is described, but also same applies for the EE calculation. A comprehensive review of the EE/ECO_2 calculation for geotechnical structures was undertaken by Arup and Cambridge University for the Highways Agency (http://www.highways.gov.uk/knowledge_compendium/11AB8738BADB4EA8B58406DB8C0154AD.aspx).

Figure 1 shows a process flow chart involved in the life of a typical geotechnical structure, from the extraction of raw materials to its decommissioning and disposal, following the same assessment process as Chau *et al.* (2007). Due to the difficulties in establishing energy or carbon contributed by the post-construction activities, the present work has only addressed the processes up to and including completion of construction (blue boxes on the chart),

From: Radford T. A. (ed.) 2012. *Earthworks in Europe*. Geological Society, London, Engineering Geology Special Publications, **26**, 175–180, http://dx.doi.org/10.1144/EGSP26.19

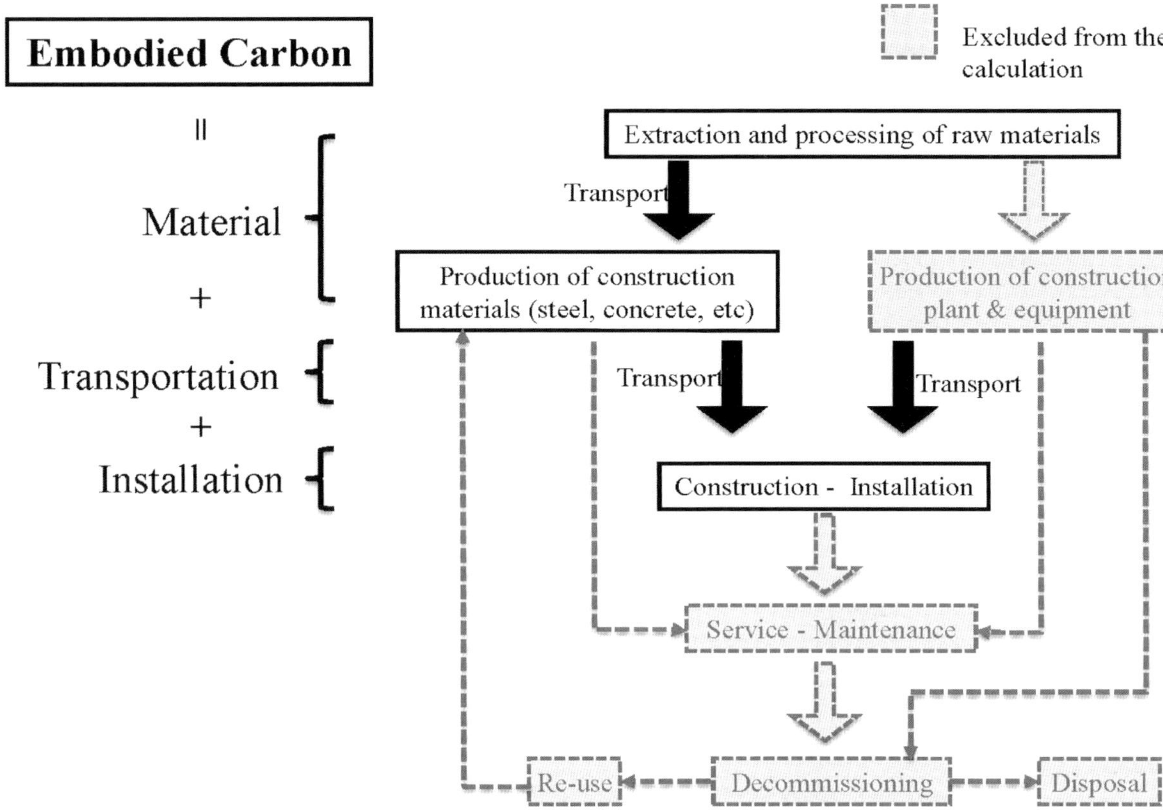

Fig. 1. Flow chart of the construction process for a structure.

without consideration of post-construction activities. Arup/Cambridge produced a simple spreadsheet with a Bill of Quantities (BoQ) format for calculating cost and EE/ECO_2 of geotechnical structures. The calculation is based on the differentiation between the carbon embodied in the (a) material, (b) transportation and (c) installation (site activities) associated with each of the quantities of the construction of the geotechnical structure (Fig. 1).

Materials: CO_2 emitted for the production of the construction materials; this includes the extraction of raw materials and their processing into construction materials. The carbon emission factors (CEF) adopted for this study are in accordance with the Inventory of Carbon and Energy (ICE) of University of Bath (Hammond & Jones 2006).

Transportation: CO_2 emitted during transportation of material and plant to and from site. The CEF used for diesel and electricity consumption are as per NAEI (2007). The amount of fuel consumed during transportation to and from site is carried out in accordance with the methodology described in the Design Manual for Roads and Bridges (DMRB) approach (Highways Agency 2007).

Installation: CO_2 emitted from construction plant during operation on site. The amount of fuel and/or electricity consumed by the machinery and equipment during construction on site is quantified using information mainly based on Japanese guidelines (Nansai *et al.* 2002) in the absence of UK-specific information. The contribution from the service and maintenance of plant are not included in the calculation. Similarly, transportation of labour to and from site is excluded. The sum of the three components of carbon for each quantity is its ECO_2 to the end of construction (excluding maintenance during operation and decommissioning).

The BoQ format of the tool makes it easy to identify the main culprits for the energy/emissions of the geotechnical structure, the first step in an effort to reduce environmental impact. This can serve as a decision support tool for option selection.

Example: construction of a new highway in the UK

The BoQ approach to carbon calculation is identical to that of the embodied energy and can be extended to general infrastructure projects. Arup has developed a carbon calculator

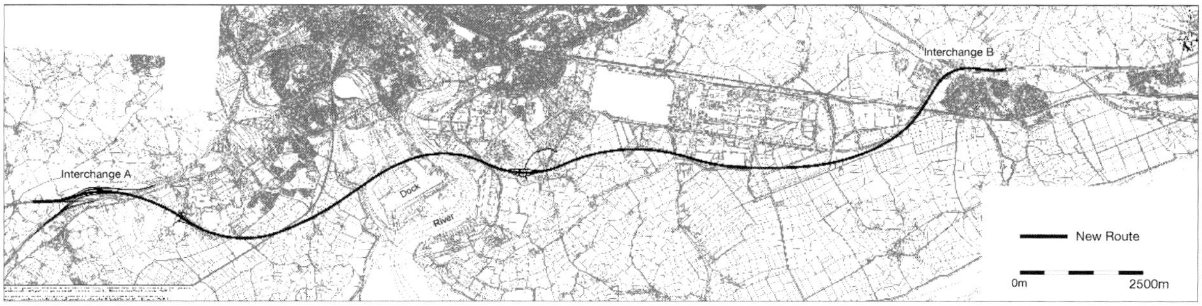

Fig. 2. UK highway example: 24 km long; dual 3-lane motorway; 3 major interchanges; 29 structures.

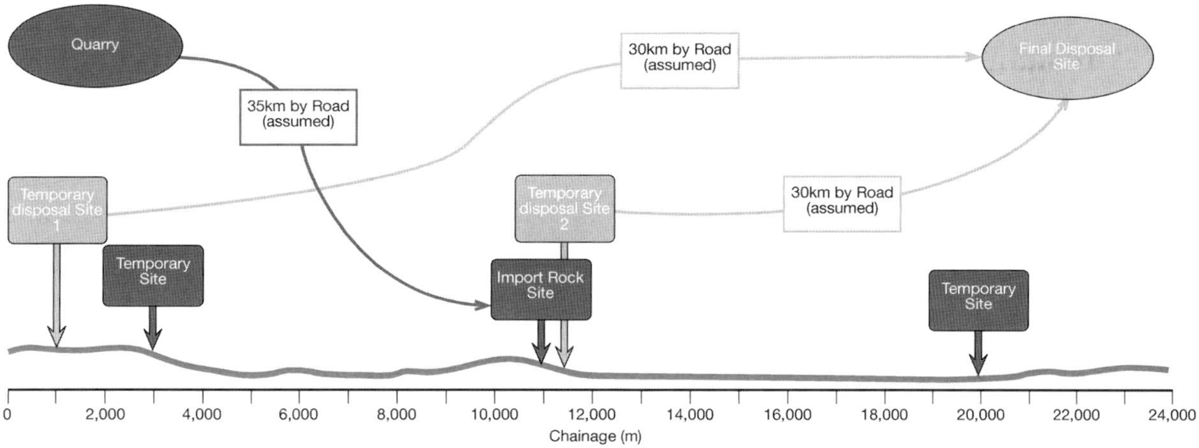

Fig. 3. Assumptions made for the earthworks calculation.

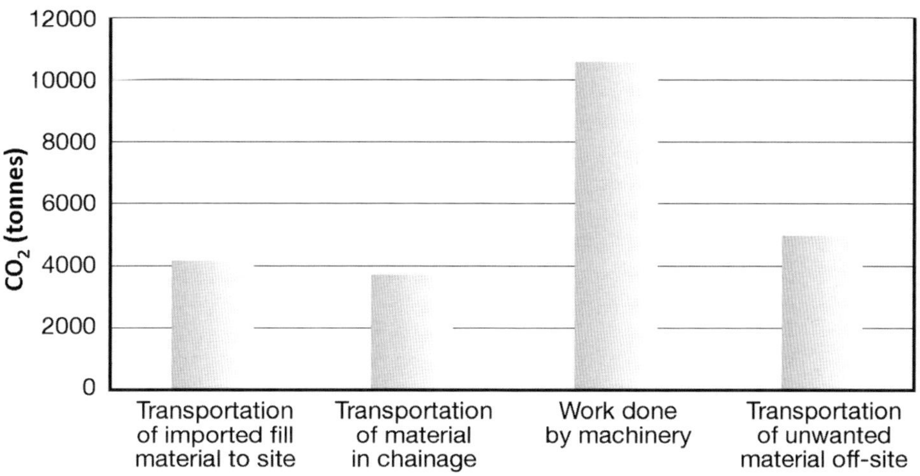

Fig. 4. Earthworks carbon emissions.

for infrastructure projects, based on the same principle of BoQ breakdown as summarized in the previous section.

We quantified the ECO_2 of a new, 24 km long highway construction (Fig. 2), comparing the CO_2 contribution from the construction of structures, pavement and earthworks. The carbon emission calculations for this project were divided into three groups: structures and foundations, pavement, and earthworks. The paper briefly looks at the earthworks carbon calculation only. The emissions from the other two components are only mentioned for comparison.

In terms of embodied carbon, earthworks are a special case of construction activity, as they do not have any carbon attributed to their material. The only components of their embodied carbon are those of transportation and installation.

For the highway project example, particular effort was made to minimize transportation of unsuitable spoil off site and imported fill into the site by optimizing on-site use of arisings. In the absence of project specific information relating to numbers and relative locations of temporary stockpiles along the alignment, or distances to external quarries and landfills, the carbon calculation was based on a number of assumptions (Fig. 3):

- distance to quarry: 35 km by road;
- distance to landfill: 30 km by road;

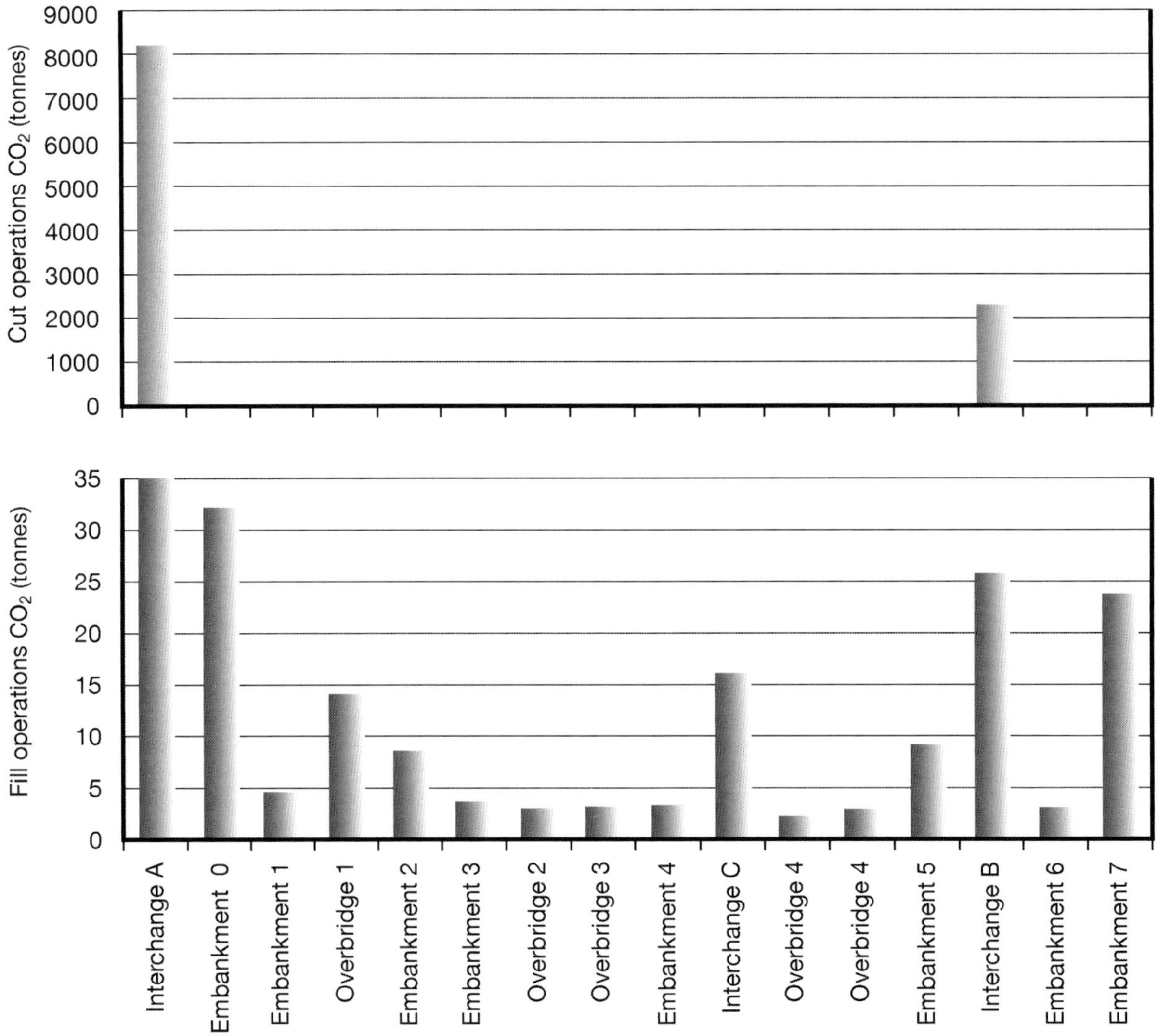

Fig. 5. Cutting operations appear to be more polluting than filling.

- 2 temporary stockpiles along the alignment; and
- plant efficiency and output emissions.

The breakdown of earthworks emissions is shown in Figures 4 and 5.

Transportation and installation (cutting, filling) result in similar quantities of CO_2 emissions, although these are dependent on the assumptions made for the calculation. Cutting operations are calculated to be significantly more polluting than filling operations, mainly due to the excavating plant emissions (Fig. 4). This is also likely to be dependent on the ground conditions and type of machinery required; its sensitivity should be investigated further. The total construction emissions calculated for the highway project were 475 800 tCO_2, of which 27 400 tCO_2 are attributed to the earthworks (Fig. 6). Earthworks emit an order of magnitude less than structures and pavement and only 6% of the total construction emissions.

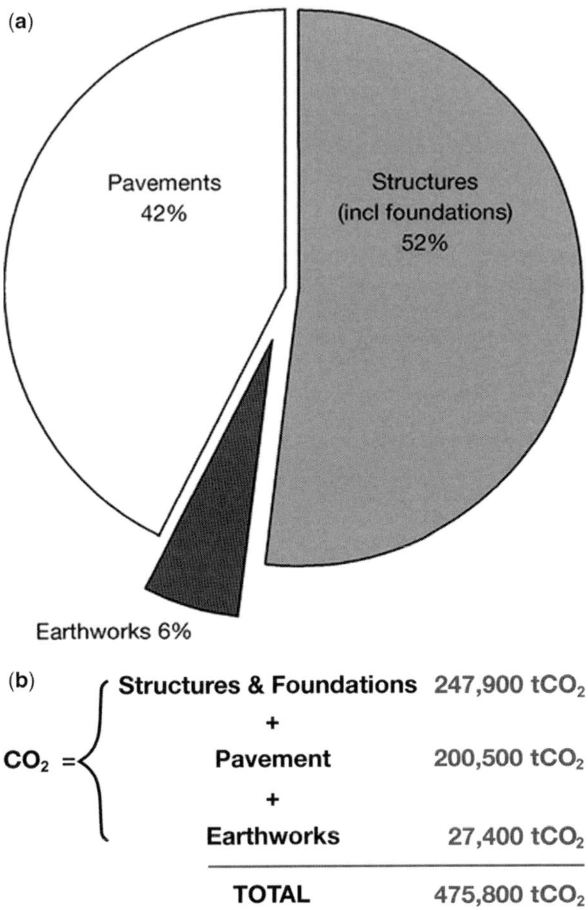

Fig. 6. Embodied carbon attributed to the three components of the highway construction example.

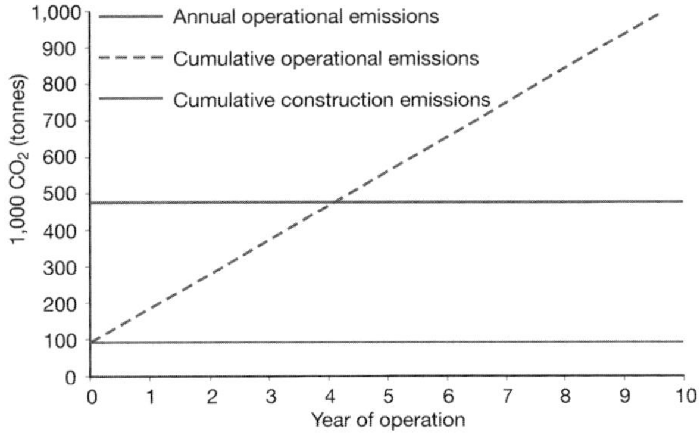

Fig. 7. Comparison of construction and operational emissions for the highway. Annual operational emissions were computed using VISSIM micro-simulation for the traffic model.

Comparison of the construction emissions with the emissions during the operational life of the highway is shown in Figure 7. Total construction emissions are equivalent to 5 years of operational emissions of the highway.

Conclusions

Carbon emissions during construction of an infrastructure project are an indicator of its environmental impact and should be used to assist in option selection. The carbon emissions during construction of a highway project example were considered, giving the opportunity to compare the contributions of the different elements of construction. For this particular example, earthworks emitted an order of magnitude less carbon than the structures and pavement.

Cutting and filling plant emissions were calculated to be of similar magnitude to the emissions from transportation of fill and spoil to and from and within the site. In addition, cutting operations are calculated to emit more CO_2 than the filling operations on this project. Both comparisons are likely to be sensitive to the input assumptions made, which should be reviewed and validated in more detail. Construction emissions for this highway example are estimated to be equivalent to 5 years of the highway operation.

References

Chau, C., Soga, K., Nicholson, D. P., O'Riordan, N. & Inui, T. 2007. Embodied Energy as an environmental Impact Indicator for Basement Wall Construction. *In*: Alshawabkeh, A. N.,

REDDY, K. R. & KHIRE, M. V. (eds) *Proceedings of GeoCongress 2008*. ASCE, USA, 867–874.

HAMMOND, G. P. & JONES, C. I. 2006. *Inventory of Carbon and Energy (ICE)*. Sustainable Energy Research Team, University of Bath. http://people.bath.ac.uk/cj219/

HIGHWAYS AGENCY. 2007. *Design Manual for Roads and Bridges*; Vol. 11: Environmental Assessment; Section 3: Environmental Assessment Techniques; Part 1: Air Quality.

NAEI. 2007. National Atmospheric Emissions Inventory/DEFRA GHG Conversion Factors, released alongside the Company Reporting Guidelines (accessed in 2007).

NANSAI, K., MORIGUNCHI, Y. & TOHNO, S. 2002. *Embodied Energy & Emission Intensity Data for Japan using Input Output Tables (3EID)*. Inventory Data for LCA. Centre for Global Environmental Research. National Institute for Environmental Studies, Japan.

Index

Figures are shown in *italic*, tables in **bold**.